AP* Q&A STATISTICS

600 QUESTIONS AND ANSWERS

Martin Sternstein, Ph.D.
Professor Emeritus
Department of Mathematics
Ithaca College
Ithaca, New York

*AP and Advanced Placement Program are registered trademarks of the College Board, which was not involved in the production of, and does not endorse, this product.

To my wife, Faith; sons, Jonathan and Jeremy; daughters-in-law, Asia and Cheryl; brother, Allan; sister-in-law, Marilyn; and grandchildren, Jaiden, Jordan, Josiah, Luna, and Jayme; all of whom keep me focused on what is truly important in life.

Special thanks to Paul Buckley, Dawn Dentato, Lee Kucera, and Kenn Pendleton for their careful reading and thoughtful suggestions, and to Linda Turner, senior editor at Barron's, for her guidance.

About the Author

Dr. Martin Sternstein, Professor Emeritus at Ithaca College, is a long-time College Board Consultant and has been a Reader and Table Leader for the AP Statistics exam for many years. He has strong interests in national educational and social issues concerning equal access to math education for all. For two years he was a Fulbright Professor in Liberia, West Africa, after which he developed a popular "Math in Africa" course, and is the only mathematician to have given a presentation at the annual Conference on African Linguistics. He also taught the first U.S. course for college credit in chess theory. Dr. Sternstein and his wife live in a house they designed and built with their own hands with the help of many of the math faculty. They have five delightful grandchildren.

© Copyright 2018 by Barron's Educational Series, Inc.

All rights reserved.
No part of this publication may be reproduced or distributed in any form or by any means without the written permission of the copyright owner.

All inquiries should be addressed to:
Barron's Educational Series, Inc.
250 Wireless Boulevard
Hauppauge, New York 11788
www.barronseduc.com

Library of Congress Control Number: 2018936456

ISBN: 978-1-4380-1189-9

PRINTED IN CANADA

9 8 7 6 5 4 3 2 1

Contents

FREE-RESPONSE QUESTIONS

ANSWERS

Introduction

This study guide is designed as a quick source of practice exercises for problems seen on the AP Statistics Exam. It is fully up-to-date concerning the ongoing changes in topic emphasis and scoring guidelines of the exam. All of the problems are at the level of the AP Exam except for a very few starred (*) exercises, which are at a more challenging level. They are included to fully test your understanding. Whereas the focus of this book is practice exercises, Barron's publishes another book, Barron's *AP Statistics*, which develops, explains, and reviews each concept.

The AP Statistics Exam consists of a 90-minute section with 40 multiple-choice problems and a 90-minute section with six free-response problems (five open-ended questions and an investigative task). In grading, the two sections are weighted equally. The multiple-choice questions are more conceptual than computational. The free-response questions require communication skills as well as statistical knowledge. (Showing just the answer to a free-response problem will not receive full credit.)

The percentage of questions on the AP Statistics Exam from each content area is approximately 25% exploratory analysis, 15% collecting and producing data, 25% probability, and 35% statistical inference.

In this book are a total of 600 questions (500 multiple-choice and 100 free-response) for you to practice. Time yourself, allowing only 2 minutes for each multiple-choice question and only 13 minutes for each free-response question. When you read through a given multiple-choice answer, think just as hard about why the wrong choices are incorrect as about why the best choice is correct.

Certainly, doing an end-of-the-year review is worthwhile. However, if you attempt, think through, and master five multiple-choice and one free-response question every day during the school year, you will be in great shape to do well on the AP Statistics Exam in May. Of course, as long as you keep up a good weekly pace, don't forget to give yourself occasional well-deserved days off! Use the following question numbers, which are listed by content, to pick which correspond to where you are in your class work.

MULTIPLE-CHOICE (QUESTIONS 1–500)

FREE-RESPONSE (Questions 501–600)

Although questions are organized into categories below, there is considerable overlap. Many, if not most, free-response questions, especially the ones under *Statistical Inference*, involve topics from throughout the course.

Exploratory Analysis (Questions 501–526)

One-variable data analysis:	Questions 501–518
Two-variable data analysis:	Questions 519–526

Collecting and Producing Data (Questions 527–536)

Probability (Questions 537–558)

Statistical Inference (Questions 559–600)

High stakes assessments can heighten anxiety and hinder learning. Frequent testing done right, however, can be an effective way to learn. Testing yourself with a few of these exercises every day will develop confidence, lead to quicker recall, and develop a deeper understanding of the material.

MULTIPLE-CHOICE QUESTIONS

MC Exploratory Analysis

CHAPTER 1

Answers for Chapter 1 are on pages 237–251.

ONE-VARIABLE DATA ANALYSIS

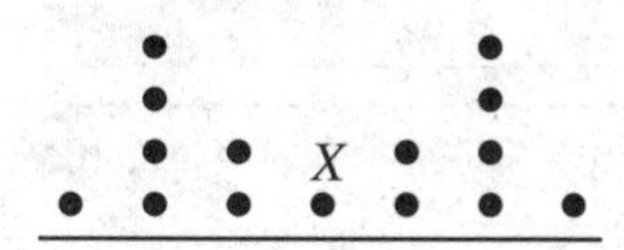

1. **One element in the dotplot above is labeled X. Which of the statements below is true?**
   (A) X has the smallest z-score, in absolute value, of any element in the set.
   (B) X has the largest z-score, in absolute value, of any element in the set.
   (C) X has the smallest value of any element in the set.
   (D) A boxplot will plot X as an isolated point.
   (E) A back-to-back stemplot will show two symmetric clusters.

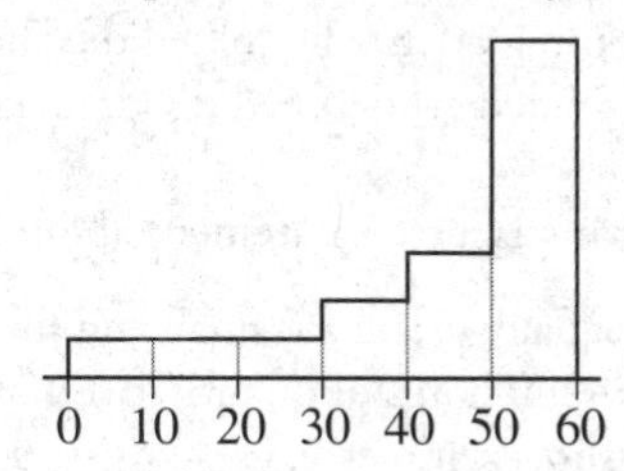

2. **The histogram shown above might correspond to which of the boxplots below?**

(A)
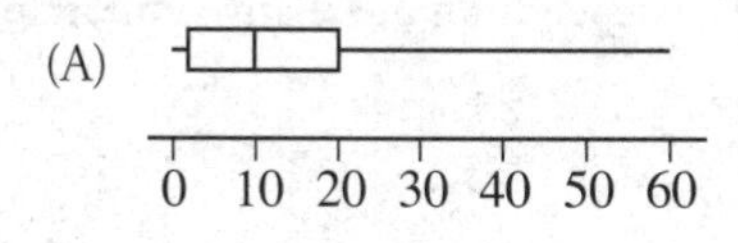

(D)
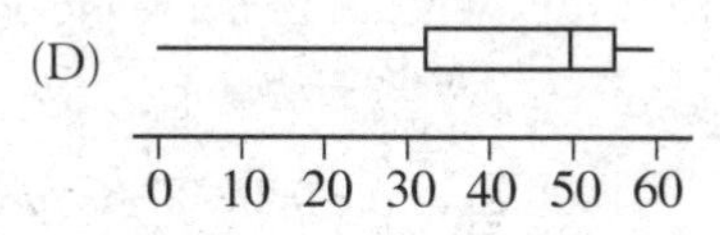

(B)
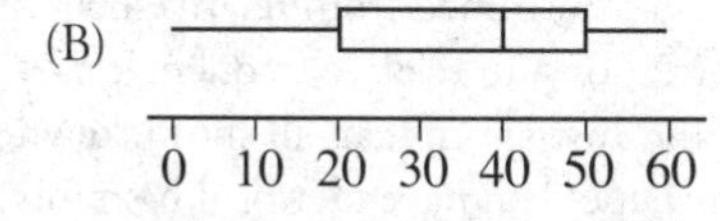

(E)
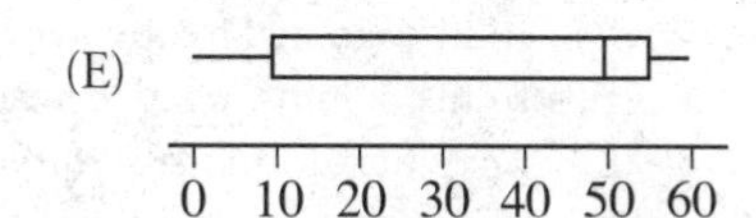

(C)
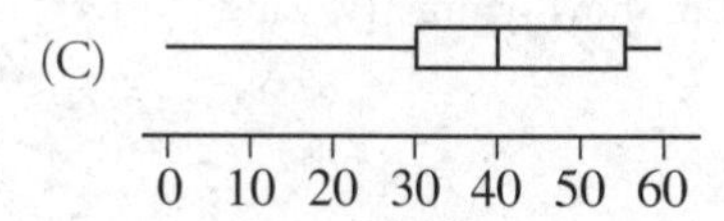

3. All students at a private middle school are required to sign up for one of the three clubs: chess, drama, or math. The graph below shows the distribution of participation by grade level.

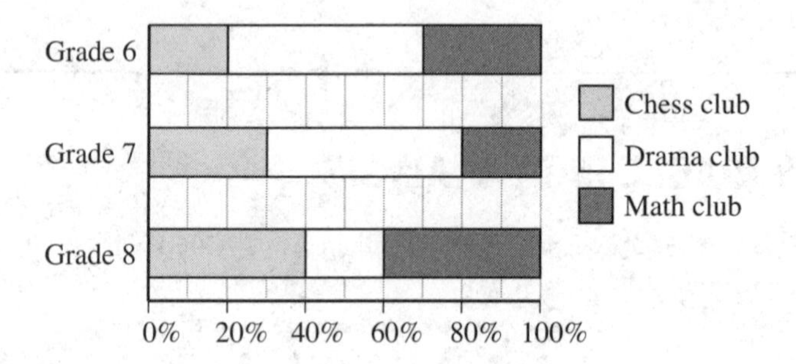

Which of the following statements is *incorrect*?

(A) 30 percent of the 6th graders joined the math club.
(B) The number of 8th graders who joined the chess club is the same as the number of 8th graders who joined the math club.
(C) The number of 6th graders who joined the drama club is the same as the number of 7th graders who joined the drama club.
(D) The percent of 8th graders who joined the math club is twice the percent of 7th graders who joined the math club.
(E) The percent of 6th graders who joined the chess club is the same as the percent of 8th graders who joined the drama club.

4. Which of the following is a true statement about stemplots?

(A) Stemplots are equally useful for small and for very large data sets.
(B) Stemplots are useful both for quantitative and for categorical data sets.
(C) Stemplots can show symmetry, gaps, clusters, and outliers.
(D) Stems should be skipped only if there is no data value for a particular stem.
(E) Whether or not to provide a key depends on the relative importance of the data.

5. Which of the following is *incorrect*?

(A) In histograms, frequencies can be determined from relative heights.
(B) In histograms, relative areas correspond to relative frequencies.
(C) Boxplots, dotplots, stemplots, and histograms can all show skewness.
(D) Sets with different distribution shapes can have identical boxplots.
(E) Both dotplots and stemplots can show symmetry, gaps, clusters, and outliers.

6. If three stations are selling gas for \$2.10 per gallon, six are selling gas for \$2.05 per gallon, and ten are selling gas for \$2.00 per gallon, what is the median price per gallon among these stations?

(A) \$2.00
(B) \$2.025
(C) \$2.0316
(D) \$2.05
(E) \$2.075

7. Suppose the average score of a national exam is 1,000 with a standard deviation of 200. If each score is increased by 50 and then each result is increased by 5 percent, what are the new mean and standard deviation?

(A) $\mu = 1{,}100$ and $\sigma = 200$
(B) $\mu = 1{,}100$ and $\sigma = 210$
(C) $\mu = 1{,}102.5$ and $\sigma = 200$
(D) $\mu = 1{,}102.5$ and $\sigma = 210$
(E) $\mu = 1{,}102.5$ and $\sigma = 212.5$

8. For which of the following variables would it be most appropriate to construct a histogram?

(A) Television brand
(B) Insect species
(C) Gender
(D) Eye color
(E) Phone call length

9. If the standard deviation of a set of observations is 0, you can conclude

(A) that a mistake in arithmetic has been made
(B) that there is no relationship among the observations
(C) that the average value is 0
(D) that all observations are the same value
(E) none of the above

10. **To which of the histograms below can the following boxplot correspond?**

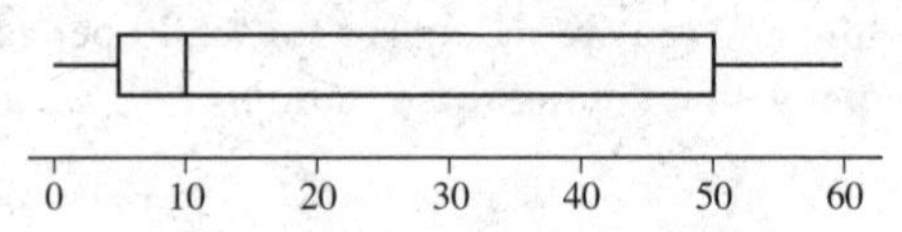

(A)

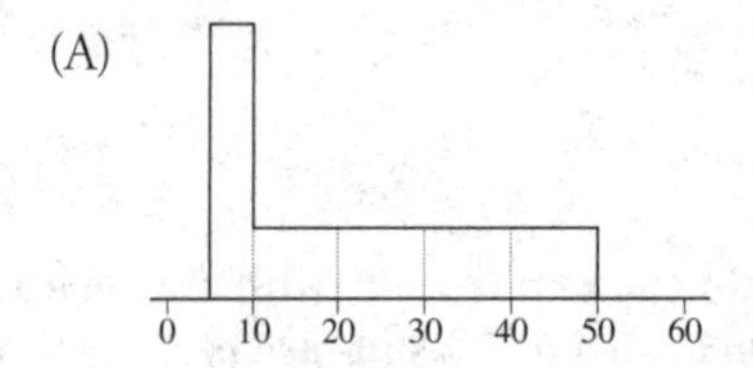

(D)

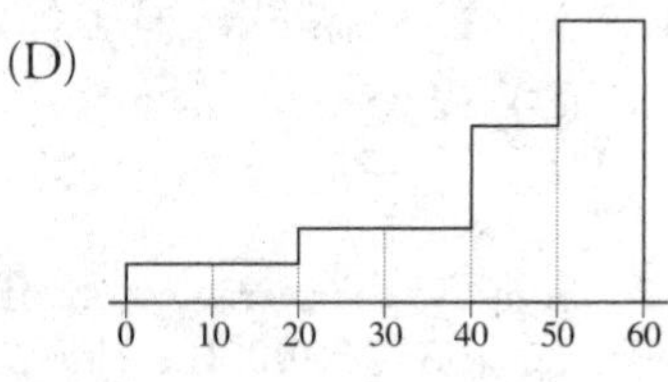

(B)

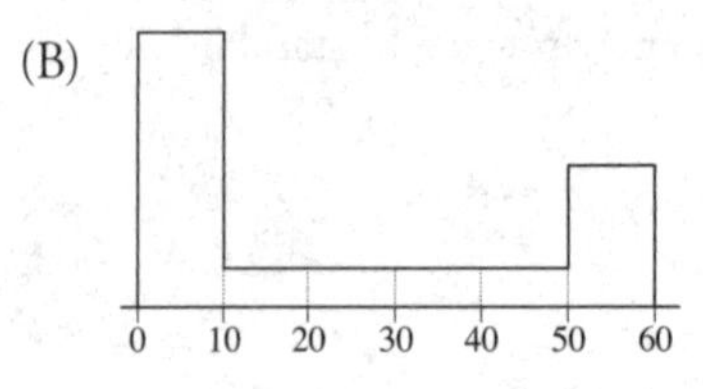

(E)

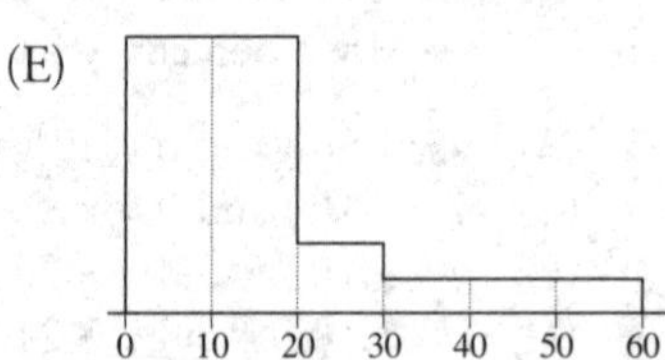

(C)

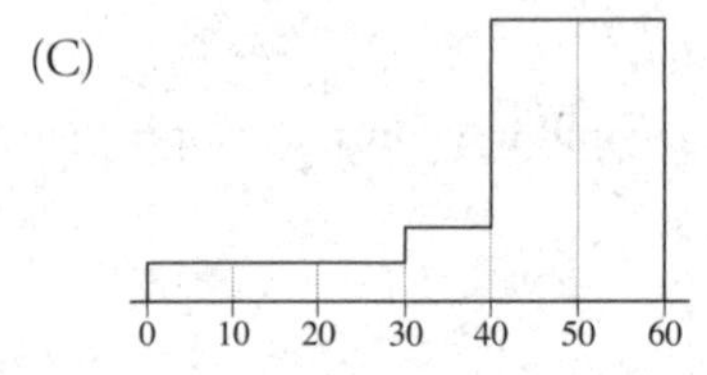

11. **Given that the median of a set is 350 and the interquartile range is 50, which of the following statements is true?**

(A) Fifty percent of the data are less than or equal to 350.
(B) Fifty percent of the data are between 325 and 375.
(C) Seventy-five percent of the data are less than or equal to 375.
(D) The mean is between 325 and 375.
(E) All of the above statements are true.

12. **The 15 tenure track professors in a university math department are paid an average of $90,000 per year, while the 10 adjunct professors are paid an average of $60,000 per year. What is the average salary among all these math professors?**

(A) $72,000
(B) $75,000
(C) $78,000
(D) $81,000
(E) $84,000

13. Which of the following statements about histograms is true?

(A) Two students working with the same set of data may come up with histograms that look different.
(B) Histograms of categorical variables can pinpoint clusters and gaps.
(C) Unlike other graphs, histogram axes do not need to be labeled.
(D) Displaying outliers is less problematic when using histograms than when using stemplots.
(E) Histograms are more widely used than stemplots or dotplots because histograms display the values of individual observations.

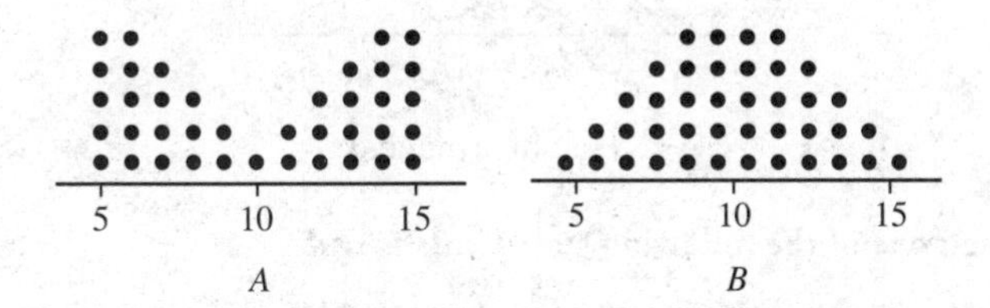

14. Which of the following statements about the above two dotplots is true?

(A) The mean of set *A* looks to be greater than the mean of set *B*.
(B) The mean of set *B* looks to be greater than the mean of set *A*.
(C) Both sets have roughly the same variance.
(D) The empirical rule applies only to set *A*.
(E) The standard deviation of set *B* is less than 4.

15. Students who apply for graduate school usually must take the GRE exam. The following boxplots were constructed from GRE quantitative scores (which take values between 130 and 170) of men and women at a state college.

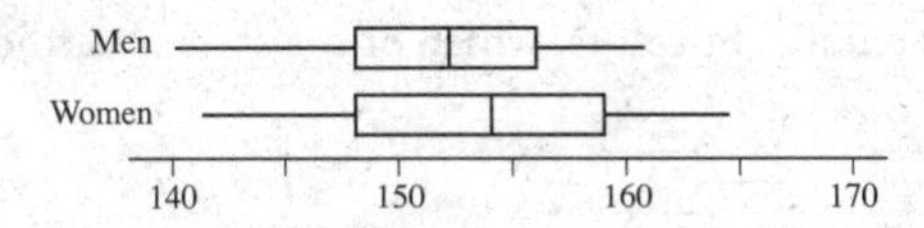

Which of the following is a possible boxplot for the combined score of all the students?

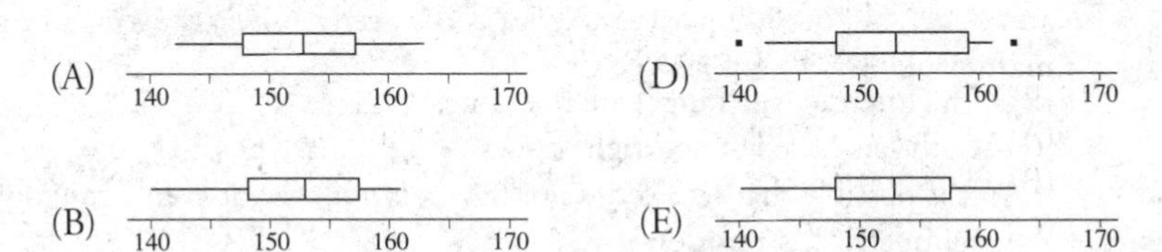

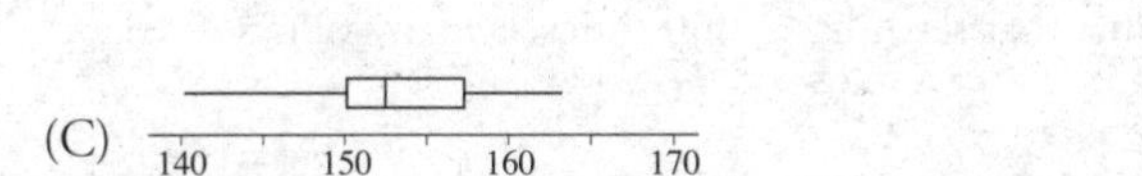

16. **The histograms below give the heights of players in two amateur basketball leagues.**

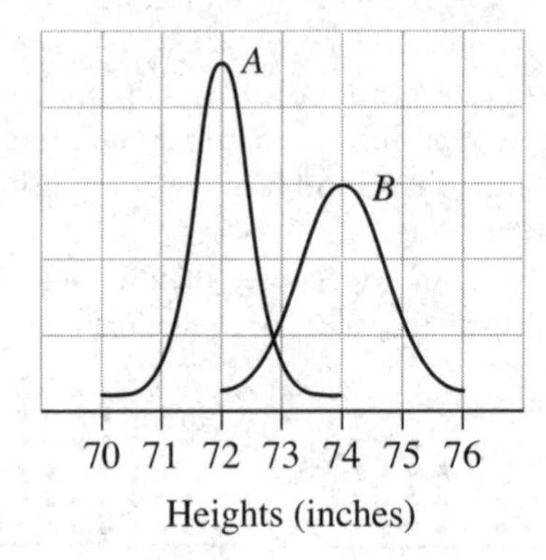

Which one of the following statements is true?

(A) The distribution of heights in League *A* has a lower mean and lower range than the distribution of heights in League *B*.
(B) The distribution of heights in League *A* has a lower mean and approximately the same range as the distribution of heights in League *B*.
(C) The distribution of heights in League *A* has a higher mean and lower range than the distribution of heights in League *B*.
(D) The distribution of heights in League *A* has a higher mean and higher range than the distribution of heights in League *B*.
(E) The distribution of heights in League *A* has a higher mean and approximately the same range as the distribution of heights in League *B*.

17. **Given these parallel boxplots, which of the statements below is *incorrect*?**

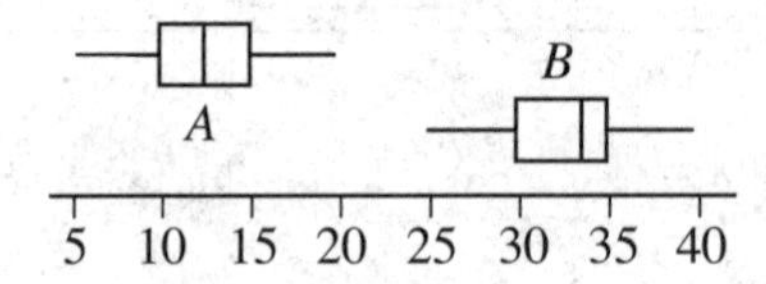

(A) The ranges are the same.
(B) The interquartile ranges are the same.
(C) Both sets have left and right skews.
(D) It is possible that set *A* does not have a symmetric or even a roughly symmetric distribution.
(E) It is possible that set *A* has 1,000 times as many values as set *B*.

18. Suppose the scores on an exam have a mean of 80 with a standard deviation of 5. If one student has a test result with a *z*-score of –1.4 and a second student has a test result with a *z*-score of 2.6, how many points higher was the second student's result than that of the first?

(A) 6
(B) 7
(C) 13
(D) 19
(E) 20

19. Each of the high school students in a random sample was given five minutes to solve as many short puzzles as possible. The results by gender are given in the dotplots below.

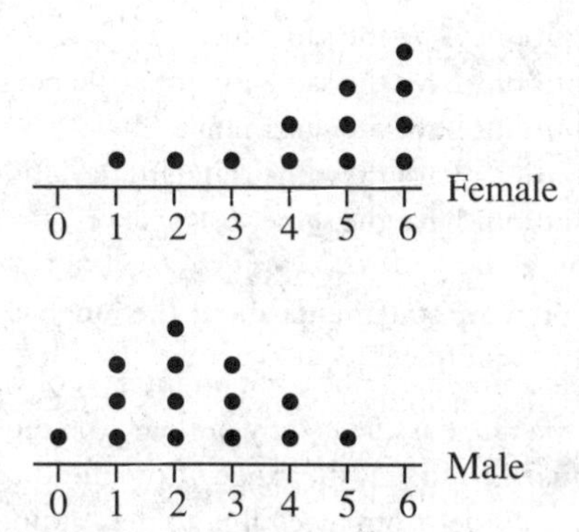

Which of the following is a true statement?

(A) One distribution is roughly symmetric; the other is skewed right.
(B) The difference in their means is less than the difference in their medians.
(C) The ranges are both 6.
(D) The standard deviations are roughly the same.
(E) The range of the entire set of results is equal to the sum of the individual ranges of the two gender specific sets.

20. A business magazine plans to compare the entry-level salaries offered to applicants with and without an MBA degree. Which of the following graphical displays is *inappropriate*?

(A) Side-by-side histograms
(B) Back-to-back stemplot
(C) Parallel boxplots
(D) Scatterplot
(E) All of the above displays are appropriate.

21. Given this back-to-back stemplot, which of the following statements is *incorrect*?

A		B
1 1	35	
6 4 3	36	1 1
7 2	37	3 4 6
9 5 2 1	38	2 7
7 1 0	39	1 2 5 9
6	40	0 1 7
	41	6

(41|6 means 416)

(A) The distributions have the same mean.
(B) The distributions have the same interquartile range.
(C) The distributions have the same range.
(D) The distributions have the same standard deviation.
(E) The distributions have the same variance.

22. Which of the following statements about the range is *incorrect*?

(A) The range is a single number, not an interval of values.
(B) Although the range is affected by outliers, the interquartile range is not.
(C) The interquartile range is the range of the middle half of the data.
(D) Changing the order from ascending to descending changes the sign of the range.
(E) The range of the sample data can never be greater than the range of the population.

23. A college statistics professor gives only two exams during the semester, a midterm and a final. The professor grades with *z*-scores. On the final exam, a student received a grade of –2.8. What is the correct interpretation of this grade?

(A) The student scored 2.8 points lower on the final than on the midterm.
(B) The student scored 2.8 points lower than the class average on the final.
(C) The student scored 2.8 standard deviations lower on the final than on the midterm.
(D) The student scored 2.8 standard deviations lower on the final than the class average on the midterm exam.
(E) The student scored 2.8 standard deviations lower on the final than the class average on the final exam.

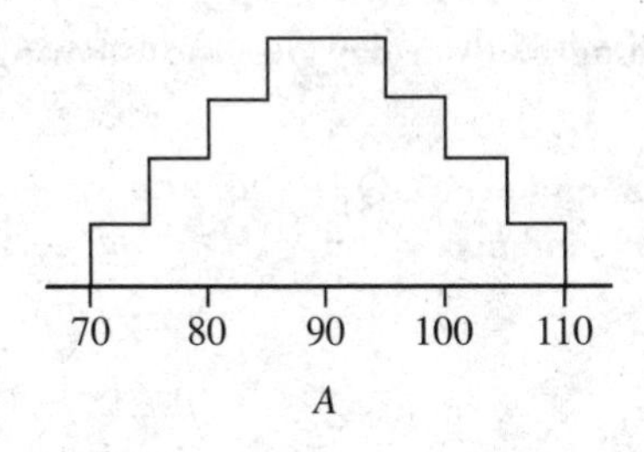

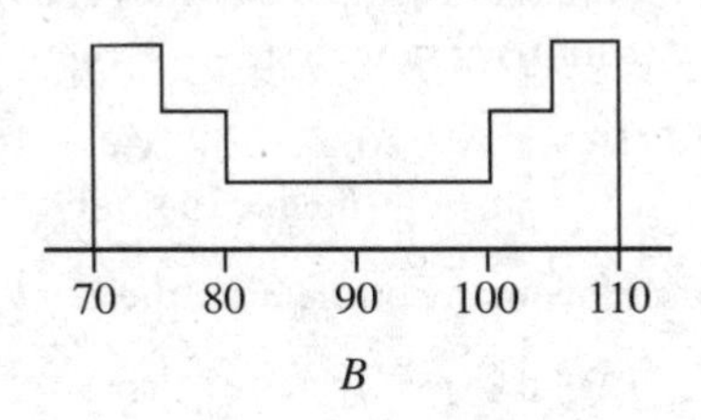

24. Given the two histograms above, which of the following statements is *incorrect*?

(A) As long as the sample sizes are at least 30, the empirical rule applies to both sets.
(B) Both sets have roughly the same ranges.
(C) The standard deviation of set A is greater than 5.
(D) The standard deviation of set B is greater than 5.
(E) A lower percentage of the values in set A are at least 10 units from its mean than is the case for set B.

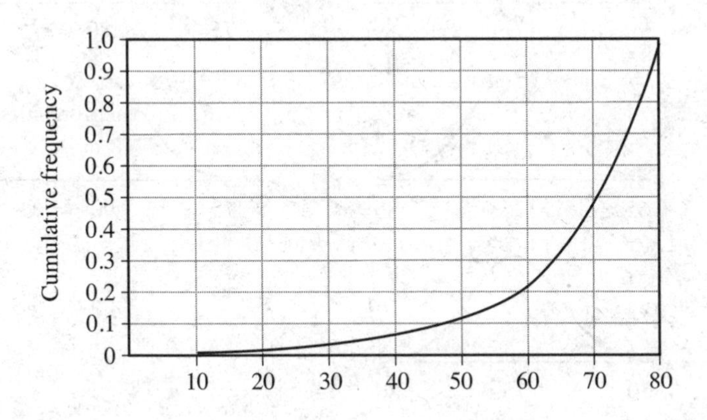

25. Given the cumulative frequency plot above, what is the value of the mean?

(A) The mean is between 75 and 80.
(B) The mean is between 70 and 75.
(C) The mean equals 70.
(D) The mean is less than 70.
(E) From such a plot, one can determine the median, but nothing can be concluded about the mean.

26. A random sample of scores at a bowling alley one day gives the following summary statistics:

$$n = 26, \bar{x} = 132.34, s = 10.18, \text{min} = 113, Q_1 = 126,$$
$$\text{med} = 134.5, Q_3 = 139, \text{and max} = 160$$

How many outliers are there?

(A) 0
(B) 1
(C) 2
(D) At least 1
(E) At least 2

27. Which of the following distributions has a mean of 75 and a standard deviation of approximately 5?

(A)
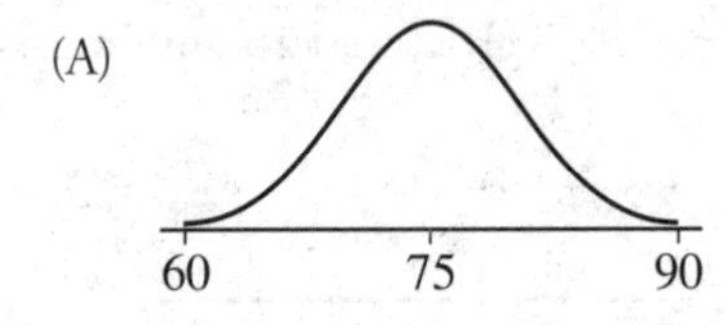

(B)
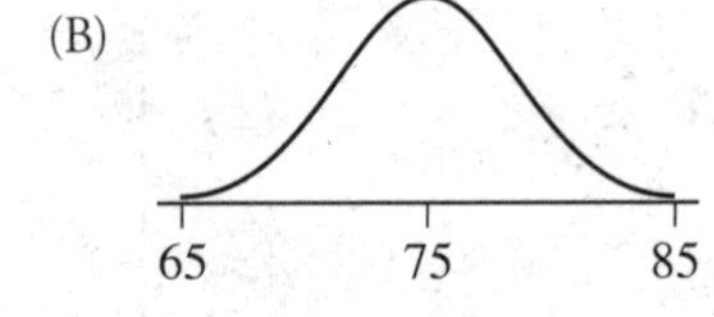

(C)
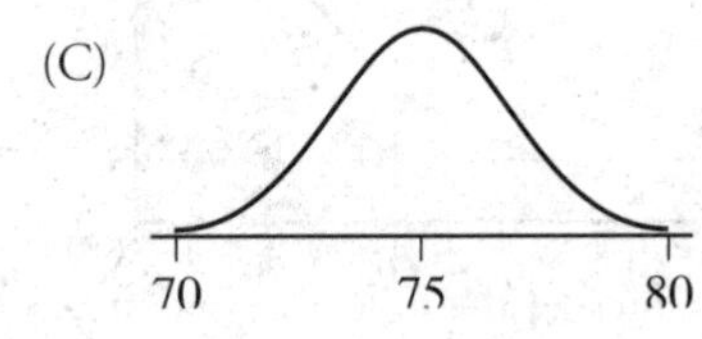

(D)
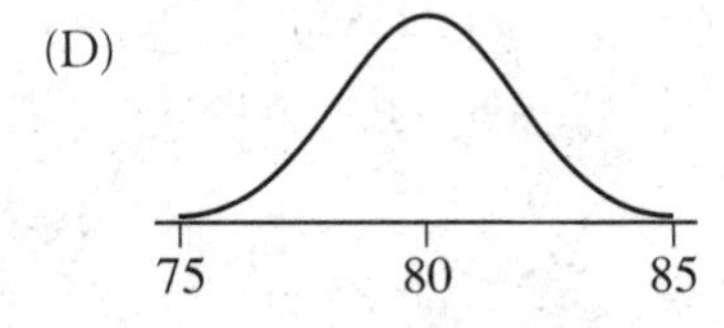

(E)
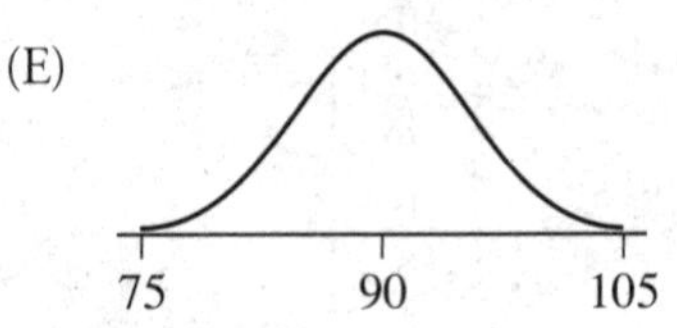

28. Assuming that weights of sumo wrestlers follow a bell-shaped distribution, how are the weights of the following wrestlers arranged in ascending order?

I. A weight with a *z*-score of 1
II. A weight with a percentile rank of 80 percent
III. A weight at the third quartile

(A) I, II, III
(B) I, III, II
(C) II, I, III
(D) III, I, II
(E) III, II, I

29. A set has 10 outliers. If every value of this set is increased by 5 percent, how many outliers will there now be?

(A) Fewer than 10
(B) 10
(C) 11
(D) More than 11
(E) It is impossible to determine without further information.

30. The distribution of the cost of a family's dine-out meal has a mean of $55.30 and a standard deviation of $4.00. If they always tip $5 plus 10 percent of the meal, what are the mean and standard deviation of the distribution of tips?

(A) Mean = $5.00 and standard deviation = $0.40
(B) Mean = $5.40 and standard deviation = $4.40
(C) Mean = $5.53 and standard deviation = $0.40
(D) Mean = $10.53 and standard deviation = $0.40
(E) Mean = $10.53 and standard deviation = $4.40

31. Male elk living in the wild have a mean weight of 720 pounds with a standard deviation of 32 pounds. Female elk living in the wild have a mean weight of 515 pounds with a standard deviation of 20 pounds. What is the weight of a male elk with the same standardized score (*z*-score) as a female elk with a weight of 500 pounds?

(A) 696 pounds
(B) 704 pounds
(C) 705 pounds
(D) 735 pounds
(E) 752 pounds

32. A psychology student measures the times each of five mice takes to solve a maze and reach the food reward. Which of the following statistics are probably most useful in describing the distribution of the five times?

(A) Mean and standard deviation
(B) Mean and variance
(C) Mean and interquartile range
(D) Median and standard deviation
(E) Median and interquartile range

33. Which of the following is a true statement?

(A) All symmetric bell-shaped curves are normal.
(B) All symmetric histograms have single peaks.
(C) All normal curves are bell-shaped and symmetric.
(D) If the mean equals the median, the distribution is symmetric.
(E) None of the above are true statements.

34. The age distribution of 343 students who attended a hip-hop concert is shown below.

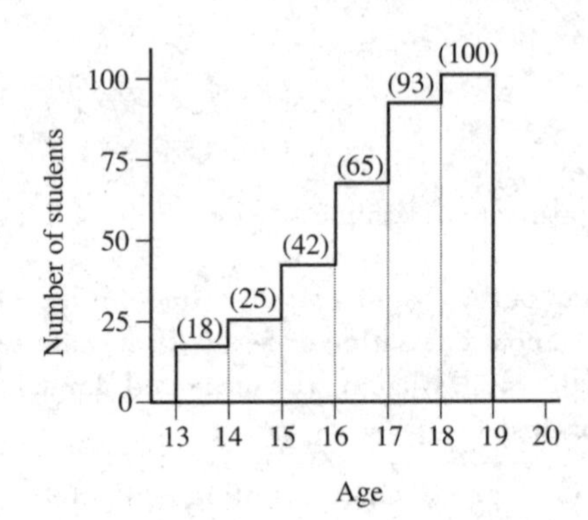

Which of the following could be the median age?

(A) 15.8
(B) 16.5
(C) 17.1
(D) 18.2
(E) 18.5

35. Given this cumulative plot of the number of Instagram pictures sent in a week by the students in an AP Art class and using the most commonly accepted definition of outliers, what numbers of Instagram pictures are considered outliers?

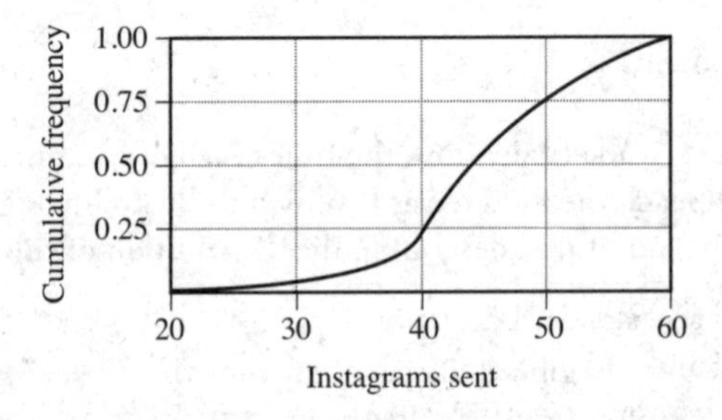

(A) Between 20 and 25
(B) Between 20 and 30
(C) Between 20 and 40
(D) Between 20 and 25, or between 55 and 60
(E) Between 20 and 30, or between 50 and 60

36. If a distribution is perfectly symmetric, which of the following must be true?

(A) The distribution is uniform.
(B) The mean and median are equal.
(C) The interquartile range is equal to twice the standard deviation.
(D) The range is twice the interquartile range.
(E) The range is equal to six times the standard deviation.

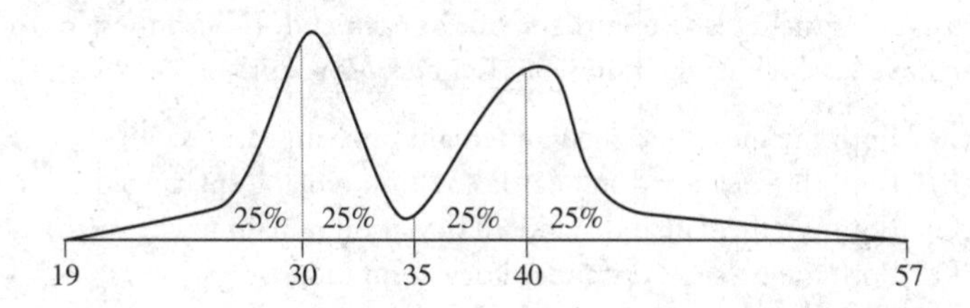

37. Given the histogram above, what values would be considered outliers?

(A) Values between 51 and 57
(B) Values between 55 and 57
(C) Values between 19 and 20, and values between 50 and 57
(D) Values between 19 and 25, and values between 45 and 57
(E) No value is an outlier.

38. On a large campus, 500 students were surveyed about their belief in an afterlife. The following table shows their responses categorized by class.

	Believe in Afterlife	**Do Not Believe in Afterlife**
Freshmen	95	30
Sophomores	87	29
Juniors	76	51
Seniors	35	97

Of the following graphical displays, which is the most appropriate to compare the proportions of students in each class who believe in an afterlife?

(A) Back-to-back stemplot
(B) Parallel boxplots
(C) Scatterplot
(D) Segmented bar chart
(E) Cumulative frequency plot

39. If quartiles $Q_1 = 25$ and $Q_3 = 45$, which of the following must be true?

(A) The median is 35.
(B) The mean is between 25 and 45.
(C) The standard deviation is at most 20.
(D) The range must be greater than 20.
(E) None must be true.

40. A data set includes two outliers, one at each end. If both these outliers are removed, which of the following is a *possible* result?

(A) Both the mean and median remain unchanged.
(B) Both the mean and standard deviation remain unchanged.
(C) Both the median and standard deviation remain unchanged.
(D) Both the mean and standard deviation increase.
(E) Both the median and standard deviation decrease.

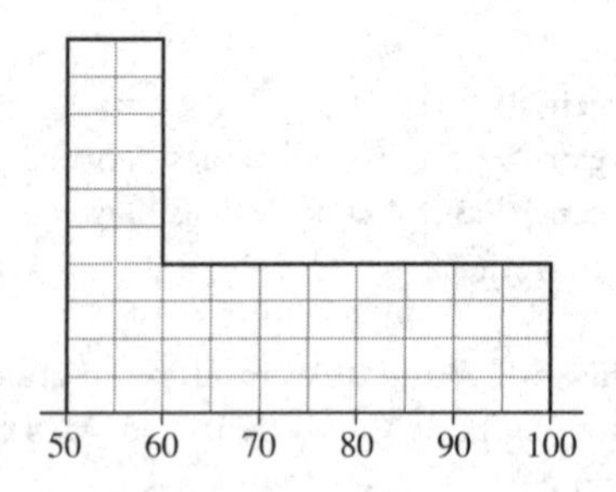

41. The histogram shown above is of 52 test scores. Which of following statements is true?

(A) The median score was 75.
(B) If 60 and below is failing, most students failed.
(C) More students scored above 80 than below 60.
(D) More students scored below the median than above the median.
(E) The mean score is probably greater than the median score.

42. The following parallel boxplots show the birth weights during one month at four hospitals.

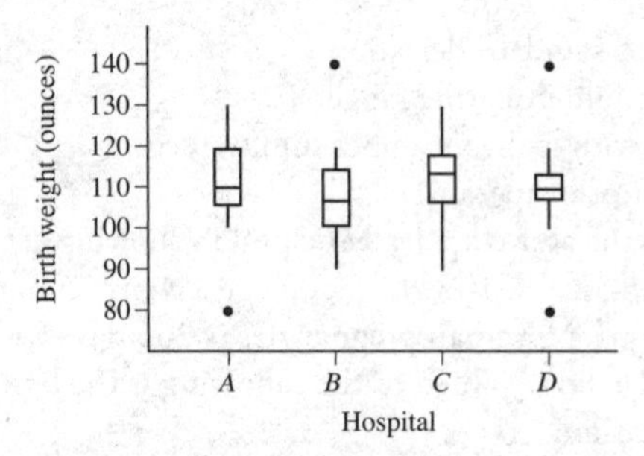

Average birth weight is considered to be seven pounds (112 ounces). In which hospital were a greater number of newborns born weighing between 102 and 122 ounces?

(A) *A*
(B) *B*
(C) *C*
(D) *D*
(E) There is insufficient information to answer.

43. Counts of the numbers of spam messages received by a company's executives during one week are summarized below.

Values Under Q_1	Q_1	Median	Q_3	Values Over Q_3
7, 9, 10	15.5	20	21.5	28, 30, 31

How many outliers are there?

(A) 0
(B) 1
(C) 2
(D) 4
(E) 6

44. Which of the following is a true statement?

(A) If the mean and median of a set are equal, then the variance is zero.
(B) If a set has variance zero, then the mean, median, and standard deviation are equal.
(C) If the population mean and median are equal, then this will also be true for any sample that is randomly selected from the population.
(D) If the population has variance zero, then the variance of any sample is also zero.
(E) If the variance of a randomly selected sample is zero, then the population mean and median are equal.

45. **When there are multiple gaps and clusters, which of the following is the best choice to give an overall picture of the distribution?**

(A) Mean and standard deviation
(B) Mean and interquartile range
(C) Boxplot with its five-number summary
(D) Stemplot or histogram
(E) None of the above are really helpful in showing gaps and clusters.

46. **The mean length of human pregnancies is 266 days with a standard deviation of 16 days. Which of the following is the best description of the standard deviation?**

(A) Approximately the median time between the lengths of individual pregnancies and 266 days
(B) Approximately the mean time between the lengths of individual pregnancies and 266 days
(C) Approximately the square root of the mean time between the lengths of individual pregnancies and 266 days
(D) One-third the time between the maximum individual length of pregnancies and 266
(E) One-sixth the time between the maximum and minimum individual lengths of pregnancies

47. **In which of the following histograms is the mean greater than the median?**

(A)
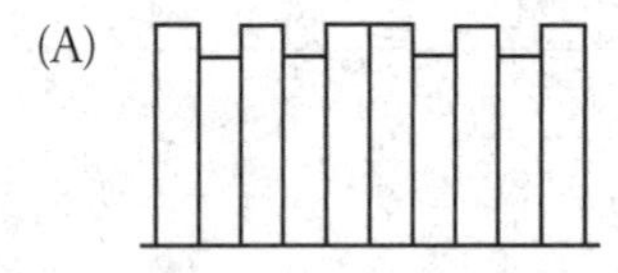

(D)
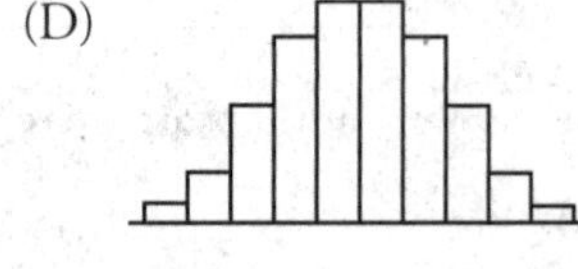

(B)
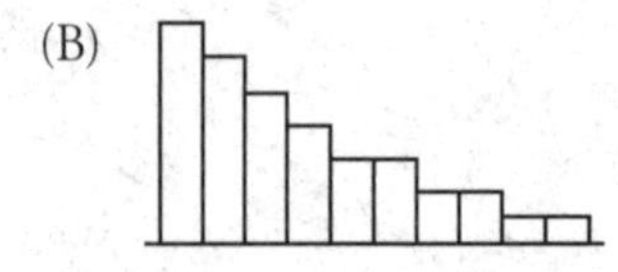

(E)
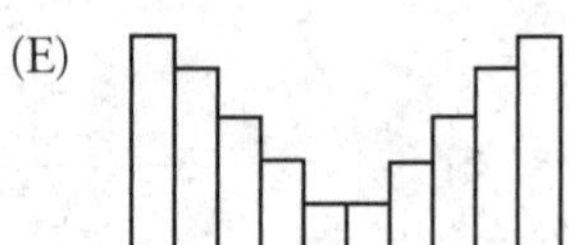

(C)

48. **Ms. D teaches three sections of AP Statistics with a total of 77 students. Her students did exceptionally well on the final exam. Some summary statistics and a histogram of the exam scores are shown below.**

Mean	Standard Deviation	Q_1	Q_3
93.4	2.6	92	95

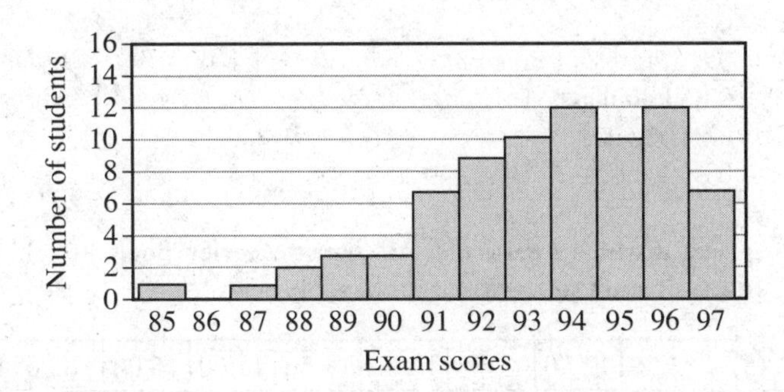

Using the standard definition of outlier, which of the following can be concluded?

(A) The mean is greater than the median, and there is only one outlier.
(B) The mean is greater than the median, and there are two outliers.
(C) The mean is less than the median, and there are no outliers.
(D) The mean is less than the median, and there is only one outlier.
(E) The mean is less than the median, and there are two outliers.

49. **The students from each of two schools were asked how many Nobel Peace Prize winners they could name. The results are shown in the following two histograms.**

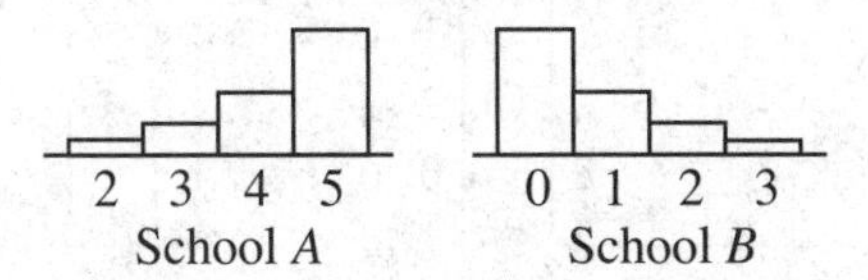

Which school has the greater median with regard to the number of Nobel Peace Prize winners students could name, and which school has the greater mean?

(A) Greater median: *A* Greater mean: *A*
(B) Greater median: *A* Greater mean: *B*
(C) Greater median: *B* Greater mean: *A*
(D) Greater median: *B* Greater mean: *B*
(E) Greater median: *B* Equal means

50. In 1999, NASA lost the $125 million Mars Climate Orbiter spacecraft after a 286-day journey to Mars. Miscalculations due to the use of English units instead of metric units apparently sent the craft off course. Suppose a particular measurement from a set of measurements in miles has a *z*-score of 0.85. If all the measurements are converted to kilometers, what will now be the *z*-score of the particular measurement? (Remember that 1 kilometer equals 0.621 miles.)

(A) 0.85
(B) 0.85 miles
(C) 0.85 kilometers
(D) (0.85)(0.621)
(E) 0.85 + 0.621

51. The prices at which a particular AP Statistics review book are sold at 92 online sites are summarized in the table below.

Price ($)	12.25	12.50	12.75	13.00	13.50	14.00	15.00	16.00	19.00
Frequency	5	11	24	21	13	10	5	2	1

Which of the following best describes the distribution of prices?

(A) Roughly normal
(B) Skewed left
(C) Skewed right
(D) Skewed both left and right
(E) Bimodal

A		*B*
7 5 5 3 2	50	2 7
9 6 5 2 0 0	51	0 3 9
7 4 1	52	1 2 7 8
6 2	53	2 3 5 9
8	54	0 6 6
3	55	1 4

55|1 = 551

52. Given the back-to-back stemplot above, which of the following statements is true?

(A) The empirical rule applies to both sets *A* and *B*.
(B) The median of each set is approximately $\frac{500 + 550}{2}$.
(C) The ranges of the two sets are equal.
(D) The variances of the two sets are approximately the same.
(E) In one set, the mean and median should be about the same. In the other set, the mean appears to be greater than the median.

53. The following histogram displays the top speeds of several roller coasters.

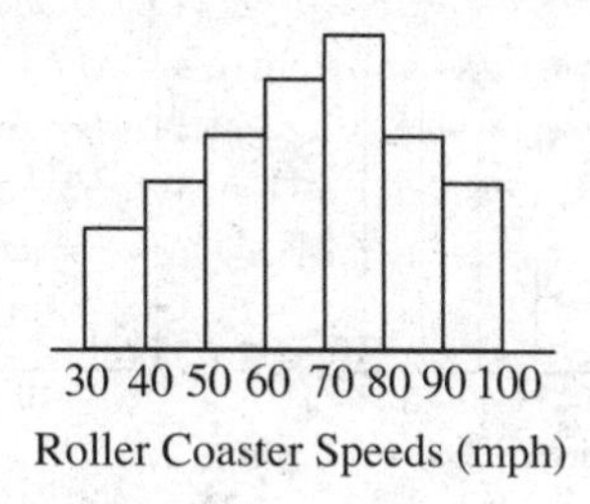

Which of the following statements must be true?

(A) The minimum and maximum top speeds are 30 mph and 100 mph, respectively.
(B) The median top speed is 60.5 mph.
(C) At least 100 more roller coasters have top speeds between 70 and 80 mph than between 30 and 40 mph.
(D) More roller coasters have top speeds between 70 and 80 then have top speeds over 80 mph.
(E) The same number of roller coasters have top speeds between 50 and 60 mph as have top speeds between 80 and 90 mph.

54. Summary statistics for the number of campsites at public parks in a large western state are as follows.

Minimum	Q_1	Median	Q_3	Maximum	Mean	Standard Deviation
0	35	61	95	300	73.2	58

A check of the data shows that 5 had been subtracted from every value to bring the minimum to 0 for some calculation purpose. If 5 is added back to every value, what will now be the interquartile range and the standard deviation?

(A) IQR = 60 and SD = 58
(B) IQR = 60 and SD = 63
(C) IQR = 65 and SD = 58
(D) IQR = 65 and SD = 63
(E) IQR = 62.5 and SD = 60.5

55. Below are histograms of two data sets.

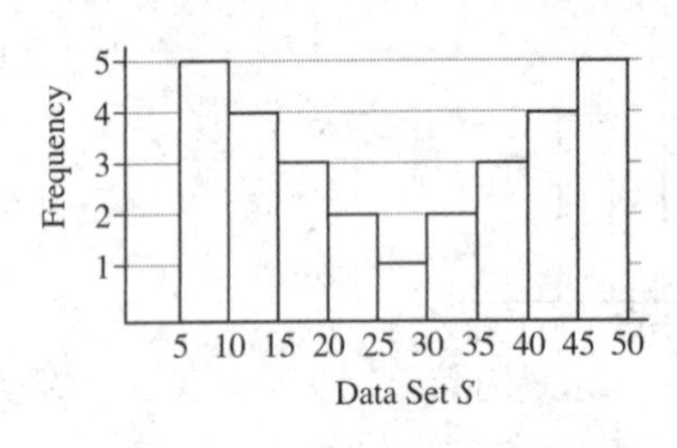

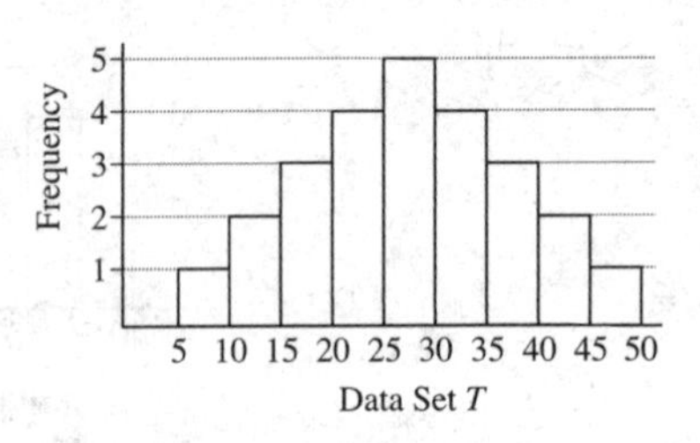

Which of the following statements *must* be true?

(A) The medians of the two data sets must be equal.
(B) The means of the two data sets must be equal.
(C) The ranges of the two data sets must be equal.
(D) The standard deviation of data set S is less than the standard deviation of data set T.
(E) The standard deviation of data set S is greater than the standard deviation of data set T.

56. A study of phone, e-mail, and texting communication in three time periods results in the following segmented bar chart.

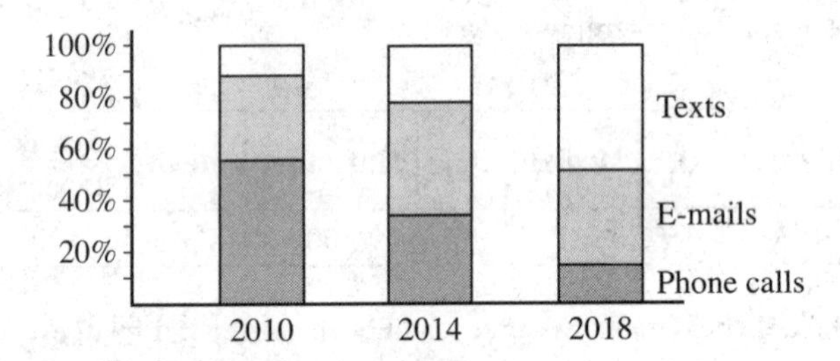

Which of the following is greatest?

(A) The number of texts and e-mails in 2010.
(B) The number of e-mails in 2014.
(C) The number of texts in 2018.
(D) The above are all equal.
(E) It is impossible to answer this question without knowing the actual numbers of individual communications involved.

57. **A normal distribution with mean 25 and standard deviation 5 is standardized by subtracting the mean from each value and dividing the difference by the standard deviation. What are the mean and standard deviation of the resulting distribution?**

(A) Mean = 0, Standard deviation = 1
(B) Mean = 5, Standard deviation = 1
(C) Mean = 5, Standard deviation = 5
(D) Mean = 25, Standard deviation = 1
(E) Mean = 25, Standard deviation = 5

58. **A barber measures hair lengths for a random sample of 100 of his customers and gathers data on the growth rates of their hair. The data are plotted in the histogram below.**

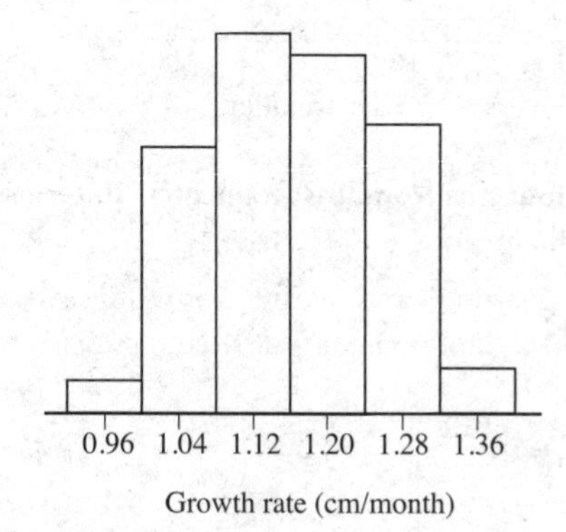

Which of the values below is closest to the standard deviation of the 100 values?

(A) 0.008
(B) 0.01
(C) 0.03
(D) 0.08
(E) 1.16

59. **A teacher had two AP Statistics classes. The mean score on the AP Exam for one class was 3.2, while the mean score for the other class, which had more students, was 3.8. What must be true about the mean score of all of the teacher's AP students?**

(A) The mean score for all students must be 3.5.
(B) It's possible that the mean score of all students is 3.5, but it doesn't have to be.
(C) The mean score for all students can be anything between 3.2 and 3.8.
(D) The mean score for all students is less than 3.5 but greater than 3.2.
(E) The mean score for all students is greater than 3.5 but less than 3.8.

60. **Consider the three distributions below.**

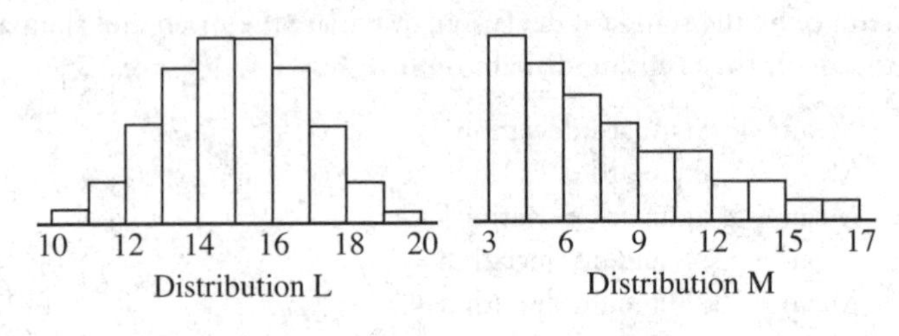

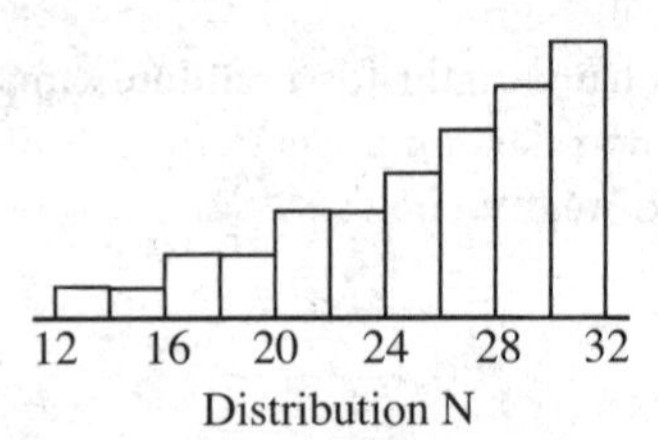

Order the distributions from least to greatest in terms of the proportion of data less than the mean.

(A) *M, L, N*
(B) *N, L, M*
(C) *M, N, L*
(D) *N, M, L*
(E) *L, N, M*

61. **The following statistics summarize the fuel economy in miles per gallon (mpg) for 50 different car models in 2018.**

Minimum	Q_1	Median	Q_3	Maximum	Mean	Standard Deviation
13.5	21	24.5	30	48.5	26	7.32

Which of the following is a true statement?

(A) 50 percent of the mpg values are between 26 and 48.5.
(B) 25 percent of the mpg values are above 21.
(C) The distribution of mpg values appears to be roughly bell-shaped.
(D) The interquartile range of mpg values is 5.
(E) There is at least one outlier.

62. College students' expenses for academic supplies have a mean of $1,100 and a standard deviation of $290. Their expenses for entertainment have a mean of $950 and a standard deviation of $250. Assuming the expenses for academic supplies and for entertainment are independent, what are the mean and standard deviation of the total yearly expenditures by college students for these two items?

(A) Mean = $1,025 and standard deviation = $270
(B) Mean = $1,025 and standard deviation = $383
(C) Mean = $2,050 and standard deviation = $270
(D) Mean = $2,050 and standard deviation = $383
(E) Mean = $2,050 and standard deviation = $540

63. Random samples are taken from three populations, and the resulting parallel boxplots of the data follows.

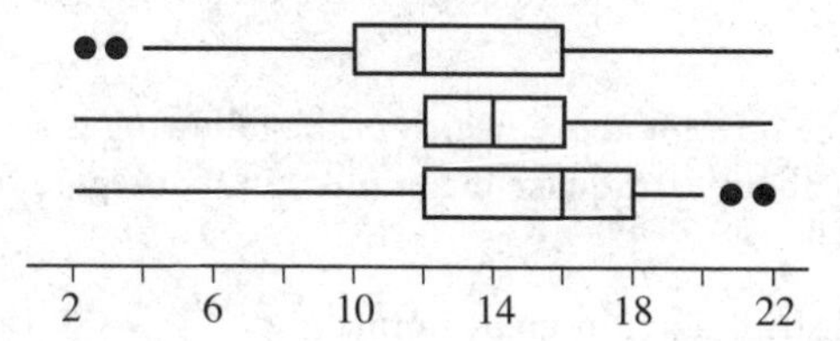

Which of the following statements is true?

(A) One of the sample distributions is roughly a normal distribution.
(B) All three sample distributions have the same interquartile range.
(C) All three sample distributions have the same range.
(D) All three sample medians are between 11 and 15.
(E) All three of the distributions can reasonably be assumed to be of samples from populations with roughly normal distributions.

64. A histogram of the educational level (in number of years of schooling) of the adult population of the United States would probably have which of the following characteristics?

(A) A roughly normal population
(B) Skewness to the right
(C) Symmetry
(D) A gap around 12 years
(E) Clusters around 8, 12, and 16 years

65. When a set of data has suspect outliers, which of the following are preferred measures of central tendency and variability?

(A) Mean and standard deviation
(B) Mean and variance
(C) Mean and range
(D) Median and range
(E) Median and interquartile range

66. Data were collected as to the magnitudes of earthquakes in a certain region over a long time period.

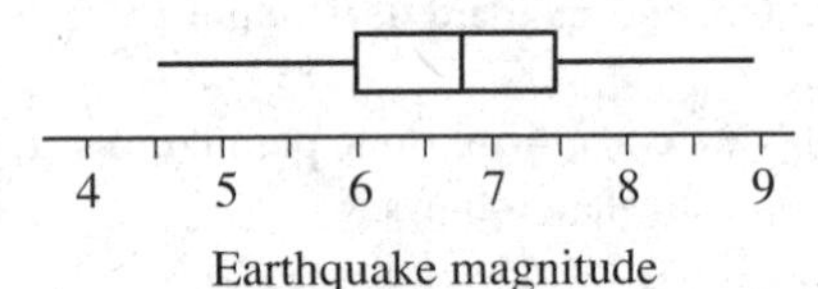

Based on the boxplot above, which of the following is a reasonable conclusion about earthquake magnitudes in that region during that time period?

(A) The distribution is roughly normal.
(B) The distribution is symmetric.
(C) There were a greater number of earthquakes with magnitudes below 6 than between 6 and 6.75.
(D) There were the same number of earthquakes between the median and the third quartile magnitudes as there were between the third quartile and the maximum magnitudes.
(E) Over half the earthquakes had magnitudes between 6 and 7.5.

67. Suppose the height distribution of male college basketball players is approximately normal and varies from 71 to 81 inches, while the height distribution of female college basketball players is approximately normal and varies from 65 to 72 inches. A random stratified sample of 500 male and 500 female college basketball players is then combined into one large sample of size 1,000. What is the distribution of this sample?

(A) Roughly normal, varying from 65 to 81 inches and centered at 71.5 inches
(B) Roughly uniform
(C) Unimodal and symmetric but not normal
(D) Skewed
(E) Bimodal

***68. A company reports mean salaries cross-classified by position and gender.**

	Male	**Female**
Junior Executives	$45,000	$48,000
Senior Executives	$93,000	$95,000
Overall	?	?

Is it possible that males have an overall higher mean salary than females?

(A) No, because $\frac{45{,}000 + 93{,}000}{2} < \frac{48{,}000 + 95{,}000}{2}$

(B) No, because female junior executives have a higher mean salary than male junior executives *and* female senior executives have a higher mean salary than male senior executives

(C) No, because $n(48{,}000 - 45{,}000) + m(95{,}000 - 93{,}000) > 0$ for all n and m

(D) Yes, depending on the number of males and females who are junior and senior executives

(E) Yes, because salaries are a numerical, not a categorical, variable

TWO-VARIABLE DATA ANALYSIS

69. Large student debt is a growing problem for college graduates, with many graduates seemingly unable ever to get out of debt. Suppose the average college debt for graduates is $120,000 with a standard deviation of $40,000, and suppose the average stress level of college graduates on a 1–100 scale is 30 with a standard deviation of 10. If the correlation between the stress level and college debt is 0.6, what is the least squares linear regression equation for stress level among college graduates based on their amount of debt (in $1,000)?

(A) Predicted stress level = 12 + 0.15(College debt in $1,000)

(B) Predicted stress level = 113.5 + 0.15(College debt in $1,000)

(C) Predicted stress level = 30 + 2.4(College debt in $1,000)

(D) Predicted stress level = 0.25(College debt in $1,000)

(E) Predicted stress level = 28.2 + 0.015(College debt in $1,000)

70. **Suppose every student who attended Saturday review classes scored exactly 10 points higher on a final exam than he/she did on a midterm exam. What would be the correlation between scores on the final exam and scores on the midterm exam for students attending the review classes?**

(A) 0
(B) Somewhat positive
(C) 0.5
(D) Nearly 1
(E) Exactly 1

71. **Paper lengths and final grades are obtained from a random sample of term papers handed in to a particular high school English teacher. The resulting regression equation is shown.**

$$\widehat{\text{Grade}} = 52.34 + 1.24(\text{Length}) \text{ with } r = 0.18$$

What percentage of the variation in final grades can be explained by the linear regression model of Grade on Length?

(A) 0.24 percent
(B) 1.24 percent
(C) 3.24 percent
(D) 18 percent
(E) 24 percent

72. **In a study of winning percentages in home games versus average home attendance for Big Ten football teams, the resulting regression line is shown in the following equation.**

$$\widehat{\text{Winning }\%} = 39 + 0.00025(\text{Attendance})$$

What is the residual if a team has a winning percentage of 55% with an average attendance of 75,000?

(A) –3.025
(B) –2.75
(C) 2.75
(D) 3.025
(E) 57.75

73. **Which of the following is an *incorrect* statement about the least square regression line?**

(A) It always passes through the point $(\bar{x}, \bar{y})$.
(B) The slope always has the same sign as the correlation.
(C) It minimizes the sum of the squares of the residuals.
(D) The slope is always between –1 and +1.
(E) The correlation of y versus x is the same as the correlation of x versus y.

74. **A linear regression analysis relating teachers' salaries to years of experience yields:**

$$\hat{y} = 37.15 + 2.405x$$

where x is years of experience and y is salary (in \$1,000). Which of the following is the most proper conclusion?

(A) A starting teacher will earn \$37,150, while one with 70 years of experience should earn \$205,500.
(B) Starting teachers average \$37,150 with bonuses of \$2,405 every year.
(C) There is a cause-and-effect relationship between teachers' salaries and experience with each extra year of experience corresponding to an extra \$2,405 in salary.
(D) Starting salaries for teachers average \$37,150, and each year of experience is associated with an average increase of \$2,405.
(E) There is a high correlation between teachers' salaries and years of experience.

75. **What is the correct regression output for the scatterplot below?**

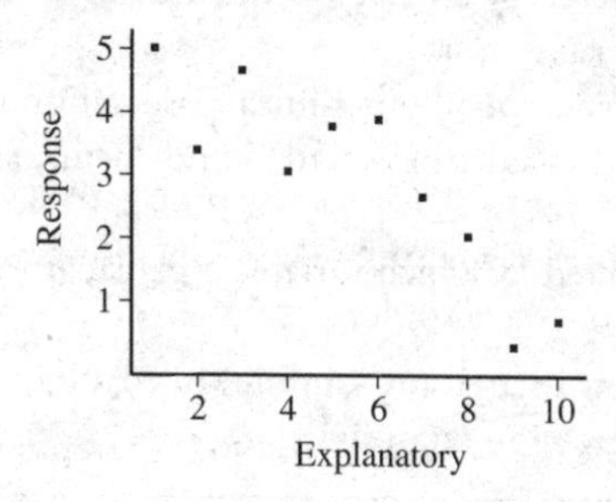

(A)

Variable	Coef	s.e.	t	p
Constant	5.45333	0.5487	9.94	0.0001
Explanatory	−0.451515	0.08844	-5.11	0.0009

s = 0.8033 R-sq = 76.5% R-sq(adj) = 73.6%

(B)

Variable	Coef	s.e.	t	p
Constant	−0.451515	0.5487	9.94	0.0001
Explanatory	5.45333	0.08844	-5.11	0.0009

s = 0.8033 R-sq = 76.5% R-sq(adj) = 73.6%

(C)

Variable	Coef	s.e.	t	p
Constant	5.45333	0.5487	9.94	0.0001
Explanatory	0.451515	0.08844	5.11	0.0009

s = 0.8033 R-sq = 36.5% R-sq(adj) = 33.6%

(D)

Variable	Coef	s.e.	t	p
Constant	5.45333	0.5487	9.94	0.0001
Explanatory	−0.451515	0.08844	-5.11	0.0009

s = 0.8033 R-sq = 36.5% R-sq(adj) = 33.6%

(E)

Variable	Coef	s.e.	t	p
Constant	5.45333	0.5487	9.94	0.0001
Explanatory	0.451515	0.08844	5.11	0.0009

s = 0.8033 R-sq = 76.5% R-sq(adj) = 73.6%

76. **Suppose a correlation is positive. Given two points from the scatterplot, which of the following is/are possible?**

    I. **The first point has a smaller x-value and a smaller y-value than the second point.**
    II. **The first point has a larger x-value and a larger y-value than the second point.**
    III. **The first point has a smaller x-value and a larger y-value than the second point.**

    (A) I only
    (B) II only
    (C) III only
    (D) I and II only
    (E) I, II, and III

77. **Which of the following statements about the correlation coefficient r is *incorrect*?**

    (A) It is not affected by changes in the measurement units of the variables.
    (B) It itself never has units.
    (C) It gives information about a linear association, not about causation.
    (D) It always takes values between –1 and +1 unless the association is nonlinear.
    (E) It is not affected by which variable is explanatory and which is response.

78. **The value of a new luxury automobile (in $1,000) versus age (in years) has the following regression analysis.**

Regression Analysis: Value versus Age

Variable	Coef	s.e. Coeff	t	p
Constant	82.9068	3.209	25.8	0.000
Age	–4.78869	0.6054	-7.91	0.000

s = 4.693 R-sq = 88.7% R-sq(adj) = 87.2%

Source	df	SS	MS	F
Regression	1	1378.18	1378.18	62.6
Residual	8	176.216	22.027	

What is the meaning of –4.78869 in the computer output?

(A) The value of the automobile will decrease by 4.78869 thousand dollars for every increase of 1 year in the age of the automobile.
(B) On average, there is a predicted decrease in the value of $1,000 for every increase of 4.78869 years in the age of the automobile.
(C) On average, there is a predicted decrease of 4.78869 thousand dollars in the value for every increase of 1 year in the age of the automobile.
(D) Approximately 4.79 percent of the variability in value can be explained by the linear model of value versus age.
(E) The standard deviation of the residuals is approximately 4.79.

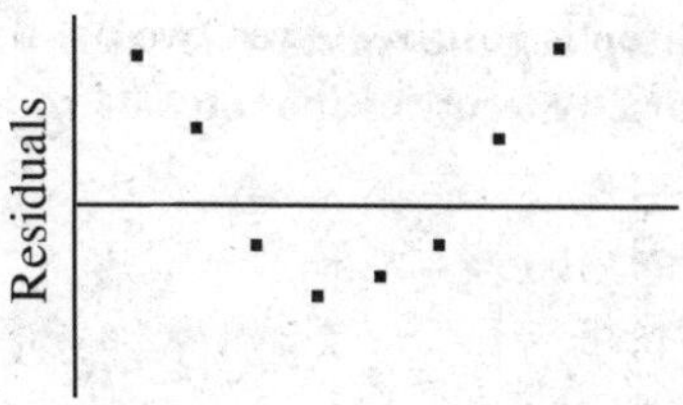

Residuals

79. **Which of the scatterplots below could have resulted in the *residual* plot shown above? (The *y*-scales are not the same in the scatterplots as in the residual plot.)**

(A)

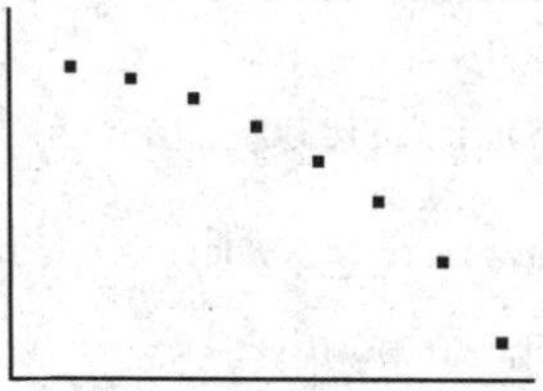

(B)

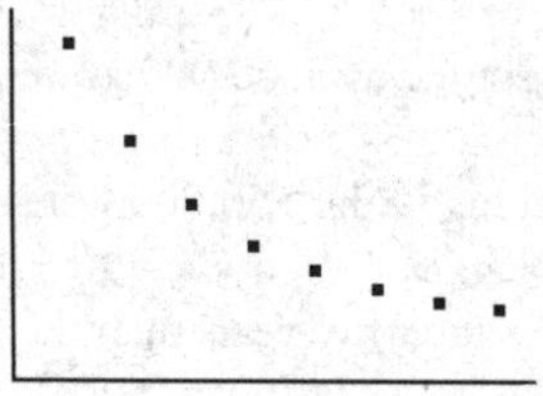

(C)

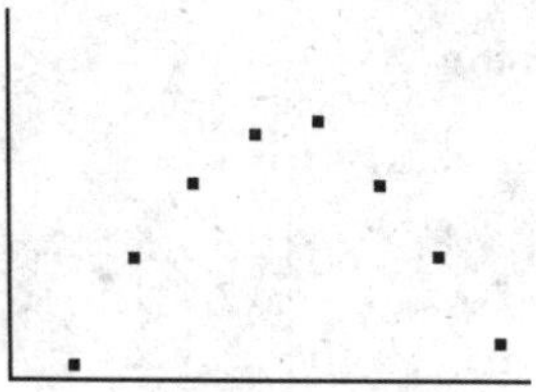

(D)

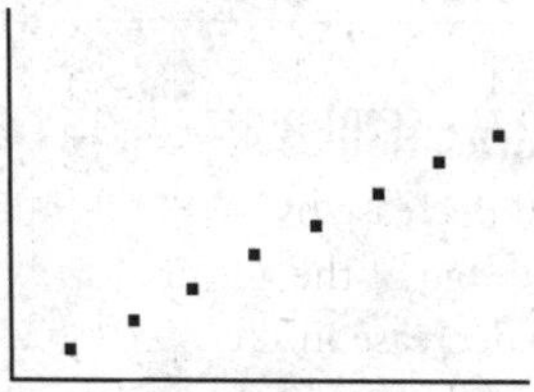

(E) None of these could result in the given residual plot.

80. Consider the scatterplot below of starting salaries and salaries after 5 years of employment for 20 executives at a software company.

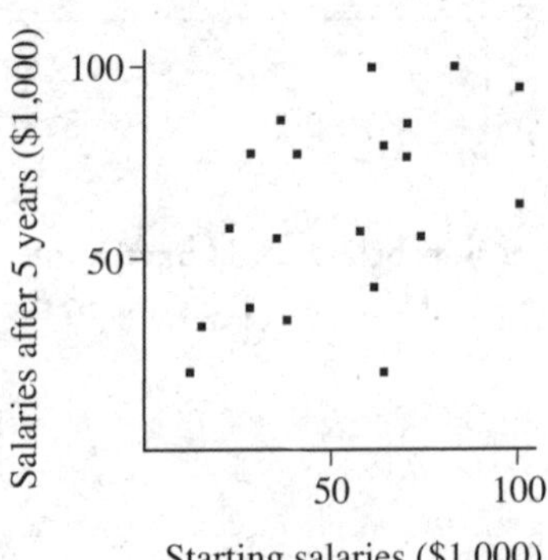

Which of the statements below is *incorrect*?

(A) Both the correlation and the slope are positive.
(B) More executives started making under $50,000 than were making under $50,000 after five years.
(C) The same number of executives started at $100,000 as were making $100,000 after five years.
(D) Most executives who started making under $50,000 were still making under $50,000 after five years.
(E) Some executives were making less after five years than they made initially.

81. Consider the following three scatterplots.

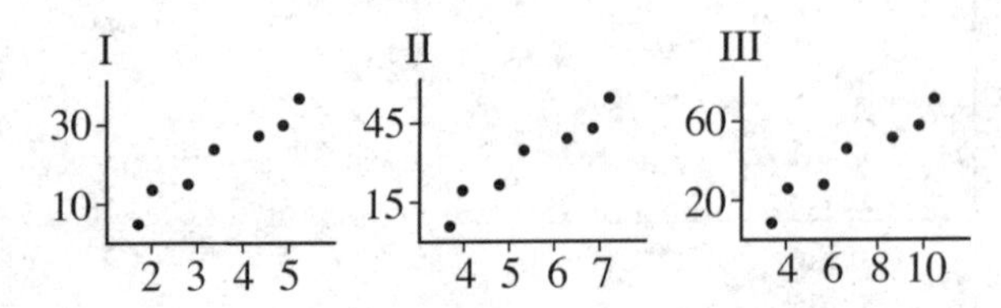

Which scatterplot has the greatest correlation coefficient *r*?

(A) I
(B) II
(C) III
(D) They all have the same correlation coefficient.
(E) The question cannot be answered without additional information.

82. **Which of the following statements about influential points is *incorrect*?**

(A) Removal of an influential point sharply affects the regression line.
(B) Looking at a residual plot is a helpful way of identifying influential points.
(C) Determining a regression model with and without a point is an excellent way of identifying influential points.
(D) Outliers in the x-direction are more likely to be influential points than are outliers in the y-direction.
(E) None of the above are incorrect statements.

83. **A regression analysis on the relationship between average annual profits of financial companies (in millions of dollars) and the yearly number of major security breaches gives the following computer output.**

```
Dependent variable is: Profits
s = 10.17  R-sq = 46.5%  R-sq(adj) = 40.5%
Source      df    SS         MS         F
Regression  1     807.309    807.309    7.81
Residual    9     930.327    103.37
Variable    Coef       s.e.Coeff  t       p
Constant    420.727    5.735      73.4    0.0001
Breaches    –2.70909   0.9694     -2.79   0.0209
```

If the analysis was rerun using the number of breaches as the dependent variable instead of profits, what would be the correlation coefficient?

(A) –0.318
(B) 0.465
(C) –0.465
(D) 0.682
(E) –0.682

84. **Suppose the regression line for a set of data, $\hat{y} = a + 5x$, passes through the point (1, 7). If $\bar{x}$ and $\bar{y}$ are the sample means of the x- and y-values, respectively, what is $\bar{y}$?**

(A) $\bar{x}$
(B) $5\bar{x}$
(C) $6 + \bar{x}$
(D) $2 + 5\bar{x}$
(E) $7 + 5\bar{x}$

85. The number of students who were victims of bullying at a high school during the years 2010–2017 is fitted with a least squares regression line. The graph of the residuals and some computer output is as follows.

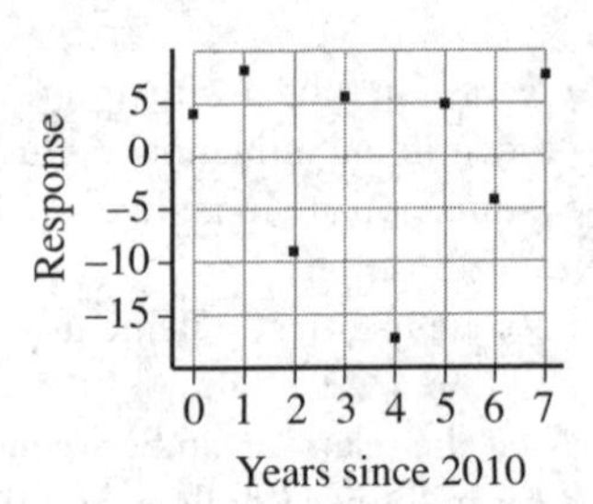


```
Dependent variable is: Students
s = 9.758  R-sq = 93.4%  R-sq(adj) = 92.4%
Variable    Coeff        s.e.    t      p
Constant    11           6.299   1.75   0.1313
Years       13.9286      1.506   9.25   0.0001
```

How many students were victims of bullying at this high school in the year 2013?

(A) 47
(B) 48
(C) 52
(D) 53
(E) 58

86. A simple random sample of 25 AP Statistics students provides the following statistics.

Number of hours studying for this class each day: $\bar{x} = 1.8$, $s_x = 0.5$
Average grade for this class: $\bar{y} = 86.5$, $s_y = 4.2$
Correlation $r = 0.45$

Based on this data, what is the resulting linear regression equation?

(A) $\widehat{\text{Grade}} = 52.9 + 18.67(\text{Hours})$

(B) $\widehat{\text{Grade}} = 79.7 + 3.78(\text{Hours})$

(C) $\widehat{\text{Grade}} = 83.1 + 1.89(\text{Hours})$

(D) $\widehat{\text{Grade}} = 85.7 + 0.45(\text{Hours})$

(E) $\widehat{\text{Grade}} = 86.4 + 0.0536(\text{Hours})$

87. A biologist studying robins collected data on the weight and the egg production for 50 female birds. The correlation is found to be 0.7. Which of the following is a true statement?

(A) On average, a 70 percent increase in weight results in a 49 percent increase in egg production.
(B) On average, a 70 percent increase in weight results in a 100 percent increase in egg production.
(C) Seventy percent of a robin's egg production can be explained by the bird's weight.
(D) Seventy percent of the variation in robin egg production can be accounted for by the linear regression model of egg production on bird weight.
(E) Greater egg production tends to be associated with greater bird weight.

88. One model for global warming is given by $\hat{y} = 315 + 1.52x$ with $r = 0.92$, where y is CO_2 atmospheric concentration in ppm and x is years since 1960. What is the correct interpretation of the slope?

(A) On average, atmospheric CO_2 concentration has been rising by 1.52 ppm per year since 1960.
(B) The baseline atmospheric CO_2 concentration is 315 ppm.
(C) The regression model explains 84.64 percent of the variation of atmospheric CO_2 concentration over the years since 1960.
(D) The regression model explains 92 percent of the variation of atmospheric CO_2 concentration over the years since 1960.
(E) Atmospheric CO_2 concentration will top 500 ppm in the year 2082.

89. Consider the points (–2, 3), (0, 7), (1, 9), (3, 12), and (10, *n*). What should *n* be so that the correlation between the *x* and *y* values is $r = 1$?

(A) 25
(B) 26
(C) 27
(D) A value different from any of the above
(E) No value for n can make $r = 1$.

***90. Suppose the data from two scatterplots, *A* and *B*, result in *identical* least squares regression lines with positive slopes. Which of the following statements is true?**

(A) The correlation in *A* equals the correlation in *B*.
(B) The sum of the squares of the residuals in *A* equals the sum of the squares of the residuals in *B*.
(C) If the sum of the squares of the residuals in *A* is greater than the sum of the squares of the residuals in *B*, then the correlation in *A* will be greater than the correlation in *B*.
(D) If the sum of the squares of the residuals in *A* is greater than the sum of the squares of the residuals in *B*, then the correlation in *A* will be less than the correlation in *B*.
(E) None of the above are true statements.

91. Two drivers, one for Uber and one for Lyft, compare their weekly miles driven for 17 consecutive weeks. The following scatterplot is the result, where the units for both axes are miles.

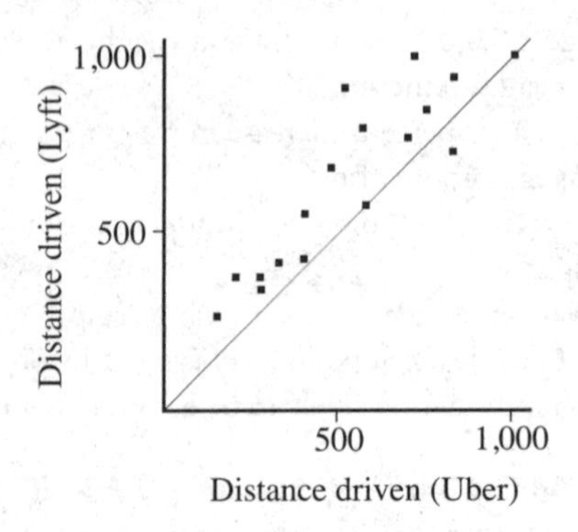

Which statement below gives the best comparison between the weekly miles driven by the two drivers?

(A) In all 17 weeks, the Lyft driver drove more miles than the Uber driver.
(B) In all 17 weeks, the Uber driver drove more miles than the Lyft driver.
(C) In all but 1 week, the Lyft driver drove more miles than the Uber driver.
(D) In all but 1 week, the Uber driver drove more miles than the Lyft driver.
(E) In all but 3 weeks, the Lyft driver drove more miles than the Uber driver.

92. Suppose a study finds that the correlation coefficient relating outside temperature in degrees Fahrenheit during the winter to the amount of natural gas a house consumes in cubic feet per day is $r = -1$. Which of the following is a proper conclusion?

(A) Lower outside winter temperatures cause a greater usage of natural gas.
(B) Higher outside winter temperatures cause a greater usage of natural gas.
(C) There is a 100% cause-and-effect relationship between outside winter temperatures and natural gas usage.
(D) There is a very strong association between outside winter temperatures and natural gas usage.
(E) None of the above are proper conclusions.

Variable	Coef	s.e.	t	p
Constant	10.3001	0.5668	18.2	0.0001
Explanatory	-1.73391	0.197	-8.8	0.0001

s = 0.9093 R-sq = 90.6% R-sq(adj) = 89.5%

93. What is the correct scatterplot for the computer output shown above?

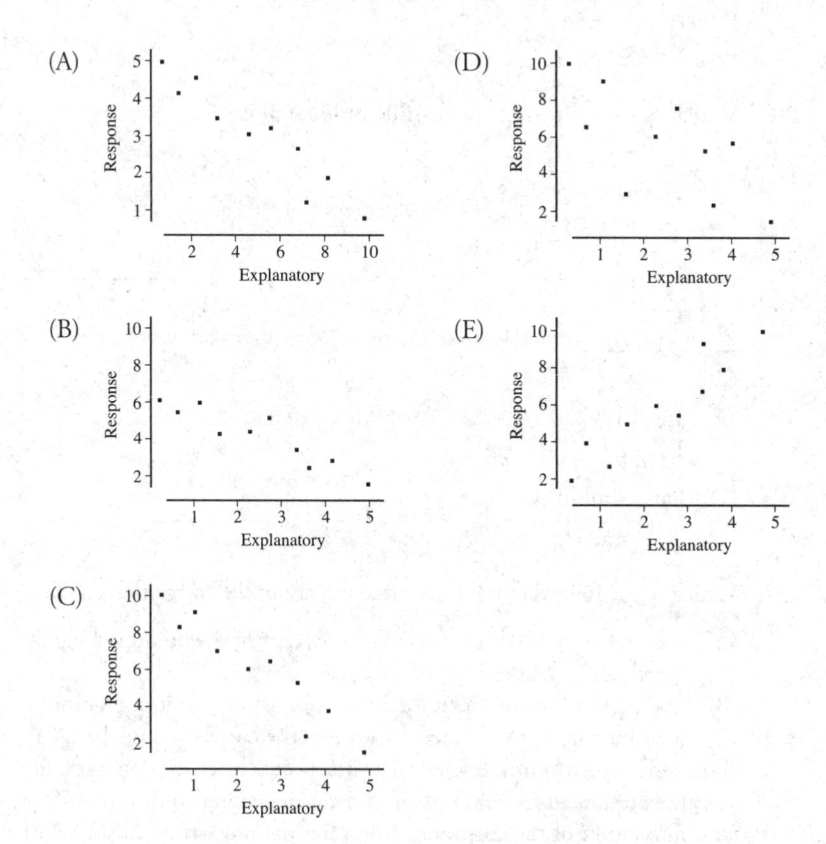

94. **Suppose the correlation between two variables is –0.28. If each of the *y*-values is multiplied by –2, which of the following is true about the new scatterplot?**

(A) It slopes up to the right, and the correlation is –0.28.
(B) It slopes up to the right, and the correlation is +0.28.
(C) It slopes down to the right, and the correlation is –0.28.
(D) It slopes down to the right, and the correlation is –0.56.
(E) It slopes up to the right, and the correlation is +0.56.

***95. Consider n pairs of numbers. Suppose $\bar{x} = 3$, $s_x = 2$, $\bar{y} = 5$, and $s_y = 4$. Which of the following could be the least squares regression line?**

(A) $\hat{y} = 11 - 2x$
(B) $\hat{y} = -4 + 3x$
(C) $\hat{y} = 17 - 4x$
(D) $\hat{y} = -10 + 5x$
(E) $\hat{y} = 23 - 6x$

***96. Which of the following are possible residual plots?**

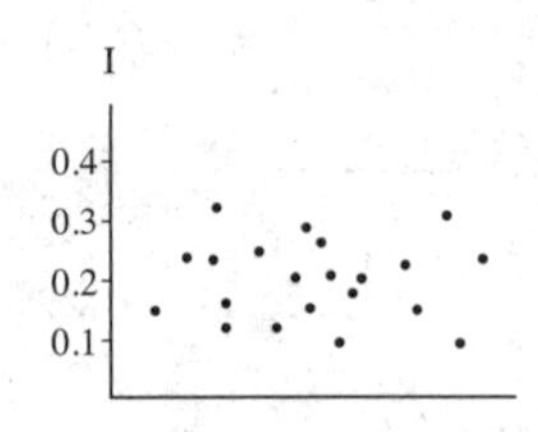

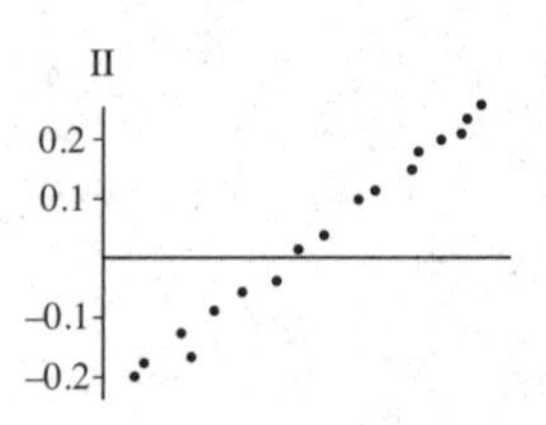

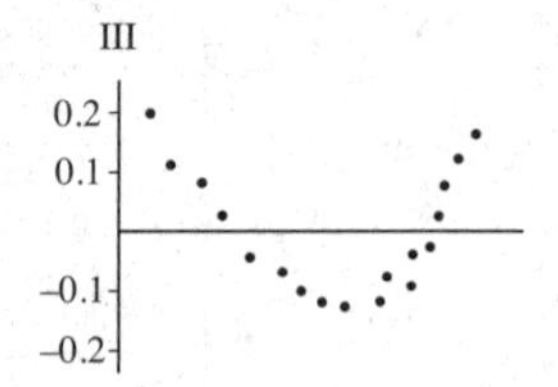

(A) I only
(B) II only
(C) III only
(D) I and II only
(E) I, II, and III

97. Which of the following is a true statement about the correlation coefficient r?

(A) A correlation of 0.4 means that 40 percent of the points are highly correlated.
(B) The unit of measurement for correlation is the y-unit per x-unit.
(C) Multiplying every y-value by –1 leaves the correlation unchanged.
(D) Perfect correlation, that is, when the points lie exactly on a straight line, results in $r = 0$.
(E) The square of the correlation measures the proportion of the y-variance that is predictable from the linear regression model.

98. Which of the following statements about correlation r is *incorrect*?

(A) The correlation and the slope of the regression line always have the same sign.
(B) Correlation r measures the strength and direction of only linear association.
(C) Outliers can greatly affect the value of r.
(D) A correlation of –0.48 and a correlation of +0.48 show the same degree of clustering around the regression line.
(E) A correlation of 0.48 indicates a relationship that is three times as linear as one for which the correlation is 0.16.

99. A scatterplot of a town's population versus time indicates a possible exponential relationship. A linear regression on y = log(population in 1,000s) against x = years since 2015 gives $\hat{y} = 2.4 + 0.03x$ with $r = 0.72$. Which of the following is a valid conclusion?

(A) On average, population goes up 0.03 thousand per year.
(B) The predicted population for year 2025 is approximately 501 thousand.
(C) 51.84 percent of the variation in population can be explained by variation in time.
(D) 72 percent of the variation in population can be explained by variation in time.
(E) None of the above are valid conclusions.

100. A scatterplot of log *X* and log *Y* shows a strong negative correlation close to –1. Which of the following is a true statement?

(A) The variables *X* and *Y* also have a correlation close to –1.
(B) A scatterplot of the variables *X* and *Y* shows a nonlinear pattern.
(C) The residual plot of the variables *X* and *Y* shows a random pattern.
(D) The residual plot of the variables log *X* and log *Y* shows a strong nonrandom pattern.
(E) None of the above are true.

101. A random sample of 12 cigarettes of different brands is selected. The tar and nicotine content of each is measured and graphed, resulting in the scatterplot below.

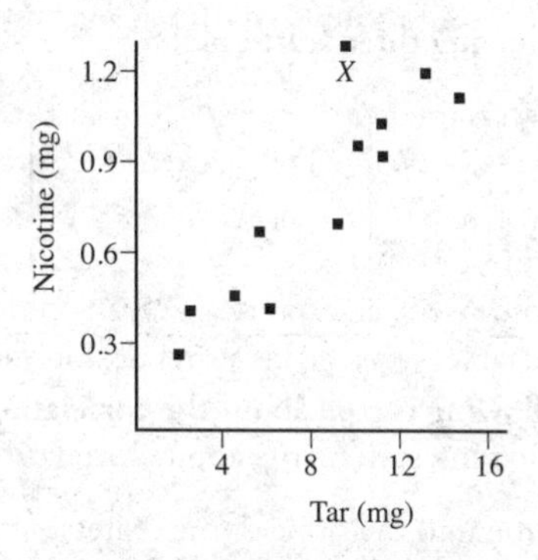

If the point labeled *X* is removed, which of the following statements would be true about the least squares regression line and the correlation coefficient?

(A) Both the slope and the correlation would remain the same.
(B) Both the slope and the correlation would increase.
(C) Both the slope and the correlation would decrease.
(D) The slope would increase, and the correlation would decrease.
(E) The slope would decrease, and the correlation would increase.

102. Ten overweight people went on a new diet, and each lost exactly 7 pounds. What is the correlation between weights before the diet and weights after the diet?

(A) 0
(B) 0.5
(C) 1
(D) This cannot be answered without seeing the actual data.
(E) This cannot be answered without knowing if the people were randomly picked from the population of all overweight people.

***103. A scatterplot with one point labeled *X* is shown below.**

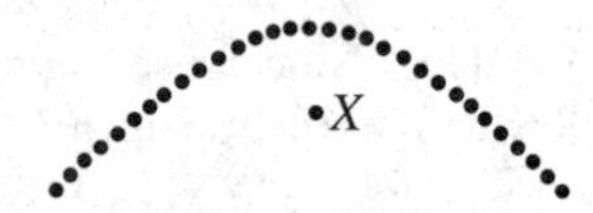

With regard to the least square regression line, which of the following statements is true?

(A) *X* is an influential point.
(B) *X* is an outlier.
(C) *X* has the largest residual, in absolute value, of any point on the scatterplot.
(D) There will be no pattern in the residual plot.
(E) None of the above are true statements.

104. Consider the following three scatterplots.

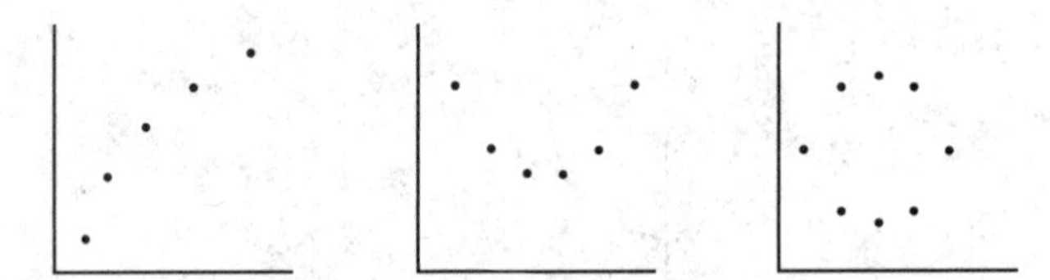

Which of the following is true about the correlations for the three scatterplots?

(A) None are close to 0.
(B) One is approximately 0, one is negative, and one is positive.
(C) One is approximately 0, and both of the others are positive.
(D) Two are approximately 0, and the other is +1.
(E) Two are approximately 0, and the other is close to +1.

105. In a random sample of eight fast food hamburgers, the association between calories and fat is summarized in the computer output below.

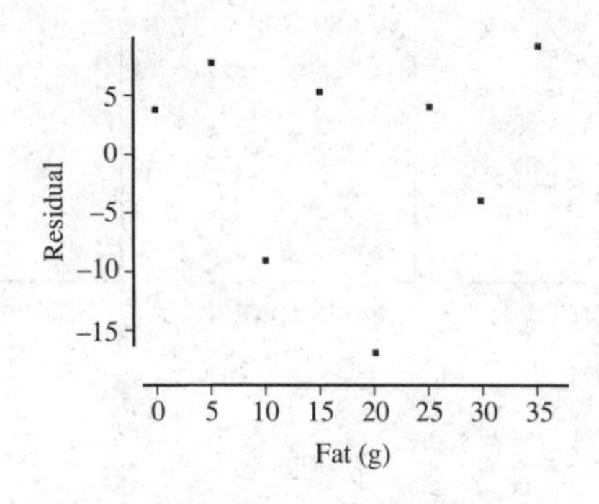


```
Dependent variable is: Calories
s = 9.758  R-sq = 93.4%  R-sq(adj) = 92.4%
Variable    Coeff       s.e.     t      p
Constant    211         6.299    1.75   0.1313
Fat         13.9286     1.506    9.25   0.0001
```

Which of the following is the best approximation of the actual calories for the hamburger in the sample that had 15 grams of fat?

(A) 410
(B) 415
(C) 420
(D) 425
(E) 430

106. A recent study on distance (in feet) drivers can see versus age (in years) of drivers resulted in a linear regression model:

Predicted distance a driver can see = 576.7 – 3.01(Age)

Which of the following is the best interpretation of the slope in context?

(A) For each year a driver's age increases, he or she sees 3.01 feet less.
(B) For each 3.01 years a driver's age increases, he or she sees 1 foot less.
(C) For each year a driver's age increases, he or she sees 3.01 feet less up to a distance of 576.7 feet.
(D) For each 3.01 years a driver's age increases, he or she sees 1 foot less on average.
(E) For each year a driver's age increases, he or she sees 3.01 feet less on average.

107. Given the following three scatterplots with correlations r_1, r_2, and r_3, respectively, what is the proper ordering of their correlations?

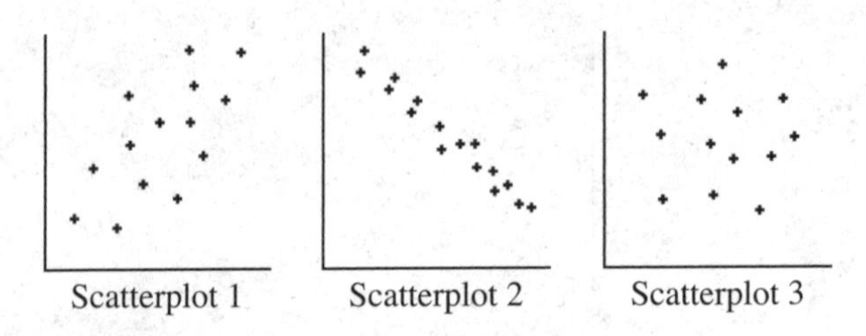

(A) $r_3 < r_1 < r_2$
(B) $r_2 < r_1 < r_3$
(C) $r_2 < r_3 < r_1 < |r_2|$
(D) $r_1 < r_3 < |r_2|$
(E) $r_3 < r_2 < r_1 < |r_2|$

*108. Consider the three points (3, 50), (4, 38), and (5, 14). Given any straight line, we can calculate the sum of the squares of the three vertical distances from these points to the line. What is the smallest possible value of this sum?

(A) 4.9
(B) 8
(C) 9.8
(D) 24
(E) 40

109. Four pairs of data are used in determining the regression line $\hat{y} = 8 + 2x$. If the four values of the independent variable are 17, 29, 35, and 43, what is the mean of the four values of the dependent variable?

(A) 31
(B) 62
(C) 70
(D) 256
(E) The mean cannot be determined from the given information.

110. Suppose the correlation between two variables is $r = 0.16$. What will the new correlation be if every value of the x-variable is tripled, 0.04 is added to all values of the y-variable, and the two variables are then interchanged?

(A) 0.16
(B) 0.20
(C) 0.48
(D) 0.52
(E) 0.60

111. Which of the following statements about the correlation r is true?

(A) When $r = 0$, there is no relationship between the variables.
(B) When $r = 0.3$, 30 percent of the variables are closely related.
(C) When $r = 1$, there is a perfect cause-and-effect relationship between the variables.
(D) A correlation close to 1 means that a linear model will give the best fit to the data.
(E) All of the above statements are false.

***112. Which of the following statements about residuals in a linear regression model is *incorrect*?**

(A) The mean of the residuals is always zero.
(B) The sum of the residuals is always zero.
(C) The regression line for a residual plot is a horizontal line.
(D) The standard deviation of the residuals gives a measure of how the points in the scatterplot are spread around the regression line.
(E) A residual equals the predicted y minus the observed y.

113. Data on ages (in months) and eBay prices for used cellphones result in a regression line.

$$\widehat{\text{Price}} = 500 - 22.5(\text{Age})$$

Given that 56.25% of the variation in price is explained by the linear regression model of Price on Age, what is the value of the correlation coefficient r?

(A) –0.75
(B) –0.5625
(C) 0.5625
(D) 0.75
(E) There is insufficient information to answer this question.

114. Studies have shown a strong, positive, linear relationship between temperature and the frequency of a cricket's chirps. What does "strong" mean in this context?

(A) The temperature is related to the frequency of a cricket's chirps.
(B) For each unit increase in the frequency of cricket chirps, the predicted temperature changes by a constant amount.
(C) The greater the frequency of cricket chirps, the greater the temperature, on average.
(D) The actual temperatures are very close to the temperatures predicted by the linear model.
(E) The variability in temperatures is related to the variability in the frequency of cricket chirps.

115. A study of family incomes and SAT scores reports a correlation of $r = +1.07$. What should be concluded?

(A) Students coming from higher-income families tend to have higher SAT scores than students from lower-income families.
(B) Students coming from higher-income families have 7 percent higher SAT scores on average than students coming from lower-income families.
(C) There is less than a 10 percent relationship between family incomes and SAT scores.
(D) There is a strong positive association between family income and SAT scores, but concluding causation would be wrong.
(E) A mistake in arithmetic was made.

116. An HR officer at a high-pressure company surveys executives as to their happiness level and accesses their salaries. The data are displayed in the scatterplot below.

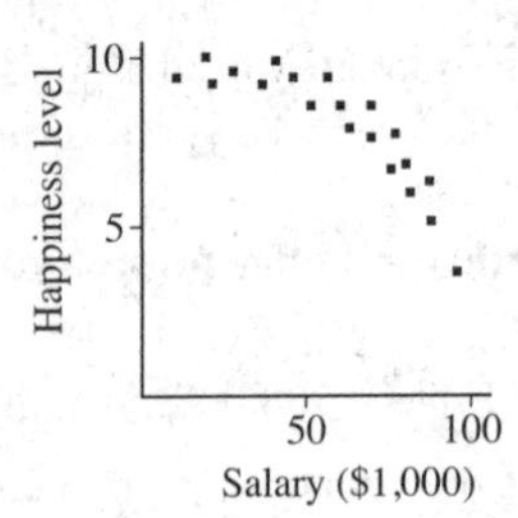

Describe the relationship between happiness level and salary.

(A) Negative, weak, linear
(B) Negative, moderate, linear
(C) Negative, strong, nonlinear
(D) Positive, strong, nonlinear
(E) Positive, weak, nonlinear

117. Data on w = weight of a car (in pounds) and s = stopping distance (in feet) when the car is traveling at 50 mph and brakes are applied results in the regression equation $\hat{s} = 35 + 0.03w$. Which of the following is the best interpretation of the intercept 35 in the equation?

(A) For each 1 pound increase in car weight, the stopping distance increases by 35 feet, on average.
(B) For each 1 foot increase in stopping distance, the car weight is 35 pounds greater, on average.
(C) The stopping distance is 35 feet when the car weight is 0, on average.
(D) The stopping distance is 0 feet when the car weight is 35 pounds, on average.
(E) The intercept, 35 feet, has no reasonable interpretation because a weight of 0 is outside the domain of car weights.

***118. A linear regression analysis of rushing yards and points scored in a random sample of 25 NFL games results in the following equation.**

$$\widehat{\text{Points scored}} = 10 + 0.05\left(\text{Rushing yards}\right)$$

A coach would like to predict rushing yards from points scored. He solves the above equation and comes up with the following.

$$\widehat{\text{Rushing yards}} = -200 + 20\left(\text{Points scored}\right)$$

Did the coach calculate correctly?

(A) Yes, because if $y = 10 + 0.05x$, solving algebraically for x gives $x = -200 + 20y$.
(B) Yes, because the x- and y-variables are interchangeable in linear regression.
(C) Yes, because $\frac{1}{0.05} = 20$, showing that the slopes are reciprocals.
(D) No, because the equations have different sets of variables.
(E) This cannot be answered without knowing the value of the correlation r.

119. A study of airfare prices versus distances gives rise to the following scatterplot.

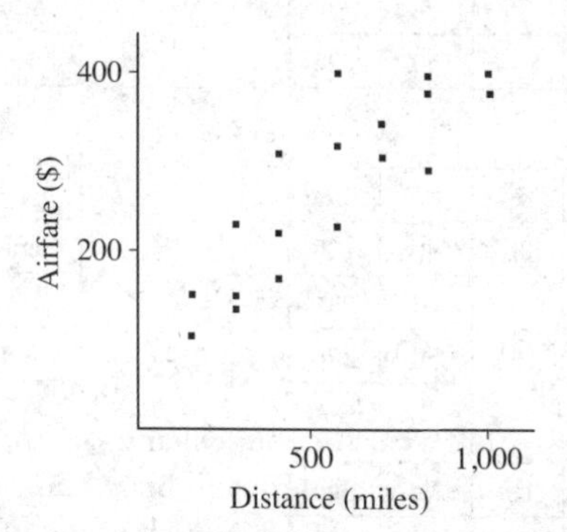

A least squares regression analysis is performed. Which of the following would be apparent from a residual plot of residuals versus distance?

(A) The variation in airfare is different across the distances.
(B) There is a positive linear relationship between the residuals and distance.
(C) The sum of the residuals is less than 0.
(D) The sum of the residuals is greater than 0.
(E) The residuals go approximately from 100 to 400.

120. In a random sample of 300 men and 200 women, 50 of the men and 40 of the women said that they brush their teeth from side to side rather than up and down. If there is no difference between the proportions of men and women who brush their teeth from side to side rather than up and down, how many men and women in the sample would be expected to brush their teeth from side to side rather than up and down?

(A) 45 men and 45 women
(B) 50 men and 40 women
(C) 54 men and 36 women
(D) 150 men and 100 women
(E) 250 men and 160 women

121. A random sample of 200 people was anonymously surveyed as to whether they sneak food into movie theaters rather than pay the high price of snacks. 120 were teenagers, and 80 were adult. 150 answered "yes," they have snuck in food, and 50 answered "no." Among those surveyed, there was independence between teenager/adult and whether or not they snuck food into movie theaters. Which of the following tables shows this conclusion?

(A)

	Yes	**No**	***Total***
Teenagers	75	45	120
Adults	75	5	80
Total	150	50	200

(B)

	Yes	**No**	***Total***
Teenagers	80	40	120
Adults	70	10	80
Total	150	50	200

(C)

	Yes	**No**	***Total***
Teenagers	90	30	120
Adults	60	20	80
Total	150	50	200

(D)

	Yes	**No**	***Total***
Teenagers	110	10	120
Adults	40	40	80
Total	150	50	200

(E)

	Yes	**No**	***Total***
Teenagers	100	20	120
Adults	50	30	80
Total	150	50	200

122. In the following table, what value of n results in a table showing perfect independence?

	Age	
	<30	≥30
Use Twitter	40	60
Do not use Twitter	50	n

(A) 30
(B) 50
(C) 70
(D) 75
(E) 100

123. A random sample of 200 movies were cross-classified by genre and rating. The following table gives the individual counts.

	Rating				
	G	**PG**	**PG-13**	**R**	***Total***
Action	5	20	15	5	45
Comedy	10	25	10	5	50
Documentary	25	10	3	2	40
Drama	20	15	17	13	65
Total	60	70	45	25	200

Of the movies in the sample that were classified as PG or PG-13, what proportion of them were either action or drama?

(A) $\frac{45 + 65}{200}$

(B) $\frac{70 + 45}{200}$

(C) $\frac{20 + 15}{70 + 45}$

(D) $\frac{20 + 15 + 15 + 17}{200}$

(E) $\frac{20 + 15 + 15 + 17}{70 + 45}$

124. For a class project, a student surveyed 125 cars in a high school parking lot. The student recorded whether each car had a student or a staff tag and whether each was an American or a foreign model car. Of the 50 American cars, 20 had staff tags. Given that whether the car has a student or a staff tag and whether it is an American or a foreign model are independent, how many of the foreign cars had student tags?

(A) 20
(B) 30
(C) 45
(D) 50
(E) 75

125. Men and women were sampled as to whether, in general, they feel "always rushed," "sometimes rushed," or "almost never rushed." A summary of those responses are given in the following table.

	Men	Women	*Total*
Always	44	95	139
Sometimes	79	69	148
Almost never	22	28	50
Total	145	192	337

Based on this table, which of the following statements is *incorrect*?

(A) Most men feel "sometimes rushed."
(B) Most women feel "always rushed."
(C) Men are more likely than women to say that they are "almost never rushed."
(D) More men than women say they are "sometimes rushed."
(E) The proportion of people who are either "sometimes rushed" or "almost never rushed" is $1 - \frac{139}{337}$.

MC Collecting and Producing Data

CHAPTER 2

Answers for Chapter 2 are on pages 252–262.

SAMPLING STRATEGIES

126. An online retailer wishes to survey its customers. The decision is made to select randomly 50 customers who pay with credit cards, 25 customers who pay with debit cards, and 10 customers who pay with checks. This procedure is an example of which type of sampling?

(A) Cluster
(B) Convenience
(C) Simple random
(D) Stratified
(E) Systematic

127. Which of the following is a true statement about sampling?

(A) Data obtained when conducting a census are always more accurate than data obtained from a sample, no matter how careful the design of the sampling method.
(B) The sampling frame is the population of interest.
(C) Sampling error implies an error, possibly very small but still an error, on the part of the surveyor.
(D) Careful analysis of a given sample will indicate whether or not it is random.
(E) The clusters in cluster sampling should all look pretty much alike.

128. Sampling error occurs

(A) when samples are too small
(B) when interviewers make mistakes resulting in bias
(C) when interviewers use judgment instead of random choice in picking the sample
(D) because a sample statistic is used to estimate a population parameter
(E) in all of the above cases

129. A town has one high school, which buses students from urban, suburban, and rural communities. Which of the following sampling techniques is most recommended in studying high school students' attitudes toward college versus trade school after high school graduation?

(A) Cluster
(B) Simple random
(C) Stratified
(D) Systematic
(E) Voluntary response

130. A company wishes to survey what people think about a new product it plans to market. The company decides to sample randomly from their large employee database as this includes phone numbers and addresses. This procedure is an example of which type of sampling?

(A) Cluster
(B) Convenience
(C) Simple random
(D) Stratified
(E) Systematic

131. Each of the 31 National Hockey League teams carries a 20-person roster. A sample of 62 players (10% of all 620 players) will be selected to undergo drug tests. To do this, each team is instructed to put the names of the 20 players into a hat and randomly draw 2 names. Will this method result in a simple random sample of the 620 hockey players?

(A) Yes, because each player has the same chance of being selected
(B) Yes, because each team is equally represented
(C) Yes, because this is an example of stratified sampling, which is a special case of simple random sampling
(D) No, because the teams are not chosen randomly
(E) No, because not each group of 62 players has the same chance of being selected

132. The United States Tennis Association (USTA) plans to survey 50 of the 1,000 top-rated tennis players in the United States. A list of the players by rank is made available, and a random number between 1 and 20 is picked. The sample consists of the person that number down the list together with every 20th person after that. This procedure is an example of which type of sampling?

(A) Cluster
(B) Convenience
(C) Simple random
(D) Stratified
(E) Systematic

133. A travel agent plans to survey passengers taking the cross-Canada train trip from Vancouver to Toronto. She is considering two sampling methods.

> **Method 1: From a boarding list, randomly choose 15 passengers traveling first class (who have roomettes for sleeping) and 35 passengers traveling coach (who have reclining chairs for sleeping).**
>
> **Method 2: Randomly select one of the cars (each holds 50 passengers), and survey all the passengers in that car.**

Identify each sampling method.

(A) Method 1 is cluster sampling, and Method 2 is stratified sampling.
(B) Method 1 is cluster sampling, and Method 2 is systematic sampling.
(C) Method 1 is stratified sampling, and Method 2 is cluster sampling.
(D) Method 1 is stratified sampling, and Method 2 is simple random sampling.
(E) Method 1 is simple random sampling, and Method 2 is cluster sampling.

134. A polling company conducts a weekly survey to determine the proportion of voters who support the direction Congress is taking. For the coming year, the company decides to double the sample size. The main benefit of this is to

(A) eliminate sampling error
(B) decrease population variability
(C) reduce undercoverage bias
(D) reduce nonresponse bias
(E) decrease the standard deviation of the sampling distribution

135. The athletic office at a large university is thinking of allowing women to try out for the baseball team and decides to survey student opinion with a stratified sample based on gender. Which of the following will achieve the desired sampling plan?

(A) Using a list of all students on campus, randomly select a starting name and every 10th name thereafter. Separate the groups by gender until the desired sample size is attained.
(B) Using a numbered list of all students on campus and a random number generator, select numbers between 1 and the size of the student body until the desired sample size is reached. A repeat of a selected number is disregarded. Separate selected students by gender.
(C) Select all students from a campus dormitory that is representative of all students on campus. Separate students by gender.
(D) Select separate random samples of females and males based on the proportions of females and males on campus.
(E) Select a random sample of students attending the next baseball game. Separate the students by gender.

136. To gauge patron use of city libraries, all patrons who check out books from randomly selected libraries will be surveyed. This procedure is an example of which type of sampling?

(A) Cluster
(B) Convenience
(C) Simple random
(D) Stratified
(E) Systematic

137. An IRS agent is given an assignment to choose and audit the tax returns of 260 companies claiming large refunds. She has an assistant list all companies whose name begins with *A*, assigns each a number, and uses a random number generator to pick 10 of these numbers and thus 10 companies. She proceeds to use the same procedure for each letter of the alphabet and combines the results into a group of 260. Which of the following is a true statement?

(A) Her procedure makes use of chance.
(B) Her procedure results in a simple random sample.
(C) Each company has an equal probability of being selected.
(D) The company named Amazon probably has a higher probability of being selected than the company named Xerox.
(E) This is an example of a systematic sample.

138. A government pollster is planning to survey randomly 50 voters in each of a state's 102 counties regarding a proposed change in the state income tax code. Why is stratification used here?

(A) Stratification helps remove bias when estimating proportions, in this case the proportion of voters favoring the change in code.
(B) Stratification helps remove bias when estimating means, in this case the mean number of voters favoring the change in code.
(C) Stratification helps remove bias when estimating variances, in this case the variance in the number of voters favoring the change in code.
(D) Stratification helps reduce sampling variability.
(E) Stratification is generally the procedure of choice when efficiency and cost is of paramount interest.

139. Which of the following statements about sampling error is *incorrect*?

(A) Sampling error is generally smaller when the sample size is larger.
(B) Sampling error concerns natural variation among samples, is always present, and can be described using probability.
(C) Sampling error is greater when working with higher confidence levels.
(D) Sampling error is unrelated to bias.
(E) Sampling error can be eliminated only if a survey is both extremely well designed and extremely well conducted.

140. A cybersecurity analyst wishes to survey her client base of 63 companies as to their vulnerability to a new computer virus. She has 63 business cards, all of the identical size, from her contacts in the companies. The analyst decides to drop all of the business cards into a small box, shake them up, and reach in to pick 5 cards for her sample. This procedure is an example of which type of sampling?

(A) Cluster
(B) Convenience
(C) Simple random
(D) Stratified
(E) Systematic

141. The Environmental Protection Agency (EPA) is concerned about a region near an old industrial waste dump and plans to take soil samples to test for toxic chemicals. The EPA is considering using either vertical or horizontal strips with regard to sampling, as shown in the figures below.

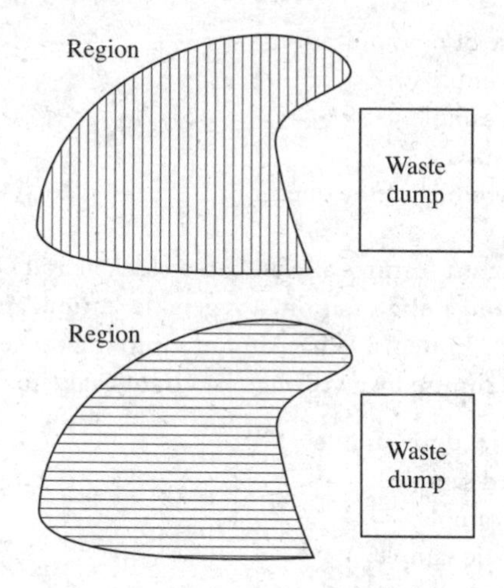

What are the two most appropriate sampling techniques?

(A) Stratified sample with vertical strips or cluster sample with vertical strips
(B) Stratified sample with vertical strips or cluster sample with horizontal strips
(C) Stratified sample with horizontal strips or cluster sample with vertical strips
(D) Stratified sample with horizontal strips or cluster sample with horizontal strips
(E) None of the above are pairs where both techniques are appropriate.

142. A study is made on whether a particular review book helps students achieve higher scores on the AP Statistics Exam. In comparing the records of 200 students, half of whom purchased the review book, it is noted that the average AP Statistics score is higher for those 100 students who purchased the book. Which of the following are true statements?

I. Although this study indicates a relation, it does not prove causation.
II. There could well be a confounding variable responsible for the seeming relationship.
III. Self-selection here makes drawing a conclusion difficult.

(A) I only
(B) I and II only
(C) I and III only
(D) II and III only
(E) I, II, and III

143. Which of the following would best help avoid response bias?

(A) Proper use of randomization
(B) A larger sample size
(C) A smaller sample size
(D) A longer survey
(E) Careful wording of the questions

144. A buffet restaurant features a salad bar, a chicken and beef entree station, a seafood station, a pizza station, a vegetable station, and a dessert bar, with a total of 235 items. What kind of sampling procedure would be recommended for use by a visiting restaurant evaluator?

(A) A simple random sample
(B) A stratified sample
(C) A cluster sample
(D) A systematic sample
(E) A convenience sample

145. A dietary supplement manufacturer wants to test consistency of the krill oil content of omega-3 supplements produced in one factory. The company decides to select one bottle of capsules randomly from each day's production run at the factory and analyze every capsule in this bottle as to krill oil content. What type of sampling is this?

(A) Cluster
(B) Convenience
(C) Simple random
(D) Stratified
(E) Systematic

146. A study is made to determine whether taking multiple AP classes in high school eventually leads to a higher 4-year graduation rate from college. In comparing the college records of 250 students, half of whom took multiple AP courses in high school, it is determined that a higher proportion of the students who took the AP courses graduated from college in four years than did students who had not taken these courses. Based on this study, students are encouraged to take multiple AP courses. Which of the following is an *incorrect* statement?

(A) This an observational study, not an experiment.
(B) Self-selection here makes drawing conclusions such as encouraging more AP enrollment difficult.
(C) Although this study indicates a relation, it does prove causation.
(D) There could well be a confounding variable responsible for the seeming relationship.
(E) A more meaningful study would be to compare a simple random sample (SRS) from each of the two groups of 125 students.

147. A survey with regard to using park funds for building a new skateboard facility resulted in the following table.

Age	**21–30**	**31–40**	**41–50**	**51–60**	**61–70**	**71–80**	***Total***
For	35	22	18	15	5	1	96
Against	20	33	37	40	50	54	234

Which of the following sampling strategies was most likely used?

(A) Simple random
(B) Cluster
(C) Stratified
(D) Systematic
(E) Proportional

148. To survey the opinions of people attending a Lincoln Center ballet, a surveyor plans to select every 30th theatergoer as he or she exits at the conclusion. Will this method result in a simple random sample?

(A) Yes, because each theatergoer has the same chance of being selected
(B) Yes, but only if everyone leaves by the same exit
(C) Yes, because the 29 out of 30 theatergoers who are not selected will form a control group
(D) Yes, because this is an example of systematic sampling, which is a special case of simple random sampling
(E) No, because not every sample of the intended size has an equal chance of being selected

149. To conduct a survey on television viewing preferences, a researcher opens a telephone book to a random page, closes his eyes, puts his finger down on the page, and then reads off the next 80 names. Which of the following is *not* a true statement?

(A) The survey incorporates chance.
(B) The procedure will not give a simple random sample.
(C) The procedure results in a systematic sample.
(D) The procedure could easily result in selection bias.
(E) The use of a phone book will result in undercoverage bias.

150. The annual AP national conference is attended by a large number of AP teachers from all AP subject areas. At one of these conferences, a random sample of AP Statistics teachers at the conference was selected and surveyed as to the proportion of their students who go on to do well in college. To what population can the result be generalized?

(A) All students
(B) All students taking an AP course
(C) All students taking AP Statistics
(D) Students of all AP teachers who attended the conference
(E) Students of all AP Statistics teachers who attended the conference

151. Sampling error is

(A) the result of bias
(B) the mean of a sample statistic
(C) the standard deviation of a sample statistic
(D) the standard error of a sample statistic
(E) the difference between a population parameter and an estimate of that parameter

BIAS IN SAMPLING

152. A typical method of telephone survey is to call a random number (with a random number generator). If there is no answer, call another number. If someone answers, check if the person is at least 18. If not, call another number; if yes, proceed with the survey. There are many concerns with this methodology. Which of the following is *not* a concern?

(A) People living in large households may be underrepresented.
(B) People with multiple phones may be overrepresented.
(C) People with unlisted numbers may be underrepresented.
(D) People with children under 18 years old may be underrepresented.
(E) People with caller ID and recording machines may be underrepresented.

153. Two surveys using different sampling methods were both aimed at determining the extent of illegal drug use on a college campus. The first method was able to obtain responses from only a small percentage of those selected to be surveyed. The second method received a very high rate of response; however, the answers to the question on drug use indicated a much smaller use of drugs on campus than the health counselors knew to be true. What type of bias did each sampling method illustrate?

(A) The first method illustrates response bias; the second method illustrates nonresponse bias.
(B) The first method illustrates nonresponse bias; the second method illustrates response bias.
(C) The first method illustrates selection bias; the second method illustrates nonresponse bias.
(D) The first method illustrates selection bias; the second method illustrates response bias.
(E) The first method illustrates nonresponse bias; the second method illustrates selection bias.

154. 42 percent of Americans have blood type A. In a random sample of 200 Americans, 40 percent had blood type A. What is the most likely explanation for this difference between the observed and expected percents?

(A) Selection bias
(B) Confounding between those who have and do not have blood type A
(C) Experimental design error
(D) Incomplete sampling frame
(E) Sampling variability

155. Pollsters are becoming more and more concerned about the accuracy of telephone surveys because of nonresponse bias. Which of the following best explains their concern?

(A) The attitude of the surveyor might cause the respondent to hang up.
(B) The wording of the questions might influence the way people respond.
(C) If people are uncomfortable with the questions, they might not answer truthfully.
(D) Not everyone has phones, so certain groups of people—for example, the homeless—will be missed in telephone surveys.
(E) With caller ID and answering machines, many people simply choose to not participate in telephone surveys.

156. A school district with 25 elementary schools is considering having the school day begin and end one hour later. They are considering four methods of surveying parents.

    Method 1: Randomly select one of the 25 schools, and contact every parent with children in that school by mail.
    Method 2: Send a survey home with every elementary school student, and ask parents to fill it out and return it within a week.
    Method 3: Randomly select 10 parents from each elementary school, send them a survey, follow up with a phone call if they don't return the survey, and follow up with a home visit if they don't return the survey or answer the phone.
    Method 4: Make a public announcement on a local television news program asking people to post their opinions to an online website.

    How many of the above methods clearly illustrate some form of bias?

    (A) 0
    (B) 1
    (C) 2
    (D) 3
    (E) 4

157. An assembly line produces 1,000 circuit boards a day. As a quality check, a sample of 50 boards are checked. One day unbeknownst to everyone, the first 600 boards off the assembly line are fine but the last 400 are defective. Which sampling method would have best picked up on the fraction of defective boards produced that day?

    (A) A simple random sample, randomly picking 50 of the 1,000 boards
    (B) A systematic sample, randomly picking one of the first 20 boards and then testing it and every 20th board from that point on
    (C) A convenience sample, randomly picking one of the first 951 boards and then testing it and the following 49 boards
    (D) Go through the boards one at a time, flipping a coin for each board, and testing the first 50 boards for which the coin comes up "heads"
    (E) All of the above methods involve randomization, so all are equally likely to have picked up on the fraction of defective boards produced that day

158. A home economist is interested in estimating the proportion of people who make their beds in the morning, and she plans to interview a random sample of 500 people. If instead she increases the sample size to 750, what effect will this have on bias and on the variance of the estimator?

(A) The bias and the variance will remain the same.
(B) Both the bias and the variance will decrease.
(C) The bias will decrease, while the variance will remain the same.
(D) The bias will remain the same, while the variance will decrease.
(E) The bias will increase, while the variance will decrease.

159. What fault do all of these sampling designs have in common?

I. The *Wall Street Journal* plans to make a prediction for the Republican nominees for open Senate seats based on a survey of its readers.
II. An Internet site asks viewers to vote on their choice for "Television Series of the Year."
III. A college teacher randomly picks a sample of his students and interviews them concerning how much they feel they have learned during the academic semester.

(A) None of the designs satisfactorily controls for sampling error.
(B) All the designs confuse association with cause and effect.
(C) All the designs have errors that can lead to strong bias.
(D) All the designs make improper use of stratification.
(E) None of the designs makes use of chance in selecting a sample.

160. A researcher plans a study to examine the depth of belief in faith healing (the power of prayer in curing illnesses) among the adult population. She interviews a simple random sample of 50 adults leaving church one Sunday morning. All but two of them agree to participate in the survey, consisting of a series of neutrally worded questions. Which of the following is a true statement?

(A) Selection bias makes this a poorly designed survey.
(B) The high response rate makes this a well-designed survey.
(C) The use of neutral wording makes this a well-designed survey.
(D) The proper use of chance as evidenced by the simple random sample and neutral wording makes this a well-designed survey.
(E) The large sample, $n \geq 30$, helps makes this a well-designed survey.

161. Which of the following statements is *incorrect*?

(A) Convenience samples often lead to undercoverage bias.
(B) There is no way to fix the results if a biased sampling method was employed.
(C) Questionnaires with nonneutral wording are likely to have response bias.
(D) Voluntary response samples often underrepresent people with strong opinions.
(E) Nonresponse bias should be avoided because those who do not respond might have different views from those who do respond.

162. Two wordings for a questionnaire on whether Guamanians want independence from the U.S. are as follows.

I. Would you vote for sovereignty for Guam?
II. Would you support a Guam separate from the United States?

One of these questions showed 32 percent support for independence while the other showed 38 percent support. Which question produced which result and why?

(A) The first question showed the 38 percent support because of lack of randomization in choice of subjects as evidenced by the wording of the questions.
(B) The first question showed the 32 percent support because of a placebo effect.
(C) The first question showed the 38 percent support due to lack of blocking.
(D) The first question showed the 32 percent support because of response bias due to the wording of the question.
(E) The first question showed the 38 percent support because of response bias due to the wording of the question.

163. When conducting surveys, *bias* refers to which of the following?

(A) Lack of a control group
(B) Difficulty in concluding cause and effect
(C) Confounding variables
(D) An example of sampling error
(E) A tendency to favor the selection of certain members of a population

164. To find out a town's average family size, a researcher interviews a random sample of parents at a school's PTA meeting. The average family size in the 50-family sample is 3.72. Is this estimate probably too low or too high?

(A) Too low, because of undercoverage bias
(B) Too low, because convenience samples underestimate average results
(C) Too high, because of undercoverage bias
(D) Too high, because convenience samples overestimate average results
(E) Too high, because voluntary response samples overestimate average results

165. Which of the following is a true statement about sampling?

(A) There is no such thing as a "bad sample."
(B) Sampling techniques that use probability techniques effectively eliminate bias.
(C) Sampling techniques that allow the surveyor to choose participants with care and precision go a long way in controlling bias.
(D) If bias is present in a sampling procedure, it can be overcome by dramatically increasing the sample size.
(E) When choosing a sample, sample size is more important than the fraction of the population that is surveyed.

166. In general, for a survey to yield usable results,

(A) sampling error must be avoided
(B) a sample size of at least $n = 30$ is necessary
(C) researchers must carefully choose people who they think are representative of the population
(D) researchers must be careful in the way questions are worded
(E) a census should be strived for as it is the only truly accurate methodology

167. A researcher planning a survey of high school mathematics teachers in Illinois has faculty lists for each of the 97 high school districts in the state. The procedure is to obtain a simple random sample of teachers from each of these districts rather than grouping all the districts together and obtaining a sample from the entire group. Which of the following is *not* a true statement about the resulting stratified sample?

(A) It is more susceptible to bias than a simple random sample.
(B) It is easier and more cost-effective than a simple random sample.
(C) It gives comparative information that a simple random sample wouldn't give.
(D) It recognizes that opinions of teachers in rural districts may differ from those in urban districts.
(E) All of the above are true statements.

168. A national magazine aimed at college students asks its subscribers if they would have chosen their current college if they had to make the choice over again. Of the 12,000 or so responses, 75 percent said no. What does this show?

(A) The survey is meaningless because of voluntary response bias.
(B) No meaningful conclusion is possible without knowing something more about the characteristics of the subscribers.
(C) The survey would have been more meaningful if it had picked a random sample of the 12,000 subscribers who responded.
(D) The survey would have been more meaningful if it had used a control group.
(E) This was a legitimate sample, randomly drawn from the subscribers and of sufficient size to allow the conclusion that most of the subscribers would have chosen a different college if they had to make the choice over again.

169. Internet surveys have been increasing dramatically. Which of the following is *not* true about such surveys?

(A) They allow for the use of multimedia elements not available to other survey modes.
(B) There is less response bias than associated with interviewer-administered modes.
(C) There is less nonresponse and undercoverage bias.
(D) They have a low cost compared with most other types of surveys.
(E) It is convenient for respondents to take the surveys at their own time and own pace.

EXPERIMENTS AND OBSERVATIONAL STUDIES

170. An advantage to using observational studies as opposed to experiments is that

(A) observational studies are generally cheaper to conduct
(B) concluding cause and effect is generally easier from observational studies
(C) observational studies are generally not subject to bias
(D) observational studies involve the use of randomization
(E) observational studies can make use of stratification

171. Two studies are run to compare the number of doctor visits of low-income families enrolled in the Children's Health Insurance Program (CHIP) to those receiving cash subsidies for health care. The first study interviews 75 families who have been in each government program for at least five years, while the second randomly assigns 75 families to each program and interviews them after five years. Which of the following is a true statement?

(A) Both studies are observational studies because of the time period involved.
(B) Both studies are observational studies because there are no control groups.
(C) The first study is an observational study, while the second is an experiment.
(D) The first study is an experiment, while the second is an observational study.
(E) Both studies are experiments because in each, families are receiving treatments (due to either CHIP or cash).

172. Which of the following is most useful in establishing cause-and-effect relationships?

(A) A complete census
(B) A least squares regression line showing high correlation
(C) A simple random sample (SRS)
(D) A well-designed, well-conducted observational study incorporating chance to ensure a representative sample
(E) A controlled experiment

173. Some medical investigations have indicated that nondrinkers are more likely to develop dementia than drinkers. Assuming these investigations were carefully carried out, what is the most reasonable conclusion?

(A) These were probably observational studies, and so no conclusion about drinking providing protection against dementia is proper.
(B) Given that these investigations were carefully carried out, they probably were experiments. Thus they do show that drinking provides protection against dementia (although it may have other undesired outcomes).
(C) Without information about the use or nonuse of randomization, no conclusion about cause and effect is possible.
(D) Without the use of blinding—for example, having all participants drink something (alcohol or a harmless placebo)—no conclusion is proper.
(E) No matter how carefully the investigations were carried out, no conclusion is possible without knowing the sample sizes.

174. A critical difference between experiments and observational studies is that

(A) experiments are free to choose subjects from an entire population while an observational study considers only a random sample
(B) tests of significance can be used on data collected from experiments but not on data from observational studies
(C) observational studies make use of randomization, while experiments do not
(D) experiments are generally more cost-effective and time-effective than observational studies
(E) an experiment often suggests a causal relationship, while an observational study only suggests an association

175. Which of the following is a true statement?

(A) The purposes behind random sampling and random assignment are different.
(B) In well-designed observational studies, responses are systematically influenced during the collection of data.
(C) In well-designed experiments, the treatments result in responses that are as similar as possible.
(D) A well-designed experiment always has a single treatment but may test that treatment at different levels.
(E) Randomized block design refers to deciding randomly which blocks receive which treatments.

176. In which of the following studies are *cause-and-effect* conclusions (rather than simple *association* conclusions) probably reasonable?

(A) Studies showing that drivers who speed more than 25 percent above the speed limit have higher mortality rates in accidents than those who do not speed.
(B) Studies showing that college students who regularly use illegal drugs have lower GPAs than students who don't.
(C) Studies showing that teenagers using a particular skin cream tend to have fewer skin problems than teenagers not using that cream.
(D) Studies noting weights and ages at death tend to show that underweight people live longer than overweight people.
(E) Studies noting that children born to parents who are nonsmokers and nondrinkers tend to engage in less risky behavior as adults.

177. Suppose you wish to compare the average age of math/science teachers to the average age of English/social studies teachers in your high school. Which is the most appropriate technique for gathering the needed data?

(A) Census
(B) Sample survey
(C) Experiment
(D) Observational study
(E) None of these methods is appropriate.

178. Does "an apple a day keep the doctor away?" A study is proposed to pick a random sample of 100 healthy adults, divide them into two groups based on who eats apples regularly and who doesn't, follow them for five years, and calculate the mean number of doctor visits of each group. Which of the following is a correct statement?

(A) This is an observational study because no treatment is applied.
(B) This is an experiment, but one design fault is that the two groups might be different sizes.
(C) This is an experiment where eating apples is the treatment but there is no control.
(D) This is an experiment, but no causation can be concluded because there was no random assignment.
(E) This is an experiment, but no causation can be concluded because of confounding variables.

179. Personnel entering the armed forces undergo extensive medical testing, and these records are a valuable source of data. Medical researchers compared vitamin D blood level records of military personnel who later developed multiple sclerosis (MS) with the vitamin D blood level records of military personnel who never developed MS. What can be said about this study?

(A) This is an observational study, and the subjects are military personnel. Two variables are considered, vitamin D blood levels and whether or not the subject develops MS.
(B) This is an experiment where the treatments are the different levels of vitamin D in the blood.
(C) This is an experiment where the response variable is whether or not the subject develops MS.
(D) This is an experiment where the subjects who do not develop MS act as a control group.
(E) This is an experiment. However, causation cannot be concluded because there is no random assignment.

180. In a study of medical patches to help prevent the urge to smoke, 500 heavy smokers (at least two packs per day) were randomly sorted into two groups. The smokers in one group were given a new nicotine patch, while smokers in the other group were given a similar-looking patch containing no medication. The subjects did not know who received the drug. In the weeks to follow, all 500 smokers showed a similar reduction in the urge to smoke. This is an example of which of the following?

(A) The effect of a treatment unit
(B) The placebo effect
(C) The control group effect
(D) Sampling error
(E) Voluntary response bias

181. One hundred patients suffering from severe back pain are randomly selected from hospital records. Half the patients are told to listen to classical music and sit in the dark the next time they experience severe pain. The remaining patients are told to use neither of these possible remedies. Participants then report back as to relief, if any. Faults of this experimental design include all of the following *except* which response?

(A) Lack of randomization
(B) Confounding variables
(C) Lack of blinding
(D) Unclear factor levels
(E) Measurement of response variable

182. Which of the following are most important in minimizing the placebo effect?

(A) Randomization and blinding
(B) Randomization and a control
(C) Replication and randomization
(D) Replication and blinding
(E) Blinding and a control

183. Which of the following is a true statement about blocking?

(A) The paired (matched pairs) comparison design is a special case of blocking.
(B) Blocking is a useful procedure when there are certain attributes, which are not under study, that may affect the outcomes.
(C) Blocking is to experiment design as stratification is to sampling design.
(D) By controlling certain variables, blocking can make conclusions more specific.
(E) All of the above are true statements about blocking.

184. Two procedures are to be compared in the treatment of autism: the use of antipsychotic medication versus family therapy. The experimental design is to create homogeneous blocks with respect to the severity of the developmental disorder. How should randomization be used for a randomized block design?

(A) Within each block, randomly pick half the patients to receive each procedure.
(B) Randomly pick half of all patients to receive each procedure, but then analyze the results separately by blocks.
(C) Randomly choose which blocks will receive which procedure.
(D) Randomly choose half the blocks to receive each procedure for a given time period. Then for the same time period, switch the procedure in each block and compare the results.
(E) For ethical reasons, allow families to choose which procedure they prefer taking, but then randomly assign patients to the blocks.

185. Before playing a soccer game, players rested for 2, 4, or 6 hours. Half of each group was given an energy drink before starting the game. Determine the number of factors, levels for each factor, and number of treatments.

(A) 1 factor with 2 levels; 5 treatments
(B) 2 factors, one with 1 level and one with 2 levels; 3 treatments
(C) 2 factors, one with 2 levels and one with 3 levels; 5 treatments
(D) 2 factors, one with 2 levels and one with 3 levels; 6 treatments
(E) 3 factors, each with 2 levels; 6 treatments

186. A medical researcher is running an experiment on the treatment of carotid artery stenosis with an invasive procedure (stenting or surgery) and a blood thinner (Plavix or aspirin). She has 40 volunteers, half men and half women, all suffering from symptomatic carotid artery stenosis. The researcher randomly selects 10 men and 10 women for treatment with stenting and Plavix, while the remaining volunteers are treated with surgery and aspirin. Of the following, which is the most important observation about this procedure?

(A) The variables—invasive procedure and blood thinner—are confounded.
(B) The variables—gender and invasive procedure—are confounded.
(C) The variables—gender and blood thinner—are confounded.
(D) No variables are confounded.
(E) There is a hidden confounding variable.

187. A sports product agency measures average feet per second for balls hit by high school baseball players using each of four different brand bats (Easton, Louisville Slugger, Rawlings, and DeMarini). Which of the following is true?

(A) There are four explanatory variables and one response variable.
(B) There is one explanatory variable with four levels of response.
(C) Feet per second is the only explanatory variable, but there are four response variables corresponding to the different brands.
(D) There are four levels of a single explanatory variable.
(E) Each explanatory level has an associated level of response.

188. In a random survey of 30 students at an elementary school, those with greater weight appear to have higher math levels. Of the following, which is the most important conclusion about this observation?

(A) Parents interested in their child's math level should feed their children more.
(B) Generalizing from a single elementary school and a single study is unreasonable.
(C) The sample size is too small for any reasonable conclusion.
(D) There is a confounding variable.
(E) As long as the sample was a simple random sample (SRS) from among all students at the school, a cause-and-effect conclusion is valid.

189. A sociologist and musician team wants to estimate the proportion of adults who sing in the shower. They go to a Broadway musical and a Lincoln Center concert and then randomly pick 25 adults from each event to survey. If the team members want to be able to generalize their findings, what are the appropriate sample and population?

(A) The sample is the 50 musical/concert attendees, and the population is all adults.
(B) The sample is the 50 musical/concert attendees, and the population is all adults who attend Broadway musicals or Lincoln Center concerts.
(C) The sample is the 50 musical/concert attendees, and the population is all adults who enjoy music.
(D) The sample is all adults who attend Broadway musicals or Lincoln Center concerts, and the population is all adults.
(E) The sample is all adults who attend Broadway musicals or Lincoln Center concerts, and the population is all adults who enjoy music.

190. Which of the following best explains why researchers try to guard against confounding when designing experiments?

(A) Confounding can conflict with randomization.
(B) Confounding can negate the benefits of blinding.
(C) Confounding can lead to bias.
(D) Confounding can lead to uncertainty as to which variable is causing an effect.
(E) Confounding can make it more difficult to separate subjects into treatment and control groups.

191. A psychologist takes 50 volunteers suffering from acrophobia to the top of the Empire State Building and instructs them to stay there for 12 straight hours in an effort to cure their fear of heights. He contacts them a week later, and 60 percent claim to no longer have a fear of heights. Which of the following is a true statement?

(A) Because the subjects were volunteers, their desire to be cured could be a confounding variable.
(B) With 60 percent claiming a cure after a single treatment, this is statistically significant.
(C) With this large of a sample, $n = 50 \geq 30$, the positive conclusions can be generalized to the general population of people suffering from acrophobia.
(D) Bringing the subjects to the top of the Empire State Building is a treatment, so this is an experiment and positive conclusions are justified.
(E) Because the subjects were volunteers, positive conclusions can be generalized only to other volunteers.

192. A nutritionist conducts a study on serving size. As each attendee arrived at an ice cream social, they were handed either a 12-ounce or an 18-ounce bowl depending upon whether the next digit in a random digit table was odd or even. The nutritionist then noted how much ice cream each attendee put into his or her bowl. All attendees participated. The nutritionist compared the mean helping of ice cream taken by those given 12-ounce bowls versus those given 18-ounce bowls. What method of testing is the nutritionist using?

(A) A census because all attendees participated
(B) An observational study because the nutritionist did not tell anyone how much ice cream to take but, rather, simply observed
(C) An experiment with a blocked design, blocked on size of bowl
(D) An experiment with a completely randomized design
(E) An experiment with a matched-pair design

193. A winemaker is undecided as to which of two grape varieties, *A* or *B*, to use in the production of malbec. He runs a test with a random sample of wine connoisseurs. The winemaker randomly assigns each participant to one of four groups and then has taste tests as follows.

	Taste First	**Taste Second**
Group 1	Variety *A*, Bottle labeled "10"	Variety *B*, Bottle labeled "20"
Group 2	Variety *B*, Bottle labeled "20"	Variety *A*, Bottle labeled "10"
Group 3	Variety *B*, Bottle labeled "10"	Variety *A*, Bottle labeled "20"
Group 4	Variety *A*, Bottle labeled "20"	Variety *B*, Bottle labeled "10"

Which of the following explains the use of random assignment here?

(A) The connoisseurs might be partial to whichever wine they taste first. So groups 1 and 4 will first taste the malbec made from grape variety *A*, while groups 2 and 3 will first taste the malbec made from grape variety *B*.
(B) The connoisseurs might subconsciously be partial to the bottle labeling, "10" and "20." So groups 1 and 2 will taste from bottles labeled "10" with variety *A* grapes and bottles labeled "20" with variety *B* grapes. Groups 3 and 4 will taste from bottles labeled "10" with variety *B* grapes and bottles labeled "20" with variety *A* grapes.
(C) Random assignment in this context is needed to minimize the effect of confounding variables, the most obvious of which are the order of tasting and the labels on the bottles.
(D) All three of the above statements explain the use of random assignment here.
(E) This study is really an observational study, so random assignment does not apply.

194. Twenty volunteers with chronic backache agreed to try yoga exercises for a month, at the end of which 45 percent reported a lower level of back pain. An influential yoga instructor announced on his TV program that yoga exercises are better at relieving back pain than not doing anything. Which of the following *best* explains why the results of the study do not support the yoga instructor's TV claim?

(A) The subjects were volunteers.
(B) The subjects self-reported pain levels.
(C) The sample was small.
(D) There was no control group.
(E) There was no randomization.

195. A randomized block design will be used in an experiment to test whether or not having alcoholic drinks affects drug effectiveness for people taking different medications. Which of the following relates to the composition of the blocks?

(A) Subjects should be randomly assigned to the blocks.
(B) Subjects in the same block should all receive the same treatment (alcohol or no alcohol).
(C) Subjects in the same block should be as similar as possible (all should be taking the same medication).
(D) Subjects in the same block should be as different as possible (all the medications under study should be represented by subjects in each block).
(E) Participants in each block should be blinded as to which medication they are taking.

196. A high school manager runs a study on whether players can hit a ball farther with an aluminum or a wood bat. For 10 games, as each player comes to bat, the manager reads off the next digit from a random digit table. If the digit is odd, the player uses an aluminum bat. If the digit is even, the player uses a wood bat. What best describes the manager's study?

(A) An observational study
(B) A completely randomized design
(C) A matched-pair design
(D) A randomized block design with two blocks
(E) A flawed experiment because the two types of bat can be confounded

197. What is the purpose of blocking in experimental design?

(A) To control the level of an experiment
(B) To be a first step in randomization
(C) To reduce bias
(D) To reduce variation
(E) To provide a substitute for a control group

198. A botanist tests three different strengths of a new fertilizer on the growth of tomato plants. Which of the following is a true statement?

(A) There are three explanatory variables and one response variable.
(B) There is one explanatory variable with three levels of response variable.
(C) Growth is the only explanatory variable, but there are three response variables corresponding to the three strengths.
(D) There are three levels of a single explanatory variable.
(E) Each explanatory variable has an associated level of response.

199. Which of the following is a true statement about the design of matched pair experiments?

(A) Randomization is unnecessary in matched pair designs.
(B) Blocking is one form of matched pair design.
(C) Each subject might receive both treatments.
(D) Stratification into two equal-sized strata is an example of matched pairs.
(E) Each pair of subjects receives identical treatment, and the differences in their responses are noted.

200. Some researchers believe that drinking a glass of red wine every day can raise the level of HDL, the "good" cholesterol. A study is performed by randomly selecting half of a group of volunteers to drink a glass of red wine every day while the rest are instructed not to drink any red wine. Is this an experiment or an observational study?

(A) An experiment with a single factor
(B) An observational study with comparison and randomization
(C) An experiment with a control group and blinding
(D) An observational study with no bias
(E) An experiment with blocking

MC Probability

Answers for Chapter 3 are on pages 262–277.

CHAPTER 3

BASIC PROBABILITY RULES

201. The following is from a particular region's mortality table.

Age	0	20	40	60	80
Number Surviving	10,000	9,800	9,350	8,600	5,100

What is the probability that a 20-year-old will survive to be 60?

(A) $\frac{1,400}{9,800}$

(B) $\frac{1,400}{10,000}$

(C) $\frac{8,600}{9,800}$

(D) $\frac{8,600}{10,000}$

(E) $\frac{9,800}{10,000}$

202. The pattern on the back of an insect causes an entomologist to suspect that she has found a rare subspecies of the insect. In the rare subspecies, 96 percent have the pattern. In the common subspecies, only 3 percent have the pattern. The rare subspecies accounts for 0.2 percent of the population. Which expression is used to calculate the probability that a bug having this pattern is rare?

(A) $(0.002)(0.96)$

(B) $1 - (0.002)(0.96)$

(C) $\frac{(0.002)(0.96)}{(0.998)(0.03)}$

(D) $\frac{(0.002)(0.96)}{(0.002)(0.96) + (0.998)(0.03)}$

(E) $\frac{(0.998)(0.03)}{(0.002)(0.96) + (0.998)(0.03)}$

203. A student calculates the probabilities of the four outcomes of an experiment to be 0, 0.25, 0.80, and –0.05. The proper conclusion is that

(A) the sum of the individual probabilities is 1
(B) one of the outcomes will never occur
(C) one of the outcomes will occur over 50 percent of the time
(D) all of the above are true
(E) the student made an error

204. There are two games involving flipping a fair coin. In the first game, you win a prize if you can throw between 40% and 60% heads. In the second game, you win a prize if you can throw more than 60% heads. For each game, would you rather flip the coin 15 times or 150 times?

(A) 15 times for each game
(B) 150 times for each game
(C) 15 times for the first game, and 150 times for the second game
(D) 150 times for the first game, and 15 times for the second game
(E) The outcomes of the games do not depend on the number of flips.

***205. Which of the following is *not* a probability density function?**

(A) $f(x) = 1, \quad 0 \le x \le 1$
(B) $f(x) = 0.05, \quad 0 \le x \le 20$

(C) $f(x) = \begin{cases} 0.3, & 0 \le x \le 2 \\ 0.2, & 2 < x \le 4 \end{cases}$

(D) $f(x) = 2x, \quad 0 \le x \le 1$
(E) $f(x) = 4x, \quad 0 \le x \le 0.25$

206. Suppose that during any weekend afternoon, the probabilities that you receive a text message is 0.94, an e-mail is 0.75, and both a text and an e-mail is 0.705. Are receiving a text message and receiving an e-mail independent events?

(A) Yes, because $(0.94)(0.75) = 0.705$
(B) No, because $(0.94)(0.75) = 0.705$
(C) Yes, because $0.94 > 0.75 > 0.705$
(D) No, because $0.5(0.94 + 0.75) \neq 0.705$
(E) There is insufficient information to answer this question.

207. Suppose you toss a fair coin five times and it comes up tails every time. Which of the following is a true statement?

(A) By the law of large numbers, the next toss is more likely to be heads rather than another tail.
(B) By the properties of conditional probability, the next toss is more likely to be tails given that five tosses in a row have been tails.
(C) Coins actually do have memories, and thus what comes up on the next toss is influenced by the past tosses.
(D) The law of large numbers tells how many tosses will be necessary before the percentages of heads and tails are again in balance.
(E) None of the above are true statements.

208. Suppose $P(X) = 0.36$ and $P(Y) = 0.41$. If $P(X \mid Y) = 0.27$, what is $P(Y \mid X)$?

(A) $\frac{(0.36)(0.41)}{0.27}$

(B) $\frac{(0.36)(0.27)}{0.41}$

(C) $\frac{(0.27)(0.41)}{0.36}$

(D) $\frac{0.27}{(0.36)(0.41)}$

(E) $\frac{0.27}{0.36 + 0.41}$

209. Body temperatures of healthy humans are roughly normal. If 10 percent of people have temperatures above 37.333°C and if 20 percent have temperatures below 36.781°C, what is the mean of this distribution?

(A) 36.965°C
(B) 37.000°C
(C) 37.057°C
(D) 37.149°C
(E) The mean cannot be calculated from the given information.

210. Given the probabilities $P(E) = 0.25$ and $P(E \cup F) = 0.64$, what is the probability $P(F)$ if E and F are mutually exclusive?

(A) $\frac{0.25}{0.64}$

(B) $0.64 - 0.25$

(C) $0.25 + 0.64$

(D) $\frac{0.64 - 0.25}{0.75}$

(E) $\frac{0.64 - 0.25}{1.25}$

211. Given the probabilities $P(E) = 0.35$ and $P(E \cup F) = 0.73$, what is the probability $P(F)$ if E and F are independent?

(A) $\frac{0.35}{0.73}$

(B) $0.73 - 0.35$

(C) $0.35 + 0.73$

(D) $\frac{0.73 - 0.35}{0.65}$

(E) $\frac{0.73 - 0.35}{1.35}$

212. If $P(A) = 0.32$ and $P(B) = 0.45$, what is $P(A \cup B)$ if A and B are independent?

(A) 0.144
(B) 0.626
(C) 0.770
(D) 0.856
(E) There is insufficient information to answer this question.

*213. A marksman hits exactly one of the first two clay pigeons thrown. From that point on, the probability that he makes the next shot is equal to the proportion of shots made up to that point. If he takes two more shots, what is the probability he ends up making a total of exactly two shots?

(A) $\frac{1}{4}$

(B) $\frac{1}{3}$

(C) $\frac{1}{2}$

(D) $\frac{2}{3}$

(E) $\frac{3}{4}$

214. In wake of the 2016 Ebola crisis, a rapid test was developed that tested positive in 99.5 percent of patients with Ebola but gave a false positive in 3 percent of healthy people. Suppose 0.1 percent of the population in a particular area has Ebola. If a person in this area tests positive, what is the probability he or she has Ebola?

(A) $\frac{0.001}{0.995}$

(B) $\frac{(0.001)(0.995)}{(0.001)(0.995)+(0.999)(0.03)}$

(C) $\frac{(0.999)(0.03)}{(0.001)(0.995)+(0.999)(0.03)}$

(D) $\frac{(0.995)(0.03)}{(0.001)(0.995)+(0.999)(0.03)}$

(E) $\frac{(0.001)(0.999)}{(0.001)(0.995)+(0.999)(0.03)}$

215. Given two events, E and F, such that $P(E) = 0.420$, $P(F) = 0.350$, and $P(E \cup F) = 0.623$, the two events are

(A) independent and mutually exclusive
(B) independent but not mutually exclusive
(C) mutually exclusive but not independent
(D) neither independent nor mutually exclusive
(E) There is not enough information to answer this question.

216. Which of the following is *not* a valid discrete probability distribution for the set $\{x_1, x_2, x_3\}$?

(A) $P(x_1) = 0, \quad P(x_2) = 1, \quad P(x_3) = 0$

(B) $P(x_1) = \frac{1}{3}, \quad P(x_2) = \frac{1}{3}, \quad P(x_3) = \frac{1}{3}$

(C) $P(x_1) = \frac{1}{6}, \quad P(x_2) = \frac{1}{3}, \quad P(x_3) = \frac{1}{2}$

(D) $P(x_1) = \frac{2}{5}, \quad P(x_2) = \frac{4}{5}, \quad P(x_3) = -\frac{1}{5}$

(E) All of the above are valid probability distributions.

217. **Suppose that when taking a multiple-choice exam and when students have no idea what an answer is, 45 percent will guess, 35 percent will choose answer (C), and the rest will choose the longest answer. If a student who has no idea of the answer doesn't guess, what is the probability he chooses the longest answer?**

(A) 0.20

(B) $\frac{0.20}{0.45}$

(C) $\frac{0.20}{0.55}$

(D) $\frac{0.20}{0.80}$

(E) $1 - \frac{0.20}{0.80}$

218. **Given that 51.0 percent of the U.S. population are female and that 5.9 percent of the population are over 75 years of age, can we conclude that (0.510)(0.059) = 3.01 percent of the population are women older than 75?**

(A) Yes, by the multiplication rule
(B) Yes, by conditional probabilities
(C) Yes, by the law of large numbers
(D) No, because the events are not independent
(E) No, because the events are not mutually exclusive

*219. **In a set of 10 boxes, 7 boxes each contain 3 red and 2 blue marbles, while the remaining boxes each contain 1 red and 4 blue marbles. A player randomly picks a box and then randomly picks a marble from that box. She wins if she ends up with a red marble. If she plays 4 times, what is the probability she wins exactly twice?**

(A) $4(0.48)^2(0.52)^2$
(B) $6(0.48)^2(0.52)^2$
(C) $4(0.7)^2(0.3)^2$
(D) $6(0.8)^2(0.2)^2$
(E) $4(0.32)^2(0.68)^2$

220. **Suppose E and F are independent events with $P(E) = 0.4$ and $P(E \text{ and } F) = 0.15$. Which of the following is a true statement?**

(A) $P(F) = 0.4$
(B) $P(F) = 0.6$
(C) $P(E \text{ or } F) = 0.375$
(D) $P(E \text{ or } F) = 0.625$
(E) $P(E \text{ or } F) = 0.775$

221. Suppose that for a certain midwestern city, in any given year the probability of a tornado hitting is 0.15, the probability of flooding from the river running through the city is 0.08, and the probability of both a tornado and flooding is 0.02. What is the probability of flooding given that a tornado hits?

(A) $\frac{0.02}{0.08}$

(B) $\frac{0.02}{0.15}$

(C) $\frac{0.08}{0.15}$

(D) $\frac{0.02}{(0.15)(0.08)}$

(E) $\frac{(0.02)(0.08)}{0.15}$

222. A dentist compiles the following summary data from interviews of her patients.

> 55% floss once a day, and these patients have a 0.04 probability of a cavity each year.
>
> 30% floss twice a day, and these patients have a 0.01 probability of a cavity each year.
>
> 15% don't floss, and these patients have a 0.10 probability of a cavity each year.

What is the probability that one of the dentist's patients flosses and has a cavity for any given year?

(A) $0.04 + 0.01$
(B) $0.04 + 0.01 + 0.10$
(C) $(0.55 + 0.30)(0.04 + 0.01)$
(D) $(0.55)(0.04) + (0.30)(0.01)$
(E) $\frac{(0.55)(0.04) + (0.30)(0.01)}{(0.55)(0.04) + (0.30)(0.01) + (0.15)(0.10)}$

*223. A game show offers the contestant who wins the grand prize a choice of three boxes, one containing 4 red chips and 1 white chip, one containing 2 red chips and 2 white chips, and one containing 1 red chip and 3 white chips. The contestant randomly picks a box and with eyes closed, reaches in and picks a chip. A white chip earns \$5,000 and a red chip earns \$10,000. The contestant then tosses a fair die. If it comes up "6," the prize is doubled. Does the contestant have more than an even chance of walking away with exactly \$10,000?

(A) Yes, because $\frac{23}{45} > \frac{1}{2}$

(B) No, because $\frac{22}{45} < \frac{1}{2}$

(C) Yes, because $\frac{31}{60} > \frac{1}{2}$

(D) No, because $\frac{29}{60} < \frac{1}{2}$

(E) The contestant has a 0.5 chance of walking away with exactly \$10,000.

224. Nine-tenths of adults change their jobs at least once. The reasons for changing and what they change to are as shown in the following tables.

Reason	Change in Career Interests	Higher Salary	Fired	Other Reason
Probability	0.55	0.10	0.20	0.15

Change To	Related Field	New Field	Unemployed
Probability	0.40	0.55	0.05

Assuming independence of all variables, what is the probability that an adult decides to change a job, based on higher salary, and move into a new field?

(A) 0.0495
(B) 0.055
(C) 0.585
(D) 0.595
(E) 0.65

225. Suppose two events, S and T, have the nonzero probabilities p and q, respectively. Which of the following is *impossible*?

(A) $p + q > 1$
(B) $p - q < 0$
(C) $\frac{p}{q} > 1$
(D) S and T are both independent and mutually exclusive.
(E) S and T are neither independent nor mutually exclusive.

226. A computer security company puts in bids for a state contract and for a federal contract. The company's chief financial officer believes that the probabilities of being awarded the state and federal grants are 0.32 and 0.15, respectively. If the probability of receiving both grants is 0.048 and if these three probabilities are correct, which of the following statements is true?

(A) Receiving the two grants are independent and mutually exclusive events.
(B) Receiving the two grants are independent but not mutually exclusive events.
(C) Receiving the two grants are mutually exclusive but not independent events.
(D) Receiving the two grants are not mutually exclusive and not independent events.
(E) None of the above statements can be known for sure until the probability of receiving at least one grant is known.

227. According to a college's records, 46 percent of its students live in the dorms, 65 percent have meal contracts, and 33 percent do both. For a randomly selected student, what is the probability that he/she either lives in a dorm or has a meal contract, but not both?

(A) 0.13
(B) 0.22
(C) 0.32
(D) 0.45
(E) 0.78

228. The distribution of jelly beans in a large assorted bag is shown in the following table.

Flavor	Very Cherry	Vanilla	Green Apple	Licorice	Peach
Probability	0.4	0.2	0.2	0.1	0.1

If two jelly beans are picked at random, what is the probability they are different flavors?

(A) 0.2
(B) 0.26
(C) 0.37
(D) 0.74
(E) 0.63

229. A medical researcher examines the records of cases where a child comes in to a pediatrician complaining of a sore throat and the next day a sibling comes in also complaining of a sore throat. Suppose 20 percent of children with sore throats test positive for strep throat. If a child tests positive for strep, there is a 0.7 probability that his/her sibling who comes in the next day will also test positive for strep. If a child tests negative for strep, there is a 0.1 probability that his/her sibling who comes in the next day will test positive for strep. Let S be the random variable for the number of children in such pairs who test positive for strep. What is the probability distribution for S?

(A)

x	0	1	2
$P(x)$	1/3	1/3	1/3

(B)

x	0	1	2
$P(x)$	0.5	0.25	0.25

(C)

x	0	1	2
$P(x)$	0.5	0.33	0.17

(D)

x	0	1	2
$P(x)$	0.72	0.14	0.14

(E)

x	0	1	2
$P(x)$	0.8	0.06	0.14

230. 85 percent of people wear seat belts. The probability of serious injury in an accident is 8 percent for those wearing seat belts and 36 percent for those not wearing seat belts. What is the probability of serious injury in an accident?

(A) 0.054
(B) 0.068
(C) 0.122
(D) 0.318
(E) 0.44

231. At many colleges, many nonmath majors choose to take at least one math/computer science course. Suppose 16 percent take computer science, 53 percent take statistics, and 62 percent take at least one of these two offerings. What is the probability a randomly chosen nonmath major takes both a computer science class and a statistics class?

(A) (0.16)(0.53)
(B) 0.16 + 0.53 – 0.62
(C) 0.62 – (0.16)(0.53)
(D) (0.62 – 0.16)(0.62 – 0.53)
(E) $\frac{(0.16)(0.53)}{0.62}$

*232. The following are parts of the probability distributions for the random variables X and Y.

x	$P(x)$
1	0.1
2	0.4
3	?

y	$P(y)$
1	0.2
2	?
3	0.3
4	?

If X and Y are independent and the joint probability $P(X = 3, Y = 2) = 0.1$, what is $P(Y = 4)$?

(A) 0.1
(B) 0.2
(C) 0.3
(D) 0.4
(E) 0.5

233. A research firm is successful in contacting 65 percent of the households randomly selected for telephone surveys. What is the probability that the firm is successful in contacting a household given that it was unsuccessful in contacting the previous two households?

(A) 0.65
(B) $(0.35)^2(0.65)$
(C) $3(0.35)^2(0.65)$
(D) $1 - (0.35)^2(0.65)$
(E) $1 - [(0.35)^3 + (0.35)^2(0.65)]$

BINOMIAL AND GEOMETRIC PROBABILITIES

234. Suppose only 70 percent of claims on a particular news program are accurate. If four independent claims are made during one program, what is the probability that less than half are accurate?

(A) $6(0.7)^2(0.3)^2$
(B) $(0.3)^4 + 4(0.7)(0.3)^3$
(C) $(0.3)^4 + 4(0.7)(0.3)^3 + 6(0.7)2(0.3)^2$
(D) $1 - [(0.3)^4 + 4(0.7)(0.3)^3]$
(E) $1 - [(0.7)^4 + 4(0.3)(0.7)^3]$

235. Which of the following statements is *incorrect*?

(A) The histogram of a binomial distribution with $p = 0.5$ is always symmetric no matter the value of n, the number of trials.
(B) The histogram of a binomial distribution with $p = 0.1$ is unimodal.
(C) The histogram of a binomial distribution with $p = 0.1$ is skewed to the left.
(D) The histogram of a binomial distribution with $p = 0.01$ looks more and more symmetric, the larger the value of n.
(E) The histogram of a binomial distribution with $p = 0.001$ looks more and more like a normal distribution the larger the value of n.

236. An inspection procedure at a manufacturing plant involves picking three items at random and then accepting the whole lot if at least two of the three items are in perfect condition. If in reality 91 percent of the whole lot is perfect, what is the probability that the lot will be accepted?

(A) $(0.91)^2$
(B) $(0.91)^2 + (0.91)^3$
(C) $3(0.91)^2 + (0.91)^3$
(D) $3(0.09)(0.91)^2 + (0.91)^3$
(E) $1 - (0.91)^3$

237. In a recent poll, only 6 percent of the American public say they have significant confidence in Congress. In a random sample of five people, what is the probability that at least one has significant confidence in Congress?

(A) $(0.06)(0.94)^4$
(B) $5(0.06)(0.94)^4$
(C) $5(0.06)(0.94)^4 + (0.94)^5$
(D) $1 - (0.94)^5$
(E) $1 - [5(0.06)(0.94)^4 + (0.94)^5]$

238. Suppose we have a random variable X where the probability associated with the value k is $\binom{15}{k}(0.28)^k(0.72)^{15-k}$ for $k = 0, \ldots, 15$. What is the mean of X?

(A) 0.28
(B) 0.72
(C) 4.2
(D) 10.8
(E) None of the above

239. It is estimated that 20 percent of all backseat passengers don't wear seat belts. In a random sample of 5 backseat passengers, what is the probability that at least two don't wear seat belts?

(A) $1 - (0.2)^5$
(B) $1 - [(0.8)^5 + 5(0.2)(0.8)^4]$
(C) $10(0.2)^2(.8)^3$
(D) $10(0.2)^3(0.8)^2 + 5(0.2)^4(0.8) + (0.2)^5$
(E) $10(0.2)^2(0.8)^3 + 5(0.2)(0.8)^4 + (0.8)^5$

240. You win a game if you toss a fair die with 5 coming up in exactly 1/6 of the tosses. Would you rather flip 12 times or 120 times?

(A) 12 times because $0.296 > 0.0973$
(B) 12 times because of the central limit theorem
(C) 120 times because ${}_{120}C_{20} > {}_{12}C_2$
(D) 120 times because of the law of large numbers
(E) Your chance of winning is the same with 12 or 120 flips.

241. A polygraph test (lie detector) will indicate a lie 5 percent of the time when a person is telling the truth and 98 percent of the time when a person is lying. Suppose three suspects, all innocent and telling the truth, are given polygraph tests. What is the probability that at least one of them is falsely accused of lying?

(A) $(0.02)(0.98)^2$
(B) $3(0.02)(0.98)^2$
(C) $3(0.05)(0.95)^2$
(D) $1 - (0.95)^3$
(E) $1 - (0.98)^3$

*242. Suppose we have a binomial random variable where the probability of exactly five successes is $\binom{n}{5} p^5 (0.41)^8$. What is the mean of the distribution?

(A) 2.05
(B) 2.95
(C) 3.28
(D) 5.33
(E) 7.67

243. For which of the following is a binomial an appropriate model?

(A) The number of 5s in 20 flips of an unfair die weighted so that even numbers come up twice as often as odd number
(B) The number of baskets in 20 shots where the probability of making the basket is either 0.92 or 0.34 depending upon whether it is a layup or a jump shot
(C) The number of tosses of a fair coin before tails appear on three consecutive tosses
(D) The number of rainy days in a given week
(E) The binomial is not appropriate for any of the above.

244. Suppose the Census Bureau reports a city as having 35 percent African-American residents. If 12 people are randomly selected for a jury, what is the probability that fewer than two of the chosen people are African-American?

(A) $(0.35)^{12} + 12(0.65)(0.35)^{11}$
(B) $(0.65)^{12} + (0.35)(0.65)^{11}$
(C) $1 - (0.65)^{12}$
(D) $1 - [(0.35)^{12} + 12(0.65)(0.35)^{11}]$
(E) $1 - [(0.65)^{12} + 12(0.35)(0.65)^{11}]$

245. A blind taste test is conducted with 6 volunteers. Each is presented with three small cups of water. For each subject, either two of the cups are tap water and the other is bottled water or two of the cups are bottled water and the other is tap water. Each volunteer guesses about which of his/her three cups of water is different from the other two. Suppose 4 volunteers give the correct answer. Could this have happened by chance if they all were blind guessing? What is the probability of at least 4 correct answers if the volunteers were all simply guessing?

(A) 0.0178
(B) 0.0823
(C) 0.1001
(D) 0.8999
(E) 0.9822

246. Which of the following graphs represents a binomial distribution with $n = 10$ and $p = 0.3$?

(A)

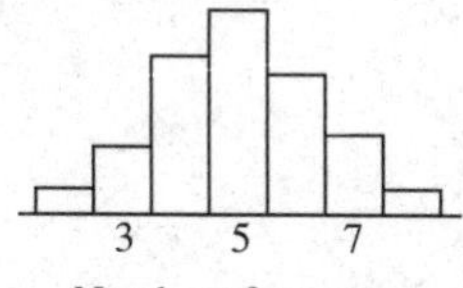

(D)

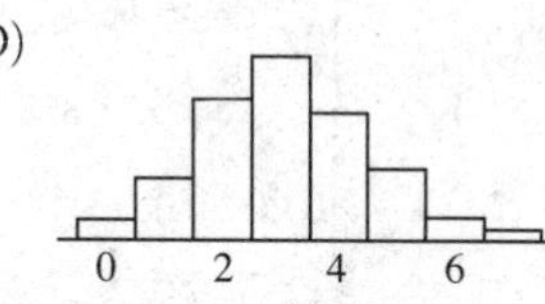

(B)

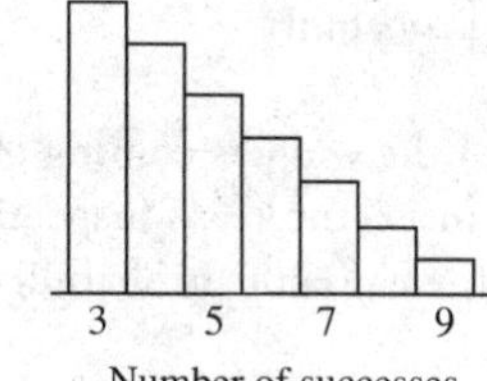

(E)

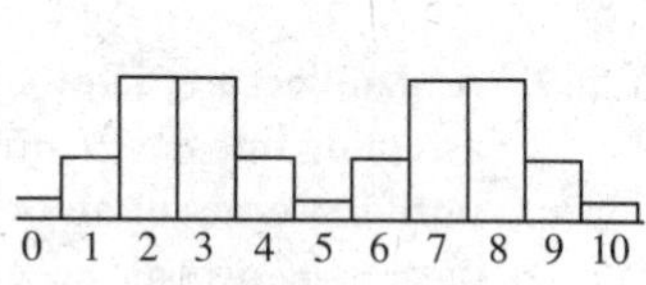

(C)

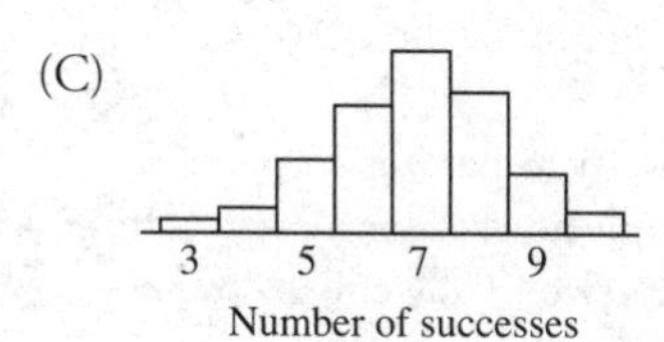

247. A rapid Zika antibody test can be performed in minutes and, if positive, can be followed up by a more time-consuming, more accurate test. Suppose the rapid test results in positive in 2 percent of the population and the more accurate test shows that 94 percent of those who test positive in the rapid test actually have Zika. Assuming that those testing negative in the rapid test do not have Zika, what is the probability that in a random sample of three people, none of them have Zika?

(A) 0.054
(B) 0.831
(C) 0.885
(D) 0.945
(E) 0.9812

248. A person has a 7 percent chance of winning the daily office lottery. What is the probability she first wins on the third day?

(A) $\binom{3}{1}(0.07)^2(0.93)$
(B) $\binom{3}{2}(0.07)(0.93)^2$
(C) $(0.07)^2(0.93)$
(D) $(0.07)(0.93)^2$
(E) None of the above gives the correct probability.

249. A manufacturer knows that 15 percent of the widgets coming off the assembly line have a minor defect. If an inspector keeps inspecting units until he comes upon one with the defect, what is the probability he will have to inspect at most four widgets?

(A) 0.1275
(B) 0.2775
(C) 0.3859
(D) 0.4780
(E) 0.5563

250. It has been estimated that 78.6 percent of all e-mail is spam. On a random day, what is the probability that the first spam e-mail you receive is the fourth e-mail of the day?

(A) $(0.214)^3$
(B) $(0.214)^3(0.786)$
(C) $(0.214)(0.786)^3$
(D) $4(0.214)^3(0.786)$
(E) $1 - (0.214)^3(0.786)$

RANDOM VARIABLES

251. Which of the following are true statements?

I. By the law of large numbers, the mean of a random variable will get closer and closer to a specific value.
II. The standard deviation of a random variable is never negative.
III. The standard deviation of a random variable is 0 only if the random variable takes a lone single value.

(A) I and II only
(B) I and III only
(C) II and III only
(D) I, II, and III
(E) None of the above gives the complete set of true responses.

252. A carnival game has three prizes with the following probabilities:

Prize ($)	0	25	50
Probability	0.8	0.16	0.04

What is the variance of the prize variable?

(A) $\frac{1}{3}\left[(0-6)^2+(25-6)^2+(50-6)^2\right]$

(B) $(25-6)^2(0.16)+(50-6)^2(0.04)$

(C) $(0-6)^2(0.8)+(25-6)^2(0.16)+(50-6)^2(0.04)$

(D) $\frac{1}{3}\left[(0-25)^2+(25-25)^2+(50-25)^2\right]$

(E) $(0-25)^2(0.8)+(25-25)^2(0.16)+(50-25)^2(0.04)$

253. An insurance company charges $750 annually for flood insurance policies for those living in designated flood zones. The policy specifies the company will pay $25,000 for major flood damage and $5,000 for minor flood damage. If the probability of a storm with minor flood damage during the year is 0.04 and of a storm with major flood damage is 0.01, how much can the insurance company expect to make on 25 policies?

(A) $ 300
(B) $ 450
(C) $ 7,500
(D) $11,250
(E) $18,750

254. **Box *A* has 3 \$20 bills and a single \$100 bill. Box *B* has 100 \$10 bills and 300 \$50 bills. Box *C* has 40 \$1 bills. You can have all of Box *C* or blindly pick one bill out of either Box *A* or Box *B*. Which choice offers the greatest expected winning?**

(A) Box *A*
(B) Box *B*
(C) Box *C*
(D) Either Box *A* or Box *B*, but not Box *C*
(E) All boxes offer the same expected winning.

255. **An Ivy League college has an acceptance rate of 8 percent. In a random sample of 50 applications, what is the expected number of applicants that will be turned down?**

(A) $50(0.08)$
(B) $50(0.92)$
(C) $50(0.08)(0.92)$
(D) $\sqrt{50(0.08)(0.92)}$
(E) $\sqrt{\dfrac{(0.08)(0.92)}{50}}$

256. **The number of condos a real estate agent sells monthly has the following probability distribution:**

Number of Condos	0	1	2	3
Probability	0.37	0.28	0.25	0.10

The agent averages a commission of \$6,500 per sale. What is the expected monthly commission from selling condos?

(A) $4 \times 6{,}500(0.37 + 0.28 + 0.25 + 0.10)$
(B) $6{,}500[0(0.37) + 1(0.28) + 2(0.25) + 3(0.10)]$
(C) $4 \times 6{,}500[0(0.37) + 1(0.28) + 2(0.25) + 3(0.10)]$
(D) $6{,}500\left(\dfrac{0.37 + 0.28 + 0.25 + 0.10}{4}\right)$
(E) $\dfrac{6{,}500}{4}[0(0.37) + 1(0.28) + 2(0.25) + 3(0.10)]$

257. A company has a choice of three investment schemes. Option I gives a sure $50,000 return on investment. Option II gives a 50% chance of returning $70,000 and a 50% chance of returning nothing. Option III gives a 10% chance of returning $90,000 and a 90% chance of returning nothing. Which option should the company choose?

(A) Option I if it wants to maximize expected return.
(B) Option II if it needs at least $60,000 to pay off an overdue loan.
(C) Option III if it needs at least $80,000 to pay off an overdue loan.
(D) All of the above answers are correct.
(E) Because of chance, it really doesn't matter which option the company chooses.

258. Two fair coins are tossed. If both land on heads, the player wins $5. If exactly one lands on heads, the player wins $2. If it costs $3 to play, what is the player's expected outcome after six games?

(A) Loss of $0.75
(B) Loss of $1.50
(C) Loss of $4.50
(D) Win of $2.25
(E) Win of $13.50

259. The number of days a juror serves on a grand jury can be considered a random variable. The table below shows the relative frequency distribution for the number of days someone serves on a federal grand jury.

Number of Days	1	2	3	4	5
Relative Frequency	0.3	0.4	0.15	0.1	0.05

Federal jurors are paid $40 per day. Based on the above distribution, what is the mean amount of pay federal jurors receive for their service?

(A) $ 2.20
(B) $ 17.60
(C) $ 40
(D) $ 88
(E) $200

260. A carnival game charges 25 cents per minute while a player is trying to win a prize. The player will receive prizes of different values. However, it is very difficult to win, and prizes are actually independent of the number of minutes played. The average number of minutes a player tries before giving up has a mean of 8 and a standard deviation of 2. The average value of the prize a player receives has a value of $1.75 with a standard deviation of $0.50. What are the expected value and standard deviation for playing this game for the player?

    (A) Expected value = –$0.25 and standard deviation = $0.56
    (B) Expected value = –$0.25 and standard deviation = $0.71
    (C) Expected value = $1.50 and standard deviation = $0.56
    (D) Expected value = $1.50 and standard deviation = $0.71
    (E) Expected value = $1.50 and standard deviation = $0.75

261. The four sides of a tetrahedron are labeled 1 through 4. For a certain carnival game, the player wins the number of points on the face-down side when the tetrahedron is tossed. Unbeknownst to the player, the tetrahedron is weighted so that the side with a 1 is twice as likely to land facedown as any of the other sides, which all have equal probabilities. What are the mean and variance of the number of points a player should expect from one toss of the polyhedron?

    (A) Mean is 2.5 and variance is 1.118
    (B) Mean is 2.5 and variance is 1.291
    (C) Mean is 2.2 and variance is 1.166
    (D) Mean is 2.2 and variance is 1.36
    (E) Mean is 2.2 and variance is 1.4

262. Given two independent random variables, X with mean 28.1 and standard deviation 3.4, and Y with mean 23.7 and standard deviation 2.9, which of the following is a true statement?

    (A) The mean of $X - Y$ is 51.8.
    (B) The median of $X - Y$ is 4.4.
    (C) The range of $X - Y$ is 51.8.
    (D) The standard deviation of $X - Y$ is 6.3.
    (E) The variance of $X - Y$ is 19.97.

263. Suppose X and Y are random variables with $\mu_x = 45$, $\sigma_x = 12$, $\mu_y = 42$, and $\sigma_y = 5$. Given that X and Y are independent, what is the standard deviation of the random variable $X - Y$?

(A) $\sqrt{7}$
(B) $\sqrt{17}$
(C) 7
(D) 13
(E) 169

264. Given a random variable X taking three possible values, x_1, x_2, x_3, which of the following is a true statement?

(A) $x_1 + x_2 + x_3 = 1$
(B) $E(X) = \frac{1}{3}\sum x_i$
(C) $\text{var}(X) = \frac{1}{3}\sum (x_i - \bar{x})^2$
(D) $E(X + c) = E(X) + c$
(E) $\text{var}(aX) = a\,\text{var}(X)$

265. A store sells metallic, printed, and glow-in-the-dark fidget spinners for $11.98, $1.47, and $5.40, respectively. If metallic, printed, and glow-in-the-dark fidget spinners represent 10 percent, 60 percent, and 30 percent, respectively, of the fidget spinner sales, what is the expected monetary value of one day's sales if 200 fidgets are sold?

(A) 11.98(0.10) + 1.47(0.60) + 5.40(0.30)
(B) 200[3(11.98 +1.47 + 5.40)]
(C) 200[11.98(0.10) + 1.47(0.60) + 5.40(0.30)]
(D) $200\left(\frac{11.98 + 1.47 + 5.40}{3}\right)$
(E) $\frac{11.98(0.10) + 1.47(0.60) + 5.40(0.30)}{200}$

266. Suppose X and Y are random variables with $E(X) = 29$, $\text{var}(X) = 7$, $E(Y) = 35$, and $\text{var}(Y) = 9$. What are the expected value and variance of the random variable $X + Y$?

(A) $E(X + Y) = 32$, $\text{var}(X + Y) = 8$
(B) $E(X + Y) = 64$, $\text{var}(X + Y) = 8$
(C) $E(X + Y) = 64$, $\text{var}(X + Y) = 16$
(D) $E(X + Y) = 64$, $\text{var}(X + Y) = \sqrt{7^2 + 9^2}$
(E) There is insufficient information to answer this question.

267. Suppose X and Y are independent random variables with $E(X) = 34$, $\text{var}(X) = 4$, $E(Y) = 23$, and $\text{var}(Y) = 3$. What is the variance of the random variable $X - Y$?

(A) $4 - 3$
(B) $4 + 3$
(C) $4^2 - 3^2$
(D) $4^2 + 3^2$
(E) $\sqrt{4^2 + 3^2}$

268. Suppose X and Y are independent random variables, both with normal distributions. If X has a mean of 45 with a standard deviation of 4 and if Y has a mean of 35 with a standard deviation of 3, what is the probability that a randomly generated value of X is greater than a randomly generated value of Y?

(A) $P(z > -2)$
(B) $P(z > -1)$
(C) $P(z > 0)$
(D) $P(z > 1)$
(E) $P(z > 2)$

269. The random variable X has a mean of 15 and a standard deviation of 10. The random variable Y is defined by $Y = 5 + 3X$. What are the mean and standard deviation of Y?

(A) The mean is 45, and the standard deviation is 30.
(B) The mean is 45, and the standard deviation is 35.
(C) The mean is 50, and the standard deviation is 10.
(D) The mean is 50, and the standard deviation is 30.
(E) The mean is 50, and the standard deviation is 35.

270. A random variable X has a mean of 25 and a standard deviation of 8. The random variable $X - Y$ has a mean of 9 and a standard deviation of 10. Assuming X and Y are independent, what are the mean and standard deviation of the random variable Y?

(A) $\mu_Y = 16$ and $\sigma_Y = 2$
(B) $\mu_Y = 16$ and $\sigma_Y = 6$
(C) $\mu_Y = 16$ and $\sigma_Y = 12.8$
(D) $\mu_Y = 34$ and $\sigma_Y = 2$
(E) $\mu_Y = 34$ and $\sigma_Y = 6$

271. Bowler A bowls an average of 225 with a standard deviation of 15. Bowler B bowls an average of 190 with a standard deviation of 20. Together they form a team with an average total score of 225 + 190 = 415. Under what conditions is the standard deviation of their total score equal to $\sqrt{15^2 + 20^2} = 25$?

(A) No conditions are required. The standard deviation is 25.
(B) No matter what conditions are given, the standard deviation is different from 25.
(C) As long as they consistently bowl their averages, the standard deviation is 25.
(D) The standard deviation is 25 only if their scores are independent.
(E) The standard deviation is 25 only if their scores are mutually exclusive.

272. The four sides of a tetrahedron are labeled 1 through 4. Let X be the random variable whose values are the numbers that end up face down when the tetrahedron is tossed. Assume this is a fair tetrahedron, that is, all sides are equally likely to end up face down. The mean of X is 2.5, and its variance is $\frac{5}{4}$. Let Y be the random variable whose values are the differences between the face-down values of the first and second rolls of the tetrahedron if it is rolled twice. What is the standard deviation of Y?

(A) $\sqrt{\frac{5}{4}}$

(B) $\sqrt{\frac{5}{4}} + \sqrt{\frac{5}{4}}$

(C) $\sqrt{\frac{5}{4} + \frac{5}{4}}$

(D) $\sqrt{\frac{5}{4} - \frac{5}{4}}$

(E) $\frac{5}{4} + \frac{5}{4}$

273. The variance of a set of data is 5. If every value in the set is multiplied by 4 and then added to 100, what is the new variance?

(A) 5
(B) 20
(C) 80
(D) 120
(E) 180

NORMAL PROBABILITIES

274. Diet Plan *A* advertises an average monthly weight loss of 10 pounds with a standard deviation of 3 pounds. Diet Plan *B* claims an average monthly weight loss of 12 pounds with a standard deviation of 1.8 pounds. Assuming both assertions are correct and assuming roughly normal distributions, which diet plan is more likely to result in a monthly weight loss of over 15 pounds?

(A) Diet Plan *A* is more likely to result in a monthly weight loss of over 15 pounds because of its greater standard deviation.
(B) Diet Plan *B* is more likely to result in a monthly weight loss of over 15 pounds because of its greater mean.
(C) For both plans, the probability of a weight loss over 15 pounds is 0.04779.
(D) For both plans, the probability of a weight loss over 15 pounds is 0.95221.
(E) The problem cannot be solved from the information given.

275. Which of the following are true statements?

I. The area under a normal curve is always equal to 1, no matter the mean and standard deviation.
II. The smaller the standard deviation of a normal curve, the higher and narrower the graph is.
III. Normal curves with different means are centered around different numbers.

(A) I and II only
(B) I and III only
(C) II and III only
(D) I, II, and III
(E) None of the above gives the complete set of true responses.

276. A cell phone takes an average of 11 minutes to move through an assembly line. If the standard deviation is 2 minutes and if the distribution is roughly normal, what is the probability that a cell phone will take over 12 minutes to move through the assembly line?

(A) $P\left(z > \frac{12-11}{2}\right)$

(B) $P\left(z > \frac{12-11}{\frac{2}{\sqrt{2}}}\right)$

(C) $P\left(z > \frac{11-12}{2}\right)$

(D) $P\left(z > \frac{11-12}{\frac{2}{\sqrt{2}}}\right)$

(E) $2P\left(t > \frac{11-12}{2}\right)$

277. A bowler's scores are approximately normally distributed with a mean of 210. What is the standard deviation if 30 percent of her scores are below 200?

(A) $\dfrac{200-210}{-0.5244}$

(B) $\dfrac{210-200}{-0.5244}$

(C) $\dfrac{200-210}{0.5244}$

(D) $\dfrac{200-210}{0.4756}$

(E) $\dfrac{210-200}{0.4756}$

278. Which of the following is a true statement?

(A) The area under the standard normal curve between 0 and 2 is twice the area between 0 and 1.
(B) The area under the standard normal curve between 0 and 2 is half the area between –2 and 2.
(C) For the standard normal curve, the interquartile range is approximately 3.
(D) For the standard normal curve, approximately 1 out of 1,000 values are greater than 10.
(E) The 68-95-99.7 rule applies only to normal curves where the mean and standard deviation are known.

279. The starting national average salary for a computer security specialist is $58,760. Assuming a roughly normal distribution and a standard deviation of $6,500, what is the probability that a randomly chosen computer security specialist will start with a salary between $50,000 and $60,000?

(A) $P\left(\dfrac{50{,}000}{6{,}500} < z < \dfrac{60{,}000}{6{,}500}\right)$

(B) $2P\left(z > \dfrac{60{,}000-50{,}000}{6{,}500}\right)$

(C) $P\left(\dfrac{58{,}760-50{,}000}{6{,}500} < z < \dfrac{58{,}760-60{,}000}{6{,}500}\right)$

(D) $P\left(\dfrac{58{,}760-50{,}000}{\frac{6{,}500}{\sqrt{n}}} < z < \dfrac{58{,}760-60{,}000}{\frac{6{,}500}{\sqrt{n}}}\right)$

(E) $P\left(\dfrac{50{,}000-58{,}760}{6{,}500} < z < \dfrac{60{,}000-58{,}760}{6{,}500}\right)$

280. Populations $P1$ and $P2$ are roughly normally distributed and have identical means. However, the standard deviation of $P1$ is half the standard deviation of $P2$. What can be said about the percentage of observations falling within one standard deviation of the mean for each population?

(A) The percentage for $P1$ is twice the percentage for $P2$.
(B) The percentage for $P1$ is greater, but not twice as great, as the percentage for $P2$.
(C) The percentage for $P2$ is twice the percentage for $P1$.
(D) The percentage for $P2$ is greater, but not twice as great, as the percentage for $P1$.
(E) The percentages are identical.

***281. A college library determines that books are checked out for an average of 8 days with a standard deviation of 2 days. Assuming a roughly normal distribution, what is the shortest time interval for which two-thirds of the books are out?**

(A) 0 to 8.9 days
(B) 6.1 to 7.1 days
(C) 6.1 to 8 days
(D) 6.1 to 9.9 days
(E) 7.1 to 14 days

282. A set of employee commuting distances from work has a roughly normal distribution with a mean of 12 miles and a standard deviation of 3.5 miles. If a randomly selected employee commutes over 10 miles, what is the probability she commutes under 15 miles?

(A) 0.521
(B) 0.545
(C) 0.647
(D) 0.716
(E) 0.727

283. Which of the following statements is *incorrect*?

(A) In all normal distributions, the mean and median are equal.
(B) Bell-shaped curves may not have normal distributions.
(C) Virtually all the area under a normal curve is within three standard deviations of the mean, no matter the particular mean and standard deviation.
(D) A normal distribution is completely determined by two numbers, its mean and its standard deviation.
(E) Standardized scores (z-scores) always have a normal distribution no matter the original distribution.

284. A trucking firm determines that the miles per gallon (mpg) achieved by trucks in its fleet are roughly normally distributed with a standard deviation of 2.5 mpg. What is the mean mpg if 75 percent of the trucks achieve better than 13.2 mpg?

(A) $13.2 + 0.3255(2.5)$
(B) $13.2 + 0.6745(2.5)$
(C) $13.2 + 0.7500(2.5)$
(D) $13.2 - 0.6745(2.5)$
(E) $13.2 - 0.7500(2.5)$

285. The mean yearly medical expenses (including insurance payments) for individuals in a large city is $15,300 with a standard deviation of $3,600. Assuming a roughly normal distribution, what is the probability that two randomly chosen individuals in the city both have yearly medical expenses over $20,000?

(A) $2P\left(z > \frac{20{,}000 - 15{,}300}{3{,}600}\right)$

(B) $\left[P\left(z > \frac{20{,}000 - 15{,}300}{3{,}600}\right)\right]^2$

(C) $\frac{1}{2}P\left(z > \frac{20{,}000 - 15{,}300}{3{,}600}\right)$

(D) $P\left(z > \frac{20{,}000 - 15{,}300}{\sqrt{3{,}600}}\right)$

(E) $P\left(z > \frac{20{,}000 - 15{,}300}{\frac{3{,}600}{\sqrt{2}}}\right)$

286. The monthly rental paid per person for college students living off campus has a roughly normal distribution with a mean of $275 and a standard deviation of $40. Ninety percent of the rentals are greater than what amount?

(A) $275 – 1.282($40)
(B) $275 – 1.645($40)
(C) $275 – 1.96($40)
(D) $275 + 1.282($40)
(E) $275 + 1.96($40)

287. Suppose women's foot lengths are roughly normally distributed with a mean of 20 cm and a standard deviation of 4 cm, while men's foot lengths are roughly normally distributed with a mean of 26 cm and a standard deviation of 5 cm. If a policeman measures a footprint to be 22.5 cm, is there a greater probability that it belongs to a man or to a woman?

(A) Man, because $22.5 - 20 < 26 - 22.5$
(B) Man, because $\frac{2.5}{4} < \frac{3.5}{5}$
(C) Woman, because $22.5 - 20 < 26 - 22.5$
(D) Woman, because $\frac{2.5}{4} < \frac{3.5}{5}$
(E) This cannot be answered without knowing if we can assume independence.

288. A couple is looking to purchase their first house. In the neighborhood in which they are interested, home prices are roughly normally distributed with a mean of $275,000 and a standard deviation of $35,000. They ask the realtor to show them only homes under $300,000. What percentage of the homes shown to them will be over $250,000?

(A) 0.312
(B) 0.475
(C) 0.525
(D) 0.688
(E) 0.763

289. Among 2,500 subjects signing up for a program to stop smoking, the distribution of their daily number of cigarettes was roughly normal with a mean of 27.1 and a standard deviation of 4.6. Which expression below represents the 60th percentile of the distribution?

(A) $27.1 - (0.25)(4.6)$
(B) $27.1 - (0.40)(4.6)$
(C) $27.1 - (0.60)(4.6)$
(D) $27.1 + (0.25)(4.6)$
(E) $27.1 + (0.40)(4.6)$

290. The women's 500 m skating times are roughly normally distributed with a mean of 40.44 seconds. The z-score of China's Peiyu Jin's time of 38.69 seconds is -1.944. What percent of all women's 500 m skating times are above 42 seconds?

(A) 1.56
(B) 1.75
(C) 4.15
(D) 98.25
(E) 95.85

*291. The number of miles that a highway construction team can lay on good-weather days is roughly normally distributed with a mean of 3 and a standard deviation of 0.35. The number of miles that a highway construction team can lay on bad-weather days is roughly normally distributed with a mean of 1.9 and a standard deviation of 0.4. If the probability of good weather is 0.7 (and of bad weather is 0.3), what is the probability of laying at least 2.5 miles on a random day?

(A) 0.06681
(B) 0.49511
(C) 0.64638
(D) 0.6664
(E) 0.92344

292. A distribution of scores is approximately normal with a mean of 28 and a standard deviation of 3.4. Which of the following equations should be used to find a score *x* with 20 percent of the scores above it?

(A) $\frac{x-28}{3.4} = 0.80$

(B) $\frac{x-28}{\sqrt{3.4}} = 0.80$

(C) $\frac{x-28}{3.4} = 0.84$

(D) $\frac{x-28}{\sqrt{3.4}} = 0.84$

(E) $\frac{x-28}{3.4^2} = 0.84$

293. A large data set is approximately normally distributed. What is the proper order, from smallest to largest, of l, m, and n, where

l = the value with a *z*-score of –0.6
m = the value of the first quartile
n = the value of the 30th percentile

(A) $l < m < n$
(B) $m < l < n$
(C) $m < n < l$
(D) $l < n < m$
(E) $n < l < m$

294. A city's average winter low temperature is 28°F with a standard deviation of 10°F, while the city's average summer low temperature is 66°F with a standard deviation of 6°F. In which season is it more unusual to have a day with a low temperature of 51°F?

(A) Winter, because $51 - 28 = 23$ is greater than $66 - 51 = 15$
(B) Winter, because 10 is greater than 6
(C) Winter, because 51 sounds like a more unusual temperature for winter than summer
(D) Summer, because $\left|\frac{51-66}{6}\right|$ is more than $\left|\frac{51-28}{10}\right|$
(E) Summer, because $\frac{28+66}{2} = 47 < 51$

SAMPLING DISTRIBUTIONS

295. The owner of a bagel shop, who is the father of an AP Statistics student, advertises that the price of a dozen bagels on any given day will be randomly picked using a normal distribution with a mean of $10.00 and a standard deviation of $0.50. If a customer buys a dozen bagels on each of five days, what is the probability that he will pay a total exceeding $52?

(A) $P\left(z > \frac{10.40 - 10.00}{0.50}\right)$
(B) $P\left(z > \frac{52.00 - 50.00}{0.50}\right)$
(C) $2P\left(z > \frac{52.00 - 50.00}{0.50}\right)$
(D) $P\left(z > \frac{10.40 - 10.00}{\left(\frac{0.50}{\sqrt{5}}\right)}\right)$
(E) $P\left(z > \frac{52.00 - 50.00}{\left(\frac{0.50}{\sqrt{5}}\right)}\right)$

296. 6.7 percent of women graduate college with STEM degrees. In a simple random sample (SRS) of 1,000 women, what is the probability that more than 8 percent will graduate college with a STEM degree?

(A) 0.033
(B) 0.05
(C) 0.067
(D) 0.10
(E) 0.933

297. Which of the following statements is *incorrect*?

(A) The larger the sample is, the larger the spread is in the sampling distribution.
(B) Provided that the population size is significantly greater than the sample size, the spread of a sampling distribution does not depend on the population size.
(C) Bias has to do with the center, not the spread, of a sampling distribution.
(D) Sample distribution and sampling distribution refer to different things.
(E) The larger the sample is, the closer the sample distribution generally becomes to the population distribution.

298. Suppose 15 percent of the mines in a particular Alaska region will strike gold. In a random sample of 200 mines in this region, what is the probability that more than 18 percent will strike gold?

(A) $P\left(z > \frac{0.18 - 0.15}{\sqrt{200(0.15)(0.85)}}\right)$

(B) $P\left(z > \frac{0.18 - 0.15}{\sqrt{200(0.5)(0.5)}}\right)$

(C) $P\left(z > \frac{0.18 - 0.15}{(0.15)(0.85) / \sqrt{200}}\right)$

(D) $P\left(z > \frac{0.18 - 0.15}{\sqrt{\frac{(0.15)(0.85)}{200}}}\right)$

(E) $P\left(z > \frac{0.18 - 0.15}{\sqrt{\frac{(0.5)(0.5)}{200}}}\right)$

299. Which of the following statements is *incorrect*?

(A) The sampling distribution of $\bar{x}$ has a mean equal to the population mean μ even if the sample size n is small.
(B) The sampling distribution of $\bar{x}$ has a standard deviation of $\frac{\sigma}{\sqrt{n}}$ even if the population is not normally distributed and even if n is small.
(C) The sampling distribution of $\bar{x}$ is normal, no matter what n is, if the population has a normal distribution.
(D) When n is large, the sampling distribution of $\bar{x}$ is approximately normal even if the population is not normally distributed.
(E) Even if the observations are not independent, the central limit theorem applies as long as n is large enough.

300. It is known that 73 percent of the employees at one factory are women, while 61 percent of the employees of a second factory are women. In a simple random sample (SRS) of 115 employees from the first factory and an independent SRS of 80 employees from the second, what is the probability that the difference between the percentages of women picked (first factory minus second) is more than 10 percent?

(A) $P\left(z > \dfrac{0.10 - 0.12}{\sqrt{(0.5)(0.5)\left(\dfrac{1}{115} + \dfrac{1}{80}\right)}}\right)$

(B) $P\left(z > \dfrac{0.10 - 0.12}{\sqrt{(0.67)(0.33)\left(\dfrac{1}{115} + \dfrac{1}{80}\right)}}\right)$

(C) $P\left(z > \dfrac{0.10 - 0.12}{\sqrt{\dfrac{(0.73)(0.27)}{115} + \dfrac{(0.61)(0.39)}{80}}}\right)$

(D) $P\left(z > \dfrac{0.10 - 0.12}{\dfrac{(0.73)(0.27)}{\sqrt{115}} + \dfrac{(0.61)(0.39)}{\sqrt{80}}}\right)$

(E) $P\left(z > \dfrac{0.10 - 0.12}{\sqrt{\left(\dfrac{84 + 49}{115 + 80}\right)\left(1 - \dfrac{84 + 49}{115 + 80}\right)\left(\dfrac{1}{115} + \dfrac{1}{80}\right)}}\right)$

301. A population is normally distributed with a mean of 37. Consider all samples of size 8. The variable $\dfrac{\bar{x} - 37}{\dfrac{s}{\sqrt{8}}}$

(A) has a normal distribution.
(B) has a t-distribution with $df = 8$
(C) has a t-distribution with $df = 7$
(D) has neither a normal distribution nor a t-distribution
(E) has either a normal distribution or a t-distribution depending on the characteristics of the population standard deviation

302. Which of the following statements is *incorrect*?

(A) Sample statistics are used to make inferences about population parameters.
(B) Statistics from smaller samples have more variability than those from larger samples.
(C) Parameters are fixed, while statistics vary depending on which sample is chosen.
(D) As the sample size n becomes larger, the sample distribution becomes closer to a normal distribution.
(E) All of the above are true statements.

303. Which of the following is a true statement?

(A) The sampling distribution of $\hat{p}$ has a mean that can vary from the population proportion p by approximately 1.96 standard deviations.
(B) The sampling distribution of $\hat{p}$ has a standard deviation equal to $\sqrt{np(1-p)}$.
(C) The sampling distribution of $\hat{p}$ is considered close to normal provided that $n \geq 30$.
(D) The sample proportion is a random variable with a probability distribution.
(E) All of the above are true statements.

304. Which of the following is a true statement?

(A) The sampling distribution of the difference $\bar{x}_1 - \bar{x}_2$ has a mean equal to the difference of the population means.
(B) The sampling distribution of the difference $\bar{x}_1 - \bar{x}_2$ has a standard deviation equal to the sum of the population standard deviations.
(C) The sampling distribution of the difference $\bar{x}_1 - \bar{x}_2$ has a standard deviation equal to the difference of the population standard deviations.
(D) The sample sizes must be equal in two-sample inference.
(E) As long as the sample sizes are large enough (for example, each at least 30), the samples in two-sample inference do not have to be independent.

305. The calories in butterscotch hard candies are normally distributed with a mean of 23.1 and a standard deviation of 0.6. In a random sample of 10 butterscotch candies, what is the probability that the total calories are between 228 and 232?

(A) $P\left(\frac{228-231}{0.6} < z < \frac{232-231}{0.6}\right)$

(B) $P\left(\frac{228-231}{6} < z < \frac{232-231}{6}\right)$

(C) $P\left(\frac{22.8-23.1}{0.6} < z < \frac{23.2-23.1}{0.6}\right)$

(D) $P\left(\frac{22.8-23.1}{\frac{0.6}{\sqrt{10}}} < z < \frac{23.2-23.1}{\frac{0.6}{\sqrt{10}}}\right)$

(E) $P\left(\frac{228-231}{\frac{0.6}{\sqrt{10}}} < z < \frac{232-231}{\frac{0.6}{\sqrt{10}}}\right)$

306. Which of the following are unbiased estimators for the corresponding population parameters?

I. Sample means
II. Sample proportions
III. Difference of sample means
IV. Difference of sample proportions

(A) None are unbiased.
(B) I and II only
(C) I and III only
(D) III and IV only
(E) All are unbiased.

307. Which one of these histograms represents the sampling distribution of $\hat{p}$ for $p = 0.6$ and $n = 150$?

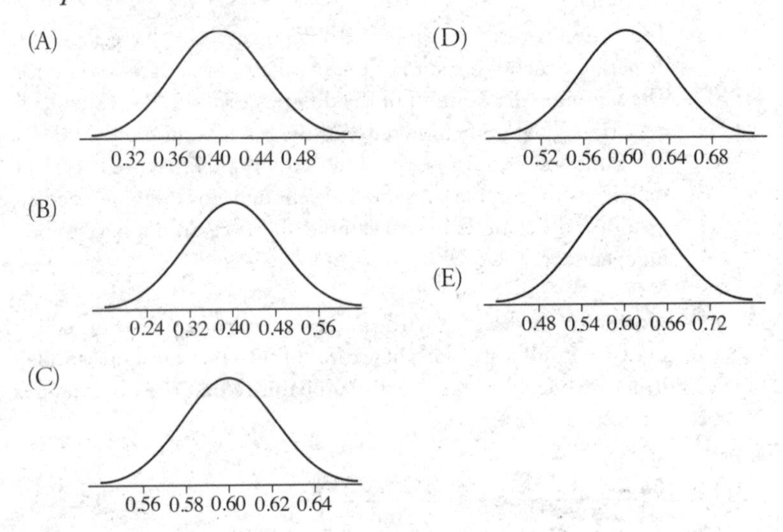

308. Which of the following statements is true?

(A) The mean of the set of sample means varies inversely as the square root of the size of the samples.
(B) The variance of the set of sample means varies directly as the size of the samples and inversely as the variance of the original population.
(C) The standard deviation of the set of sample means varies directly as the standard deviation of the original population and inversely as the square root of the size of the samples.
(D) The larger the sample size, the larger the variance of the set of sample means.
(E) One must double the sample size in order to cut the standard deviation of $\bar{x}$ in half.

309. Surveys have shown that 30 percent of people refuse to use public toilets. A random sample of 200 people is selected, and the proportion of those people who refuse to use public toilets is noted. This is repeated 100 times, and a dotplot is constructed of the 100 sample proportions. Which of the following best describes the standard deviation of the data in the dotplot?

(A) $\sqrt{100(0.3)(0.7)}$

(B) $\sqrt{200(0.3)(0.7)}$

(C) $\sqrt{300(0.3)(0.7)}$

(D) $\sqrt{\dfrac{(0.3)(0.7)}{100}}$

(E) $\sqrt{\dfrac{(0.3)(0.7)}{200}}$

310. Past surveys have shown that 9 percent of high school students and 12 percent of college students skip breakfast. A polling agency picks independent random samples of 100 high school students and of 100 college students. The agency surveys each group as to the proportion of students who skip breakfast. The difference in proportions is noted. This is repeated 400 times, and a dotplot is constructed of the 400 differences of proportions. Which of the following best describes the standard deviation of the data in the dotplot?

(A) $\sqrt{100(0.09)(0.91) + 100(0.12)(0.88)}$

(B) $\sqrt{400(0.09)(0.91) + 400(0.12)(0.88)}$

(C) $\sqrt{\dfrac{(0.09)(0.91)}{100} + \dfrac{(0.12)(0.88)}{100}}$

(D) $\sqrt{\dfrac{(0.09)(0.91)}{400} + \dfrac{(0.12)(0.88)}{400}}$

(E) $\sqrt{(0.105)(0.895)\left(\dfrac{1}{100} + \dfrac{1}{100}\right)}$

311. The distribution of weights of "18-ounce" boxes of corn flakes is given by the following histogram.

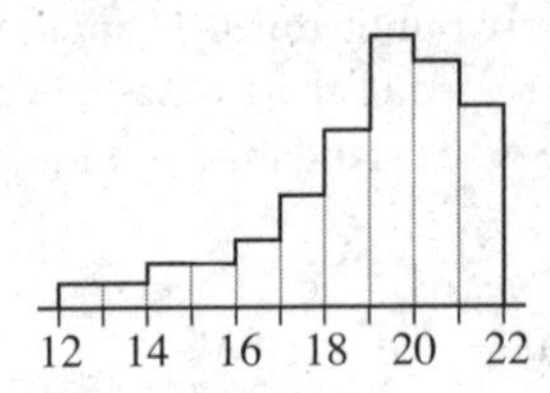

The distribution has a mean of 18.96 ounces with a standard deviation of 2.31 ounces. If 50 random samples of 12 boxes each are picked and if the mean weight of each sample is found, which of the following is most likely to represent the distribution of the sample means?

(A)

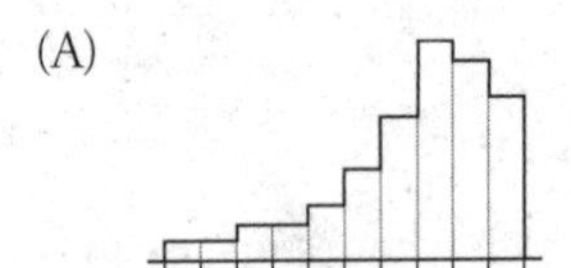

(B)

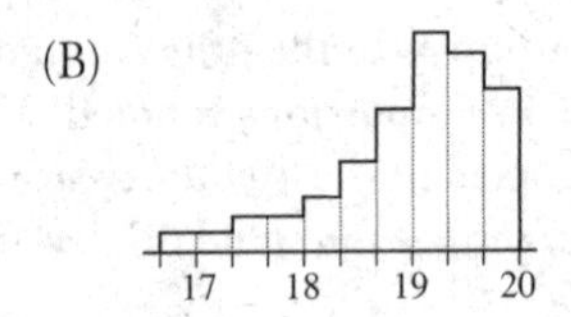

(C)

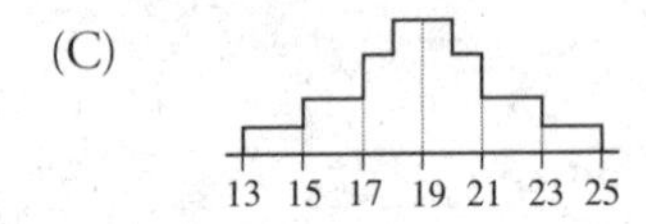

(D)

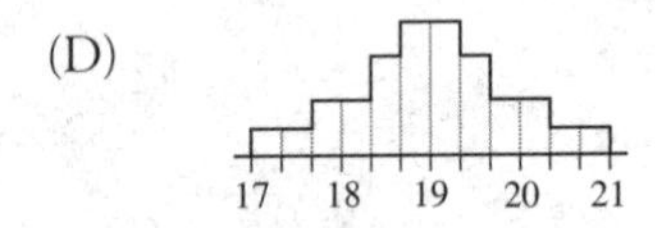

(E) None of the above gives a reasonable histogram for the sampling distribution.

312. Which of the sample sizes n and population proportions p below would result in the greatest standard deviation for the sampling distribution of $\hat{p}$?

(A) $n = 50$ and $p = 0.1$
(B) $n = 50$ and $p = 0.4$
(C) $n = 250$ and $p = 0.1$
(D) $n = 250$ and $p = 0.6$
(E) $n = 250$ and $p = 0.95$

313. The sampling distribution of the sample mean is close to the normal distribution

(A) only if the parent population is unimodal, is not badly skewed, and does not have outliers
(B) no matter the distribution of the parent population or what the value of n
(C) if n is large, no matter the distribution of the parent population
(D) if the standard deviation of the parent population is known
(E) only if both n is large and the parent population has a normal distribution

***314. Suppose that 48 percent of high school students and 44 percent of college students would answer yes to the question "Do you enjoy Drake's music?" In a simulation, a random sample of 100 high school students and 100 college students is selected. The difference in sample proportions of those who answered yes, $\hat{p}_{high\ school} - \hat{p}_{college}$, is calculated. The procedure is repeated 10,000 times. Which of the histograms below is most likely to result as the simulated sampling distribution of the difference in proportions?**

(A)

0 0.02 0.04 0.06 0.08
Difference in sample proportions

(D)

–0.36 –0.16 0.04 0.24 0.44
Difference in sample proportions

(B)

0.038 0.039 0.04 0.041 0.042
Difference in sample proportions

(E)

0.005 0.025 0.055 0.075
Difference in sample proportions

(C)

–0.205 –0.065 0.145 0.285
Difference in sample proportions

315. The body temperatures of healthy elderly adults follow a skewed left distribution with a mean of 37°C and a standard deviation of 0.4°C. A random sample of 100 healthy adults is selected, and their mean body temperature is recorded. This is repeated 500 times, and a dotplot is constructed of the 500 sample means. Which of the following best describes the dotplot?

(A) Skewed left with a mean of 37°C and a standard deviation of 0.0179°C
(B) Skewed left with a mean of 37°C and a standard deviation of 0.04°C
(C) Skewed left with a mean of 37°C and a standard deviation of 0.4°C
(D) Approximately normal with a mean of 37°C and a standard deviation of 0.0179°C
(E) Approximately normal with a mean of 37°C and a standard deviation of 0.04°C

316. Which of the following is a biased estimator?

(A) Sampling distribution of proportions
(B) Sampling distribution of means
(C) Sampling distribution of slopes
(D) Sampling distribution of maxima
(E) All of the above are unbiased.

***317. A simulation is conducted by using 25 fair dice whose faces are numbered 1 through 6, tossing them all at once, and averaging the 25 numbers showing face up. This is repeated 400 times. Which of these best describes the distribution being simulated?**

(A) A sampling distribution of a sample proportion with $\mu_{\hat{p}} = \frac{1}{6}$ and $\sigma_{\hat{p}} = \sqrt{\frac{(1/6)(5/6)}{25}}$

(B) A sampling distribution of a sample proportion with $\mu_{\hat{p}} = \frac{1}{6}$ and $\sigma_{\hat{p}} = \sqrt{\frac{(1/6)(5/6)}{400}}$

(C) A sampling distribution of a sample mean with $\mu_{\bar{x}} = 3.5$ and $\sigma_{\bar{x}} = 1.708$

(D) A sampling distribution of a sample mean with $\mu_{\bar{x}} = 3.5$ and $\sigma_{\bar{x}} = \frac{1.708}{\sqrt{25}}$

(E) A sampling distribution of a sample mean with $\mu_{\bar{x}} = 3.5$ and $\sigma_{\bar{x}} = \frac{1.708}{\sqrt{400}}$

318. Cycle times for express washing for Brand *A* dishwashers have a mean of 42.5 minutes with a standard deviation of 0.6 minutes, while times for the same cycle in Brand *B* dishwashers have a mean of 41.2 minutes with a standard deviation of 0.5 minutes. Brand *A* and Brand *B* dishwashers are each sent through an express washing 100 times. What is the probability that the mean difference (*A* minus *B*) is greater than 1 minute?

(A) $P\left(z > \dfrac{1-1.3}{\sqrt{0.6^2+0.5^2}}\right)$

(B) $P\left(z > \dfrac{1-1.3}{\sqrt{\dfrac{0.6^2}{100}+\dfrac{0.5^2}{100}}}\right)$

(C) $P\left(z > \dfrac{1-1.3}{\dfrac{0.6}{\sqrt{100}}+\dfrac{0.5}{\sqrt{100}}}\right)$

(D) $P\left(z > \dfrac{1-1.3}{\left(\dfrac{\sqrt{0.6^2+0.5^2}}{100}\right)}\right)$

(E) $P\left(z > \dfrac{1-1.13}{\left(\dfrac{\sqrt{0.6^2+0.5^2}}{100+100}\right)}\right)$

319. The mean waist size for American men is 39.7 inches with a standard deviation of 3.8 inches. In a random sample of 200 men, what is the probability that the mean waist size is less than 35 inches?

(A) $P(z < 35)$

(B) $P\left(z < \dfrac{35-39.7}{3.8}\right)$

(C) $P\left(z < \dfrac{35-39.7}{\left(\dfrac{3.8}{\sqrt{200}}\right)}\right)$

(D) $P\left(z < \dfrac{35-39.7}{\left(\dfrac{\sqrt{38}}{200}\right)}\right)$

(E) $P\left(z < \dfrac{35-39.7}{\left(\dfrac{\sqrt{3.8}}{200}\right)}\right)$

320. Based on past studies, a credit card company believes that 1.5 percent of adults receiving a new credit card offer in the mail will accept it. If the company is correct, what is the probability that less than 1 percent of adults in a random sample of 400 adults receiving the offer will accept it?

(A) $P\left(z < \dfrac{0.01-0.015}{\sqrt{400(0.01)(0.99)}}\right)$

(B) $P\left(z < \dfrac{0.01-0.015}{\sqrt{400(0.015)(0.985)}}\right)$

(C) $P\left(z < \dfrac{0.01-0.015}{\sqrt{\dfrac{(0.01)(0.99)}{400}}}\right)$

(D) $P\left(z < \dfrac{0.01-0.015}{\sqrt{\dfrac{(0.015)(0.985)}{400}}}\right)$

(E) $P\left(z < \dfrac{0.01-0.015}{\sqrt{\dfrac{(0.5)(0.5)}{400}}}\right)$

***321. Suppose that the average age of MLB players is 26.8 with a standard deviation of 3.8, while the average age of NFL players is 25.6 with a standard deviation of 2.9. In random samples of 35 MLB players and 30 NFL players, what is the probability that the mean age of the MLB players is greater than the mean age of the NFL players?**

(A) 0.8158
(B) 0.8324
(C) 0.9253
(D) 0.9584
(E) 0.9860

322. Child development scientists gathered data from 65 randomly selected babies and calculated the mean age in weeks at which the babies began to crawl. Is the sample mean $\bar{x}$ an unbiased estimator for μ, the mean age at which all babies begin to crawl?

(A) No, because some sample means do not equal the population mean
(B) No, because there is no reason to assume that the ages at which all babies begin to crawl is normally distributed
(C) No, not unless the sample size n is very large
(D) No, not unless the population standard deviation is known
(E) Yes, because for random samples, the mean of the sample means is the population mean

323. Which of the following statements is *incorrect*?

(A) Like the normal, *t*-distributions are always symmetric.
(B) Like the normal, *t*-distributions are always bell-shaped.
(C) Like the normal, *t*-distributions are always unimodal.
(D) The *t*-distributions have less spread than the normal, that is, they have less probability in the tails and more in the center than the normal.
(E) For larger values of *df*, degrees of freedom, the *t*-distributions look more like the normal distribution.

*324. **Which of the following statements about the chi-square distribution is *incorrect*?**

(A) There is a separate χ^2 curve for each *df* value
(B) The area under every χ^2 curve is the same.
(C) For small *df*, the distribution is skewed to the right; however, for large *df*, it becomes more symmetric and bell-shaped.
(D) For 1 or 2 degrees of freedom. the histogram peak occurs at 0. For 3 or more degrees of freedom, the peak is at $df - 2$.
(E) Just like for the *t*-distribution, the degrees of freedom for χ^2 distributions depend upon the sample size.

325. **Which of the following is a true statement about *t*-distributions?**

(A) The greater the number of degrees of freedom, the narrower the tails.
(B) The smaller the number of degrees of freedom, the closer the curve is to the normal curve.
(C) Thirty degrees of freedom gives the normal curve.
(D) The shape of a *t*-distribution depends on the degrees of freedom, which depend on the population size.
(E) The probability that $z > 1.96$ in a normal distribution is greater than the probability that $t > 1.96$ in a *t*-distribution with $df = 30$.

MC Statistical Inference

CHAPTER 4

Answers for Chapter 4 are on pages 277–300.

CONFIDENCE INTERVALS FOR PROPORTIONS

326. In a random sample of 1,500 adults with library cards, 525 said that their favorite books to read in bed are mysteries. What is a 95% confidence interval estimate of the proportion of all adults with library cards whose favorite books to read in bed are mysteries?

(A) $0.35 \pm 1.96\frac{(0.5)(0.5)}{\sqrt{1,500}}$

(B) $0.35 \pm 1.96\sqrt{\frac{(0.5)(0.5)}{1,500}}$

(C) $0.35 \pm 1.96\frac{(0.35)(0.65)}{\sqrt{1,500}}$

(D) $0.35 \pm 1.96\sqrt{\frac{(0.35)(0.65)}{1,500}}$

(E) $0.35 \pm 1.96\sqrt{1,500(0.35)(0.65)}$

327. A 2018 survey of 1,470 adult Americans found that 29 percent of the American adult population would be willing to pay higher taxes if the government offered "Medicare for all." Which of the following best describes what is meant by the poll having a margin of error of ±3 percent?

(A) It means that 3 percent of those surveyed refused to participate in the poll.
(B) It would not be unexpected for 3 percent of the population to agree readily to the higher taxes.
(C) Between 382 and 470 of the 1,470 adults surveyed responded that they would be willing to pay higher taxes in return for "Medicare for all."
(D) If a similar survey of 1,470 American adults was taken weekly, a 3 percent change in each week's results would not be unexpected.
(E) It is likely that between 26 percent and 32 percent of American adults would be willing to pay higher taxes in return for "Medicare for all."

328. The probability that a "walk-on" makes an intercollegiate team at a particular D-1 university is 17%. If one takes a simple random sample (SRS) of walk-ons and constructs a confidence interval estimate of the acceptance rate, which of the following statements is true?

(A) The center of the interval would be 17%.
(B) The interval would contain 17%.
(C) A 99% confidence interval estimate would contain 17%.
(D) All of the above are true statements.
(E) None of the above are true statements.

329. A telephone survey of 750 registered voters showed that 330 favored legalizing medical marijuana. At what confidence level can we say that between 40% and 48% of the electorate favor this legalization?

(A) $P(0.40 < z\, 0.48)$

(B) $P\left(\dfrac{0.40-0.44}{\sqrt{750(0.44)(0.56)}} < z < \dfrac{0.48-0.44}{\sqrt{750(0.44)(0.56)}}\right)$

(C) $P\left(\dfrac{0.40-0.44}{\sqrt{(0.5)(0.5)/750}} < z < \dfrac{0.48-0.44}{\sqrt{(0.5)(0.5)/750}}\right)$

(D) $P\left(\dfrac{0.40-0.44}{\sqrt{(0.40)(0.60)/750}} < z < \dfrac{0.48-0.44}{\sqrt{(0.48)(0.52)/750}}\right)$

(E) $P\left(\dfrac{0.40-0.44}{\sqrt{(0.44)(0.56)/750}} < z < \dfrac{0.48-0.44}{\sqrt{(0.44)(0.56)/750}}\right)$

330. In a random survey of 530 students, 28 percent said that they had experienced bullying at some time. With what degree of confidence can the pollster say that 28% ± 3% of students have experienced bullying?

(A) 12.4%
(B) 28%
(C) 87.6%
(D) 93.8%
(E) 95%

331. An online news magazine periodically gives viewers an opportunity to record electronically their agreement or disagreement with some viewpoint or commentary. On one such occasion, 250 out of 800 respondents agreed with a statement that the most practical way of meeting a potential spouse is through online dating sites. The immediate online calculation concluded that 31.25% of the viewers, with a margin of error of ±3.2%, agreed with the statement. The fine print below stated that the calculation was made with 95% confidence. What is the proper conclusion?

(A) We are 95% confident that the proportion of viewers who believe that online dating sites are the most practical way of meeting a potential spouse is between 0.280 and 0.345.
(B) Without knowing whether both np and $n(1 - p)$ are ≥ 10, the calculation is inappropriate.
(C) Without knowing whether or not the 800 respondents represent less than 10% of the entire population of viewers, the calculation is inappropriate.
(D) The z-distribution was used when the t-distribution should have been used, so the calculation is inappropriate.
(E) The data set was not a simple random sample, so the calculation is inappropriate.

332. There are 25,000 high school students in an extended metropolitan region. As each school's students came in to register for classes, guidance counselors were instructed to use a calculator to pick a random number between 1 and 100. If the number 37 was picked, the student was included in a survey. For one of the many survey questions, 86 percent of the students said they couldn't live without instant messaging. Are all conditions met for constructing a confidence interval of the proportion of this region's teenagers who believe they couldn't live without instant messaging?

(A) No, there is no guarantee that a representative random sample is chosen.
(B) No, np and $n(1 - p)$ are not both greater than 10.
(C) No, the sample size is not less than 10% of the population.
(D) No, there is no reason to assume that the population has a normal distribution.
(E) Yes, all conditions are met and a confidence interval can be constructed.

333. **For a given large sample size, which of the following gives the smallest margin of error when calculating a confidence interval for a population proportion?**

(A) 91% confidence with $\hat{p} = 0.18$
(B) 93% confidence with $\hat{p} = 0.18$
(C) 95% confidence with $\hat{p} = 0.18$
(D) 91% confidence with $\hat{p} = 0.34$
(E) 93% confidence with $\hat{p} = 0.34$

334. **We are interested in the proportion p of people who drive for Uber in a large city. Four percent of a simple random sample of 890 people say they drive for Uber. What is the midpoint for a 98% confidence interval estimate of p?**

(A) 0.01
(B) 0.49
(C) 0.5
(D) p
(E) None of these are correct.

335. **If all else is held constant, how does an interval change when constructing a 97% confidence interval rather than a 93% confidence interval?**

(A) The interval width increases by 4%.
(B) The interval width increases by 20%.
(C) The interval width increases by 35.8%.
(D) The interval width decreases by 20%.
(E) This question cannot be answered without knowing the sample size.

336. **In a random sample of men, a 95% confidence interval for the proportion who say they have ever spent a night in jail is (0.095, 0.115). In a random sample of women, the proportion who say they have ever spent a night in jail is $\hat{p} = 0.10$. If the two sample sizes are the same, how does a 95% confidence interval for the women's proportion compare to that of the men's?**

(A) The women's interval has the same width and a lower point estimate.
(B) The women's interval is narrower and has a higher point estimate.
(C) The women's interval is narrower and has a lower point estimate.
(D) The women's interval is wider and has a higher point estimate.
(E) The women's interval is wider and has a lower point estimate.

337. A restaurant evaluator reports that 22 percent of all restaurant meals include french fries. A 95% confidence interval is given by (0.202, 0.238). Which of the following is a correct interpretation of the *confidence level*?

(A) We are 95 percent confident that the true proportion of all restaurant meals that include french fries is between 0.202 and 0.238.
(B) There is a 0.95 probability that the true proportion of all restaurant meals that include french fries is between 0.202 and 0.238.
(C) 95 percent of all random samples of the same size chosen from the population result in confidence intervals that contain 0.22.
(D) 95 percent of all random samples of the same size chosen from the population result in confidence intervals that contain the true proportion of all restaurant meals that include french fries.
(E) 95 percent of all random samples of the same size chosen from the population have sample proportions in the 0.202 to 0.238 interval.

338. In a random sample of 800 married couples, 240 reported that the husband did the family laundry. What is a 90% confidence interval estimate of the proportion of all husbands who do the family laundry?

(A) $0.3 \pm 1.645\sqrt{240(0.3)(0.7)}$

(B) $0.3 \pm 1.645\sqrt{800(0.3)(0.7)}$

(C) $0.3 \pm 196\sqrt{800(0.3)(0.7)}$

(D) $0.3 \pm 1.645\sqrt{\frac{(0.3)(0.7)}{240}}$

(E) $0.3 \pm 1.645\sqrt{\frac{(0.3)(0.7)}{800}}$

339. A pollster plans to obtain a random sample of 250 adults from which he will construct a 99% confidence interval for the proportion of all adults who believe that global warming is caused by human activity. Which of the following statements must be true?

(A) The population proportion will be in the confidence interval.
(B) The probability that he will construct a confidence interval containing the sample proportion is 0.99.
(C) The probability that he will construct a confidence interval containing the population proportion is 0.99.
(D) There is a 0.99 probability that the sample proportion will equal the population proportion.
(E) None of the above are true statements.

340. In a random sample of 1,500 adults, 46 percent answered that they feel the death penalty is applied fairly by the courts. A margin of error of 2 percent is reported. Which of the following statements is appropriate?

(A) The sample is large, the sample is random, and the margin of error is relatively small. So it is reasonable to conclude that 46 percent of all adults feel the death penalty is applied fairly by the courts.
(B) Although $n = 1{,}500$ may seem large, it is actually very small in comparison to the number of all adults. So no conclusion can be drawn with any confidence.
(C) The sample proportion plus the margin of error is less than 0.50. So there is evidence that less than half of all adults feel the death penalty is applied fairly by the courts.
(D) No treatments are applied. So this is not an experiment, and cause and effect cannot be concluded.
(E) $0.02 < 0.05$, so there is sufficient evidence for any reasonable conclusion.

341. Factors involved in the creation of a confidence interval include the sample size, level of confidence, and margin of error. Which of the following is *incorrect*?

(A) For a fixed margin of error, larger samples provide greater confidence.
(B) For a given confidence level, halving the margin of error requires a sample twice as large.
(C) For a given sample size, reducing the margin of error means lower confidence.
(D) For a given sample size, higher confidence means a larger margin of error.
(E) For a fixed margin of error, smaller samples mean lower confidence.

342. In a survey of a random sample of 450 South Carolinians, 75 percent say they enjoy hot pepper sauce on their food. If the margin of error is 2.3 percent, what is the level of confidence?

(A) 67%
(B) 74%
(C) 87%
(D) 90%
(E) 93%

343. There have been various reports about restaurants and fish markets mislabeling seafood. A consumer advocacy magazine sponsored a study in which 190 pieces of seafood (out of an estimated 12,000 pieces sold each day in the region of interest) were purchased and analyzed. Laboratory results concluded that 22 percent were mislabeled. What is a 97% confidence interval for the proportion of all seafood servings/packages that were mislabeled?

(A) $0.22 \pm 2.17\sqrt{\frac{(0.22)(0.78)}{190}}$

(B) $0.22 \pm 1.88\sqrt{\frac{(0.22)(0.78)}{190}}$

(C) $0.22 \pm 2.17\sqrt{\frac{(0.5)(0.5)}{190}}$

(D) $0.22 \pm 1.88\sqrt{\frac{(0.22)(0.78)}{12{,}000}}$

(E) $0.22 \pm 2.17\sqrt{\frac{(0.22)(0.78)}{12{,}000}}$

344. Is climate change causing more severe storms? In a 2017 pre–Hurricane Irma survey, 486 out of 1,080 adults answered in the affirmative. In a 2017 post–Hurricane Irma survey, 546 out of 1,050 answered affirmatively. Establish a 90% confidence interval estimate of the difference (pre- minus post-hurricane) between the proportions of adults pre–Hurricane Irma and post–Hurricane Irma who believe climate change is causing more severe storms.

(A) $(0.45 - 0.52) \pm 1.645\sqrt{\frac{(0.45)(0.55)}{1{,}080} + \frac{(0.52)(0.48)}{1{,}050}}$

(B) $(0.45 - 0.52) \pm 1.645\left(\frac{(0.45)(0.55)}{\sqrt{1{,}080}} + \frac{(0.52)(0.48)}{\sqrt{1{,}050}}\right)$

(C) $(0.45 - 0.52) \pm 1.96\sqrt{\frac{(0.45)(0.55)}{1{,}080} + \frac{(0.52)(0.48)}{1{,}050}}$

(D) $(0.45 - 0.52) \pm 1.96\left(\frac{(0.45)(0.55)}{\sqrt{1{,}080}} + \frac{(0.52)(0.48)}{\sqrt{1{,}050}}\right)$

(E) $(0.45 - 0.52) \pm 2.576\left(\frac{\sqrt{(0.45)(0.55)}}{1{,}080} + \frac{\sqrt{(0.52)(0.48)}}{1{,}050}\right)$

345. In a random sample of 500 high school students, 350 said they had TVs in their bedrooms. In a random sample of 500 middle school students, 325 said they had TVs in their bedrooms. Which of the following represents a 99% confidence interval estimate for the difference (high school minus middle school) between the proportions of high school students and middle school students who have TVs in their bedrooms?

(A) $(0.7 - 0.65) \pm 1.96\sqrt{\frac{(0.7)(0.3)}{500} + \frac{(0.65)(0.35)}{500}}$

(B) $(0.7 - 0.65) \pm 2.326\sqrt{\frac{(0.7)(0.3)}{500} + \frac{(0.65)(0.35)}{500}}$

(C) $(0.7 - 0.65) \pm 2.576\sqrt{\frac{(0.7)(0.3)}{500} + \frac{(0.65)(0.35)}{500}}$

(D) $(0.7 - 0.65) \pm 2.326\sqrt{(0.675)(0.325)\left(\frac{1}{500} + \frac{1}{500}\right)}$

(E) $(0.7 - 0.65) \pm 2.576\sqrt{(0.675)(0.325)\left(\frac{1}{500} + \frac{1}{500}\right)}$

346. A statistician is developing a method to compare the quotient of proportions from two political polling companies. He tests this method by simulating 5,000 samples from an appropriate population and calculating a 95% confidence interval from each using his new method. If his new method works as intended, 4,750 of the confidence intervals from the simulation will capture what?

(A) The difference of sample proportions
(B) The quotient of sample proportions
(C) The quotient of sample proportions divided by an appropriate standard deviation
(D) The difference of population proportions
(E) The quotient of population proportions

347. In a survey of randomly selected Americans over the age of 65, 40 percent of 756 men and 49 percent of 825 women suffered from some form of arthritis. What is a 96% confidence interval for the difference in the proportions of senior men and senior women who have this disease, and does it indicate sufficient evidence that the proportions are not equal?

(A) The confidence interval is (–0.138, –0.042). This does not indicate sufficient evidence that the proportions aren't equal because 0 is not in the interval.
(B) The confidence interval is (–0.141, –0.039). This does not indicate sufficient evidence that the proportions aren't equal because 0 is not in the interval.
(C) The confidence interval is (–0.138, –0.042). This does indicate sufficient evidence that the proportions aren't equal because 0 is not in the interval.
(D) The confidence interval is (–0.141, –0.039). This does indicate sufficient evidence that the proportions aren't equal because 0 is not in the interval.
(E) The confidence interval is (–0.138, –0.042). This does not indicate sufficient evidence that the proportions aren't equal because the entire interval is less than 0.

CONFIDENCE INTERVALS FOR MEANS

348. Two confidence interval estimates from the same sample are (52.7, 58.3) and (51.8, 59.2). One estimate is at the 95% level, while the other is at the 99% level. Which is which?

(A) (51.8, 59.2) is the 95% level.
(B) (51.8, 59.2) is the 99% level.
(C) This question cannot be answered without knowing the sample size.
(D) This question cannot be answered without knowing the sample standard deviation.
(E) This question cannot be answered without knowing both the sample size and the sample standard deviation.

349. When making an inference about a population mean, which of the following suggests the use of *z*-scores rather than *t*-scores?

(A) The absence of outliers
(B) The population is normal
(C) The sample size is under 30
(D) The absence of strong skew
(E) The population standard deviation is known

350. Under what conditions would it be meaningful to construct a confidence interval estimate when the data consist of the entire population?

(A) If the population size is small ($n < 30$)
(B) If the population size is large ($n \geq 30$)
(C) If a higher level of confidence is desired
(D) If the population is truly random
(E) Never

351. The number of cloudy days per month (days in which the sun is never seen) in a northwestern town is noted for a random sample of 60 months with $\bar{x} = 23.4$ and $s = 3.7$. With what degree of confidence can we assert that the mean number of cloudy days per month in this town is between 22.4 and 24.4?

(A) 48%
(B) 90%
(C) 95%
(D) 96%
(E) 99%

352. Four statistics majors receiving actuarial job offers after graduation will receive the following starting salaries: $54,000, $61,000, $48,000, and $57,000. If all assumptions are met, what is a 95% confidence interval for the population mean for all statistics majors receiving actuarial job offers?

(A) $55{,}000 \pm 1.96\left(\frac{5{,}477}{\sqrt{3}}\right)$

(B) $55{,}000 \pm 2.776\left(\frac{5{,}477}{\sqrt{3}}\right)$

(C) $55{,}000 \pm 3.182\left(\frac{5{,}477}{\sqrt{3}}\right)$

(D) $55{,}000 \pm 2.776\left(\frac{5{,}477}{\sqrt{4}}\right)$

(E) $55{,}000 \pm 3.182\left(\frac{5{,}477}{\sqrt{4}}\right)$

353. Two 95% confidence interval estimates are obtained: I (34.5, 40.5) and II (36.3, 44.2).

a. If the sample sizes are the same, which has the larger standard deviation?
b. If the sample standard deviations are the same, which has the larger sample size?

(A) *a.* I, *b.* I
(B) *a.* I, *b.* II
(C) *a.* II, *b.* I
(D) *a.* II, *b.* II
(E) More information is needed to answer these questions.

354. Suppose (59, 72) is a 99% confidence interval estimate for a population mean μ. Which of the following is a true statement?

(A) Confidence level cannot be interpreted until after data are obtained.
(B) There is a 0.99 probability that μ is between 59 and 72.
(C) The probability that μ is in any particular confidence interval can be any value between 0 and 1.
(D) If 100 random samples of the given size are picked and a 99% confidence interval estimate is calculated from each, μ will be in 99 of the resulting intervals.
(E) If 99% confidence intervals are calculated from all possible samples of the given size, μ will be in 99% of these intervals.

355. A principal is informed that among the 376 seniors in her high school, the average SAT Math score is 615 with a standard deviation of 28.5. With what margin of error is the mean SAT Math score of these students known?

(A) 0
(B) 28.5
(C) $\frac{28.5}{\sqrt{615}}$
(D) $1.96\left(\frac{28.5}{\sqrt{615}}\right)$
(E) None of the above give the correct answer.

356. In a random sample of 18 high school students' backpacks, the average weight was 14.5 pounds with a standard deviation of 3.9 pounds. Assuming all conditions for inference are met, what is a 96% confidence interval for the mean weight of all high school students' backpacks?

(A) $14.5 \pm 2.054\frac{3.9}{\sqrt{18}}$

(B) $14.5 \pm 2.205\frac{3.9}{\sqrt{18}}$

(C) $14.5 \pm 2.214\frac{3.9}{\sqrt{18}}$

(D) $14.5 \pm 2.224\frac{3.9}{\sqrt{18}}$

(E) $14.5 \pm 2.235\frac{3.9}{\sqrt{18}}$

357. To determine the average spent on books and lab supplies during a year in college, a simple random sample of 50 students is interviewed, showing a mean of $970 with a standard deviation of $310. Which of the following is the best interpretation of a 95% confidence interval estimate for the average spent on books and lab supplies during a year in college?

(A) 95% of college students spend between $882 and $1,058 on books and lab supplies yearly.
(B) 95% of college students spend a mean dollar amount on books and lab supplies yearly that is between $882 and $1,058.
(C) We are 95% confident that college students spend between $882 and $1,058 on books and lab supplies yearly.
(D) We are 95% confident that college students spend a mean dollar amount between $882 and $1,058 on books and lab supplies yearly.
(E) We are 95% confident that in the chosen sample, the mean dollar amount spent on books and lab supplies yearly by college students is between $882 and $1,058.

358. A survey was conducted involving 75 of the 26,000 families living in a town. In 2017, the average amount of medical expenses paid per family in the sample was \$1,950 with a standard deviation of \$640. What is a 95% confidence interval estimate for the total medical expenses paid by all families in the town?

(A) $26{,}000[1{,}950 \pm 1.993(640)]$

(B) $1{,}950 \pm 1.993\left(\frac{640}{\sqrt{75}}\right)$

(C) $26{,}000\left[1{,}950 \pm 1.993\left(\frac{640}{\sqrt{75}}\right)\right]$

(D) $26{,}000\left[1{,}950 \pm 1.993\left(\frac{640}{\sqrt{74}}\right)\right]$

(E) $26{,}000\left[1{,}950 \pm 1.96\left(\frac{640}{\sqrt{75}}\right)\right]$

359. A home economist wants to estimate the mean temperature of the typical shower. In a random sample of 16 people, the mean temperature of their showers was 101°F with a standard deviation of 2.4°F. Assuming all conditions of inference were met, what is a 95% confidence interval estimate for the mean typical shower temperature for all people?

(A) $101 \pm 1.96\left(\frac{2.4}{4}\right)$

(B) $101 \pm 1.96\left(\frac{\sqrt{2.4}}{4}\right)$

(C) $101 \pm 2.131\left(\frac{\sqrt{2.4}}{16}\right)$

(D) $101 \pm 2.131\left(\frac{2.4}{4}\right)$

(E) $101 \pm 2.131\left(\frac{2.4}{16}\right)$

360. **A national education journal states that high school students get an average of 7.6 hours of sleep per night. A guidance counselor thinks that her school's students might have a different average and collects data from a random sample of these students. The data are used to test the hypotheses H_0: $\mu = 7.6$, H_a: $\mu \neq 7.6$. The results of the test are shown in the table below.**

Sample Mean	Standard Error	*df*	*t*	*P*
7.4	0.1091	20	−1.833	0.0817

Assuming all conditions for inference are met, which of the following gives a 90% confidence interval for μ?

(A) 7.4 ± 0.0411
(B) 7.4 ± 0.0436
(C) 7.4 ± 0.1091
(D) 7.4 ± 0.1882
(E) 7.4 ± 0.2000

361. **A random sample of 60 high school students took part in a study where they ate helpings of pizza and then, four hours later, had their LDL cholesterol levels measured. The 95% confidence interval for the mean LDL cholesterol level for all high school students four hours after eating pizza was calculated to be (103, 120). Which of the following is a true statement?**

(A) 95 percent of all high school students will have LDL cholesterol levels between 103 and 120 four hours after eating pizza.
(B) Approximately 95 percent of all random samples of size 60 of high school students will result in 95% confidence intervals containing 111.5 for the mean LDL cholesterol level of all high school students four hours after eating pizza.
(C) Approximately 95 percent of all random samples of size 60 of high school students will result in 95% confidence intervals containing the true mean population LDL cholesterol level for the mean LDL cholesterol level of all high school students four hours after eating pizza.
(D) The probability is 0.95 that a randomly selected high school student will have an LDL cholesterol level between 103 and 120 four hours after eating pizza.
(E) We are 95 percent confident that the mean LDL cholesterol level of these 60 students four hours after eating pizza is between 103 and 120.

362. A 90% confidence interval for the mean time, in minutes, for high school students to finish a video game is determined to be (35.2, 73.4). Which of the following is the best interpretation of the interval?

(A) Ten percent of players will take either less than 35.2 minutes or more than 73.4 minutes.
(B) The probability that a randomly selected player will take between 35.2 and 73.4 minutes is 0.90.
(C) For 90 percent of players, their mean finishing time is between 35.2 and 73.4 minutes.
(D) If this study was repeated many times, the resulting confidence interval will contain the true mean finishing time 90 percent of the time.
(E) We are 90 percent confident that the true mean finishing time is between 35.2 and 73.4 minutes.

363. Americans and Japanese were surveyed as to the ideal number of children a family should have. For 500 Americans, the mean was 1.95 with a standard deviation of 0.65. For 450 Japanese, the mean was 1.05 with a standard deviation of 0.40. Assuming all conditions for inference are met, what is a 90% confidence interval for the difference in population means?

(A) $0.9 \pm 1.96\sqrt{\frac{(1.95)^2}{500} + \frac{(1.05)^2}{450}}$

(B) $0.9 \pm 1.645\left(\frac{0.65}{\sqrt{500}} + \frac{0.4}{\sqrt{450}}\right)$

(C) $0.9 \pm 1.645\sqrt{\frac{(0.65)^2}{500} + \frac{(0.4)^2}{450}}$

(D) $0.9 \pm 1.96\left(\frac{0.65}{\sqrt{500}} + \frac{0.4}{\sqrt{450}}\right)$

(E) $0.9 \pm 3.29\left(\frac{0.65}{\sqrt{500}} + \frac{0.4}{\sqrt{450}}\right)$

364. A dietician plans to compare the average number of minutes spent eating each day for French and Americans. The 95% confidence interval estimate of the difference (French – Americans) is (47, 73). Which of the following is the most reasonable conclusion?

(A) The mean number of minutes daily spent eating among the French is $\frac{73}{47}$ times the mean spent by Americans.
(B) The mean number of minutes daily spent eating among the French is 73 minutes, while the mean among Americans is 47 minutes.
(C) The probability that the mean numbers of minutes spent eating are different is 0.95.
(D) The probability that the difference in minutes spent eating is greater than 47 minutes is 0.95.
(E) We should be 95 percent confident that the difference in mean minutes spent eating is between 47 and 73 minutes.

365. In a national sleep study, a random sample of adults was contacted. These subjects were asked the number of hours they slept per night and whether or not they exercised regularly. Of the 240 adults who exercised regularly, the mean number of hours of sleep per night was 8.2 with a standard deviation of 0.4. Of the 347 adults who did not exercise regularly, the mean number of hours of sleep per night was 7.9 with a standard deviation of 0.5. Which of the following is the most appropriate standard error for a confidence interval of the difference in nightly sleep hours between exercisers and nonexercisers?

(A) $\sqrt{\frac{(8.2)^2}{240}+\frac{(7.9)^2}{347}}$

(B) $\frac{8.2}{\sqrt{240}}+\frac{7.9}{\sqrt{347}}$

(C) $\sqrt{\frac{(0.4)^2}{240}+\frac{(0.5)^2}{347}}$

(D) $\frac{0.4}{\sqrt{240}}+\frac{0.5}{\sqrt{347}}$

(E) $\sqrt{\frac{(0.4)^2+(0.5)^2}{240+347}}$

366. A botanist chooses a random sample of 10 maple trees and an independent random sample of 10 elm trees. She measures the diameters of the trees and then calculates the mean and standard deviation of the diameters for each sample. Assume the distribution of diameter sizes for maples and for elms are both normal. What is the appropriate method for constructing a two-sample confidence interval of the difference in the mean diameters of maples and elms?

(A) A *z*-interval is appropriate because both distributions are normal.
(B) A *z*-interval is appropriate because both sample standard deviations are known.
(C) A *z*-interval is appropriate because the central limit theorem applies.
(D) A *t*-interval is appropriate because the population standard deviations are unknown.
(E) A *t*-interval is appropriate because the sample sizes are both below 30.

367. Suppose a two-sample *t*-test for H_0: $\mu_1 - \mu_2 = 0$ and H_a: $\mu_1 - \mu_2 > 0$ results in a *P*-value of 0.04. Which of the statements below must be true?

I. A 90% confidence interval for the difference in means will contain 0.
II. A 95% confidence interval for the difference in means will contain 0.
III. A 99% confidence interval for the difference in means will contain 0.

(A) I only
(B) I and II only
(C) I, II, and III
(D) III only
(E) II and III only

368. A nutritionist analyzes two types of hotdogs, all-beef and all-chicken. A random sample of 10 all-beef hotdogs showed a mean of 9.9 mg of fat with a standard deviation of 1.5 mg. A random sample of 8 all-chicken hotdogs showed a mean of 7.5 mg of fat with a standard deviation of 1.1 mg. What is the standard error of the difference in fat content (beef – chicken) between the sample means?

(A) $\sqrt{\frac{9.9^2}{10} + \frac{7.5^2}{8}}$

(B) $\frac{9.9}{\sqrt{10}} - \frac{7.5}{\sqrt{8}}$

(C) $\sqrt{\frac{1.5^2}{10} - \frac{1.1^2}{8}}$

(D) $\sqrt{\frac{1.5^2}{10} + \frac{1.1^2}{8}}$

(E) $\frac{1.5}{\sqrt{10}} + \frac{1.1}{\sqrt{8}}$

369. An experiment is conducted to test the effectiveness of a new sleep aid. Twenty volunteers are randomly assigned to two groups, with one group receiving the sleep aid while the other group receives a similar-looking placebo. The next day, the mean hours of sleep for each group is calculated and a two-sample *t*-test for the difference of means is performed. Which of the following is a necessary assumption?

(A) The expected number of hours of sleep for each person is at least 5.
(B) The sample sizes are greater than or equal to 10 percent of the population sizes.
(C) The 10 volunteers randomly assigned to the sleep aid group are paired with the 10 volunteers assigned to the placebo group.
(D) The distributions of hours of sleep in each population is approximately normal.
(E) Each group of 10 should have 5 successes and 5 failures.

CONFIDENCE INTERVALS FOR SLOPES

370. Below is the computer out put for a regression analysis involving starting salary (in $1,000) and college GPA in a random sample of 25 graduates.

Variable	Coef	s.e.	t	p
Constant	−0.73391	5.744	−0.128	0.9015
GPA	11.8204	1.848	6.4	0.0002
s = 3.772	R-sq = 83.6%		R-sq(adj) = 81.6%	

What is a 90% confidence interval for the slope of the regression line?

(A) $11.8204 \pm 1.645(1.848)$
(B) $11.8204 \pm 1.711(1.848)$
(C) $11.8204 \pm 1.714(1.848)$
(D) $11.8204 \pm 1.645\left(\frac{1.848}{\sqrt{25}}\right)$
(E) $11.8204 \pm 1.711\left(\frac{1.848}{\sqrt{25}}\right)$

371. An insurance adjustor is interested in the age of homes and average wind and flood damage from hurricanes. Data from 20 randomly selected homes generate the following computer output:

Variable	Coeff	s.e. of Coeff	t-ratio	prob
Constant	9717.95	7208	1.35	0.194
Age	15675.2	1678	9.34	0.000

s = 15,720 with df = 20 − 2 = 18 R-sq = 82.9% R-sq(adj) = 81.9%

Which of the following gives a 96% confidence interval for the slope of the regression line?

(A) $9{,}718 \pm 2.054$
(B) $9{,}718 \pm 2.197$
(C) $9{,}718 \pm 2.214\left(\frac{7{,}208}{\sqrt{20}}\right)$
(D) $15{,}675 \pm 2.197\left(\frac{15{,}720}{\sqrt{20}}\right)$
(E) $15{,}675 \pm 2.214(1{,}678)$

372. A study is made relating life expectancy (in days) of a laptop battery as a function of price of the battery. Data from a sample of 13 laptops generates the following computer output.

Dependent variable is: **Life**
R squared = 68.3% R squared (adjusted) = 65.5% s = 74.29

Source	Sum of Squares	Mean Square
Regression	131088	131088
Residual	60704.5	5518.59

Variable	Coefficient	s.e. Coeff	t-ratio	prob
Constant	410.997	67.79	6.06	0.000
Price	5.52979	1.135	4.87	0.000

Which of the following gives a 90% confidence interval for the slope of the regression line?

(A) $410.997 \pm 1.771(67.79)$
(B) $410.997 \pm 1.796(67.79)$
(C) $5.52979 \pm 1.645(74.29)$
(D) $5.52979 \pm 1.771(1.135)$
(E) $5.52979 \pm 1.796(1.135)$

373. A college Office of Alumni Relations gathers data and performs a linear regression analysis on donation gifts versus salary of alumni. The resulting computer output (where salary is in $1,000) is shown below.

```
Dependent variable is: Gift
s = 95.08 with 20 – 2 degrees of freedom
Variable    Coeff     s.e. of Coeff  t-ratio  prob
Constant    8.90087   161.6          0.055    0.956
Salary      10.292    0.8192         12.6     0.000
```

What is a 95% confidence interval of the slope interpreted in context?

(A) We are 95 percent confident that for each $1,000 more in salary that an alumnus earns, he/she will donate $10.29 more.
(B) We are 95 percent confident that for each $1,000 more in salary that an alumnus earns, he/she will donate between $8.57 and $12.01 more.
(C) We are 95 percent confident that for each $1,000 more in salary that an alumnus earns, it is predicted that he/she will donate between $8.57 and $12.01 more on average.
(D) 89.8 percent of the variability in gift donations is explained by the linear model.
(E) 89.8 percent of the variability in gift donations is explained by variability in salary.

374. A regression analysis of the prices of textbooks versus page lengths yields the following equation.

$$\textit{Predicted price} = -3.35 + 0.15(\textit{Pages})$$

A 92% confidence interval of the slope is (0.115, 0.185). What is a correct interpretation of this interval in context?

(A) Every additional page will raise the price $0.15.
(B) The probability is 0.92 that, on average, each additional page will raise the price $0.15.
(C) 92 percent of all random samples of textbooks will give a regression slope between 0.115 and 0.185.
(D) We are 92 percent confident that each additional page will raise the price $0.15.
(E) We are 92 percent confident that each additional page will raise the price between $0.115 and $0.185 on average.

375. The 96% confidence interval for the slope of a regression line is (–0.142, 1.036). Which of the following is a true statement?

(A) The sample slope is $b = 0$.
(B) The sample slope is $b = 0.589$.
(C) The sum of the residuals is positive.
(D) The mean of the residuals is positive.
(E) The correlation coefficient r is positive.

DETERMINING SAMPLE SIZES

376. In general, how does halving the sample size change the confidence interval size?

(A) It doubles the interval size.
(B) It halves the interval size.
(C) It multiplies the interval size by 1.414.
(D) It divides the interval size by 1.414.
(E) This question cannot be answered without knowing the sample size.

377. A hospital administrator wishes to determine the mean number of admissions per day to within ±0.15 at a 94% confidence level. What sample size should be chosen if it is known that the standard deviation is 0.56?

(A) 8
(B) 38
(C) 49
(D) 50
(E) 54

378. A survey is to be taken to estimate the proportion of voters who favor stem cell research. Among the following proposed sample sizes, which is the smallest that will still guarantee a margin of error of at most 0.045 for a 98% confidence interval?

(A) 26
(B) 200
(C) 400
(D) 600
(E) 700

379. Which of the following would result in the narrowest confidence interval?

(A) Small sample size and 93% confidence
(B) Small sample size and 97% confidence
(C) Large sample size and 93% confidence
(D) Large sample size and 97% confidence
(E) This cannot be answered without knowing an appropriate standard deviation.

380. A researcher plans to investigate the difference between the proportion of children of drug abusers and the proportion of children of nondrug abusers who experiment with drugs while in high school. Which of the following will give the sample size n that should be taken (same number for each group) to be 95 percent certain of knowing the difference to within ±0.025?

(A) $1.645\left(\frac{0.5}{\sqrt{n}}\right) \le 0.025$

(B) $1.645\left(\frac{0.5\sqrt{2}}{\sqrt{n}}\right) \le 0.025$

(C) $1.96\left(\frac{0.5}{\sqrt{n}}\right) \le 0.025$

(D) $1.96\left(\frac{0.5\sqrt{2}}{\sqrt{n}}\right) \le 0.025$

(E) $2.576\left(\frac{0.5}{\sqrt{n}}\right) \le 0.025$

381. A concerned-scientists action group wishes to learn the proportion of high school students who believe that humans are the principal cause of observed global warming. From a past study, the group knows that it will have to poll 200 people for the desired level of confidence. If the group wants to keep the same level of confidence but divide the margin of error by 3, how many people will it have to poll?

(A) 23
(B) 68
(C) 300
(D) 600
(E) 1,800

382. A guidance counselor wishes to know the difference in GPAs between students who exercise regularly and those who don't. Suppose the standard deviation of each group is known to be 0.93. Which of the following will give the sample size n that should be taken (same number for each group) to be 99 percent certain of knowing the difference to within ±0.1 on the GPA scale?

(A) $2.326\frac{0.93}{\sqrt{n}} \leq 0.1$

(B) $2.576\frac{0.93}{\sqrt{n}} \leq 0.1$

(C) $2.326\sqrt{\frac{0.93^2}{n} + \frac{0.93^2}{n}} \leq 0.1$

(D) $2.576\sqrt{\frac{0.93^2}{n} + \frac{0.93^2}{n}} \leq 0.1$

(E) $2.576\frac{0.5}{\sqrt{n}} \leq 0.1$

383. Two samples, S and T, of sizes 50 and 200, respectively, are obtained from the same population and result in the same sample proportion. In calculating a confidence interval for p from each sample, which of the following is a true statement about the margins of error obtained for each confidence interval?

(A) Since the sample proportions are equal, the margins of error are equal.
(B) The margin of error using S is 2 times the margin of error using T.
(C) The margin of error using S is 4 times the margin of error using T.
(D) The margin of error using T is 2 times the margin of error using S.
(E) The margin of error using T is 4 times the margin of error using S.

384. The presence of mercury in fish is a growing concern, especially now that people are being encouraged to eat more fish for the omega-3 content. Past studies of New Jersey coastal fish indicated mercury above 0.5 parts per million, a level that could pose a human health concern. A new study is being planned. Assuming the old study's standard deviation of 0.1 parts per million is still accurate, what size sample of fish should be caught for a 95% confidence interval for mercury with a margin of error no more than 0.008 parts per million?

(A) 25
(B) 123
(C) 423
(D) 601
(E) 15,007

385. An administrator at the National Council of Teachers is interested in the proportion of teachers who believe they will be able to retire comfortably when they reach the age of 60. Which of the following will give the minimum sample size that needs to be surveyed to be 98 percent confident of the true proportion to within ±3.5 percent?

(A) $2.054\sqrt{\frac{(0.98)(0.02)}{n}} \leq 0.035$

(B) $2.054\frac{0.5}{\sqrt{n}} \leq 0.035$

(C) $2.326\sqrt{\frac{(0.98)(0.02)}{n}} \leq 0.035$

(D) $2.326\frac{0.5}{\sqrt{n}} \leq 0.035$

(E) $2.576\sqrt{\frac{(0.035)(0.965)}{n}} \leq 0.035$

386. How fast can you download movies? We want to estimate the mean download time of new software being heavily advertised. Suppose the standard deviation of download times is known to be approximately 8 minutes. Which of the following gives the number of trial downloads we should run to have 90 percent confidence in our answer with a margin of error of at most 5 minutes?

(A) $1.645\frac{8}{\sqrt{n}} \leq 5$

(B) $1.645\frac{8(0.5)}{\sqrt{n}} \leq 5$

(C) $1.645\frac{8\sqrt{2}}{\sqrt{n}} \leq 5$

(D) $1.96\frac{8}{\sqrt{n}} \leq 5$

(E) $1.96\frac{8\sqrt{2}}{\sqrt{n}} \leq 5$

387. A cardiologist notices that 20 percent of her patients who complain of chest pain need stents. She plans to establish a confidence interval estimate at the 96 percent level with a margin of error of 3 percent. Of the following, which is the smallest sample size that she can use?

(A) 25
(B) 350
(C) 600
(D) 800
(E) 1,000

LOGIC OF SIGNIFICANCE TESTING

388. Which of the following is a true statement?

(A) A well-planned test of significance should result in a statement that either the null hypothesis is true or that it is false.
(B) The null hypothesis is one-sided and expressed using either < or > if there is interest in deviations in only one direction.
(C) When a true parameter value is further from the hypothesized value, it becomes easier to reject the alternative hypothesis.
(D) Tests of significance (hypothesis tests) are designed to measure the strength of evidence against the null hypothesis.
(E) Increasing the sample size makes it more difficult to conclude that an observed distance between observed and hypothesized values is significant.

389. Which of the following is a true statement?

(A) If the *P*-value is 0.16, the probability that the null hypothesis is correct is 0.16.
(B) The larger the *P*-value is, the more evidence there is against the null hypothesis.
(C) If the *P*-value is small enough, we can conclude that the alternative hypothesis is true.
(D) It is always wrong to use sample statistics in stating hypotheses.
(E) It is helpful to examine your data before deciding whether to use a one-sided or a two-sided hypothesis test.

390. Which of the following is a true statement?

(A) The *P*-value of a test is the probability of obtaining a result as extreme as or more extreme than the one obtained when assuming the null hypothesis is true.
(B) If the *P*-value for a test is 0.014, the probability that the null hypothesis is true is 0.014.
(C) When the null hypothesis is rejected, it is because the null hypothesis is not true.
(D) A very large *P*-value provides convincing evidence that the null hypothesis is true.
(E) The larger the *P*-value is, the greater the evidence of something statistically significant.

391. Which of the following is a true statement?

(A) If a population parameter is known, there is no reason to run a hypothesis test on that population parameter.
(B) The *P*-value can be negative or positive depending on whether the sample statistic is less than or greater than the claimed value of the population parameter in the null hypothesis.
(C) The *P*-value is based on a specific test statistic, so the *P*-value must be chosen before an experiment is conducted.
(D) If a *P*-value is larger than a specified value α, the data are statistically significant at that level.
(E) The *P*-value is a probability calculated when assuming that the alternative hypothesis is true.

392. The mean miles per gallon (mpg) of a certain model car is 32.8. Concerned that a production process change might have lowered that efficiency, an inspector tests a random sample of six cars, calculating a mean of 31.9 mpg with a *t*-score of –2.16 and a *P*-value of 0.0416. Which of the following is the most reasonable conclusion?

(A) 95.84% of the cars produced under the new process will have an mpg under 31.9.
(B) 95.84% of the cars produced under the new process will have an mpg under 32.8.
(C) 4.16% of the cars produced under the new process will have an mpg over 32.8.
(D) There is evidence at the 5% significance level to conclude that the new process is producing cars with a mean mpg under 31.9.
(E) There is evidence at the 5% significance level to conclude that the new process is producing cars with a mean mpg under 32.8.

393. Which of the following is a true statement?

(A) A *P*-value is a conditional probability.
(B) The *P*-value is the probability that the null hypothesis is true.
(C) A *P*-value is the probability the null hypothesis is true given a particular observed statistic.
(D) The *P*-value is the same as the power of a hypothesis test.
(E) A large *P*-value is evidence against the null hypothesis because it says that observed results are unlikely to occur when the null hypothesis is true.

394. A small retail store employs five men and five women. When comparing the mean years of service of these men and women, which of the following is most appropriate?

(A) A two-sample *z*-test of population means
(B) A two-sample *t*-test of population means
(C) A one-sample *z*-test on a set of differences
(D) A one-sample *z*-test on a set of differences
(E) None of the above are appropriate.

395. In a simple random sample (SRS) of 625 families who do not live near any chemical plant, 10 had children with leukemia. In an SRS of 412 families living near chemical plants, 15 had children with leukemia. A 90% confidence interval of the difference is reported to be –0.020 ± 0.017. Which of the following is a proper conclusion?

(A) The interval is invalid because it does not contain 0.
(B) The interval is invalid because probabilities cannot be negative.
(C) Families who live near chemical plants are approximately 2.0 percent more likely to have children with leukemia than are families who do not live near chemical plants.
(D) 90 percent of families living near chemical plants are approximately 2.0 percent more likely to have children with leukemia than are families who do not live near chemical plants.
(E) None of the above are proper conclusions.

396. For which of the following is it appropriate to use a *t*-distribution with 12 degrees of freedom?

I. **Constructing a confidence interval from an SRS with $n = 11$.**
II. **Doing a hypothesis test H_0: $\mu_1 - \mu_2 = 0$ and H_a: $\mu_1 - \mu_2 > 0$, where $n_1 = 12$ and $n_2 = 12$.**
III. **Doing a χ^2-test of independence where the contingency table has 4 rows and 3 columns.**

(A) I only
(B) II only
(C) III only
(D) I, II, and III
(E) None are appropriate.

397. Are generic batteries just as good as brand-name batteries? A consumer group runs a test on random samples of 20 brand-name alkaline AA batteries and 20 generic alkaline AA batteries. The group lets all the batteries run continuously and notes the number of hours when they die. Which of the following shows the proper set of hypotheses?

(A) H_0: $\bar{x}_{\text{name}} = \bar{x}_{\text{generic}}$ and H_a: $\bar{x}_{\text{name}} \neq \bar{x}_{\text{generic}}$
(B) H_0: $\bar{x}_{\text{name}} = \bar{x}_{\text{generic}}$ and H_a: $\bar{x}_{\text{name}} > \bar{x}_{\text{generic}}$
(C) H_0: $\bar{x}_{\text{name}} = \bar{x}_{\text{generic}}$ and H_a: $\bar{x}_{\text{name}} < \bar{x}_{\text{generic}}$
(D) H_0: $\bar{x}_{\text{difference}} = 0$ and H_a: $\bar{x}_{\text{difference}} \neq 0$
(E) None of the above show the proper set of hypotheses.

398. A driving instructor claims that the probability of passing the driving test on the first try after finishing his course is 0.85. A reporter believes this claim is high and plans to survey a random sample of people who took the instructor's class. Assuming the sample size will be large, what statistical test will be most appropriate?

(A) *z*-test of a proportion
(B) *z*-test for difference of two proportions
(C) chi-square test of homogeneity of proportions
(D) *t*-test of mean of a proportion
(E) *t*-test for a slope where a proportion is interpreted as a slope

399. The *P*-value for a two-sided *t*-test is 0.10. If the test had been one-sided, what would the *P*-value have been?

(A) 0.05
(B) 0.10
(C) 0.20
(D) 0.95
(E) It depends on the direction of the alternative hypothesis.

400. Medical researchers keep trying different molecular variations of a drug. They test each of these variations for effectiveness in relieving pain after minor surgeries. At the 5% significance level, 2 out of 75 of these variations show some statistical significance. Which of the following should be concluded?

(A) A sample size of 75 is not large enough for any serious medical conclusion.
(B) $\frac{2}{75} = 0.0267 < 0.05$, so there is sufficient evidence to conclude that the drug is effective in relieving pain after minor surgeries.
(C) Although no conclusion can be made about the drug in general, positive conclusions can be made about 2 of the 75 molecular variations in their effectiveness in relieving pain after minor surgeries.
(D) Pain relief is very susceptible to the placebo effect, so no conclusion is reasonable from this study without knowing whether or not random assignment and blinding with one group receiving a placebo was part of the procedure.
(E) There is insufficient evidence that any of the molecular variations are effective in relieving pain after minor surgeries because at the 5% significance level, statistically significant results in 2 out of 75 tests could simply be due to chance variability.

401. One of the conditions in performing a one-sample *t*-test is that the sample size be less than 10 percent of the population size. What is the reason for this condition?

(A) This allows for the application of the central limit theorem.
(B) This allows for using the sample standard deviation as an approximation to the population standard deviation.
(C) This guarantees an unbiased sample.
(D) If the sample is not too large when compared to the population, the dependence among observations is negligible even though we sample without replacement.
(E) This guarantees a large enough sample for the test procedure.

402. Suppose a two-sided hypothesis test results in a P-value of 0.32. If a one-sided test is conducted on the same data, which of the following would be true about the possible values of the resulting P-value?

(A) The only possible value is 0.16.
(B) The only possible value is 0.32.
(C) The only possible value is 0.64.
(D) The only possible values are 0.16 and 0.84.
(E) The only possible values are 0.32 and 0.68.

403. A politician is concerned about whether he has a problem with support from women voters. A large number of polls are taken by various polling organizations, each surveying independent random samples of 500 male voters and 500 female voters. The difference in the proportion of male support and proportion of female support is calculated for each poll. If there is actually no difference in support for the politician between male and female voters, which of the following results would be anticipated?

I. If for each poll tests are run at the 5% significance level with the hypotheses H_0: $p_M = p_W$ versus H_0: $p_M > p_W$, the P-value will be less than 0.05 for approximately 5 percent of the tests.
II. There will be approximately equal numbers of polls showing higher support levels from males and higher support levels from females.
III. If for each poll 95% confidence intervals for the difference in proportions, $p_M - p_W$, are calculated, approximately 95 percent of the confidence intervals will contain 0.

(A) I only
(B) II only
(C) III only
(D) I and II only
(E) I, II, and III

HYPOTHESIS TESTS FOR PROPORTIONS

404. A truant officer believes that the percentage of students skipping school during the World Series is even greater than the previously claimed 7 percent. She conducts a hypothesis test on a random 200 students and finds 23 guilty of truancy. Is this strong evidence against the 0.07 claim?

(A) Yes, because the P-value is 0.0062
(B) Yes, because the P-value is 2.5
(C) No, because the P-value is only 0.0062
(D) No, because the P-value is over 0.10
(E) There is insufficient information to reach a conclusion.

405. New York State law limits the decibel level of truck horns. A major truck manufacturer claims that only 35 percent of trucks on the road are even capable of a horn decibel level at or over the legal limit. However, a congressional investigator believes the true percentage is greater and runs a hypothesis test at the 5% significance level. If 57 out of a simple random sample (SRS) of 150 trucks have horns capable of decibel blasts over the legal limit, what is the appropriate test statistic?

(A) $z = \frac{0.38 - 0.35}{\sqrt{150(0.35)(1 - 0.35)}}$

(B) $z = \frac{0.38 - 0.35}{\sqrt{150(0.38)(1 - 0.38)}}$

(C) $z = \frac{0.38 - 0.35}{\sqrt{\frac{(0.35)(1 - 0.35)}{150}}}$

(D) $z = \frac{0.38 - 0.35}{\sqrt{\frac{(0.38)(1 - 0.38)}{150}}}$

(E) $z = 1.96\frac{0.38 - 0.35}{\sqrt{150(0.35)(1 - 0.35)}}$

406. It is claimed that 54 percent of lost remote controls are stuck between sofa cushions. A reporter tests this claim by checking a simple random sample (SRS) of 500 people who lost remote controls, and 280 of them reported finding the remote controls between sofa cushions. With H_0: $p = 0.54$ and H_a: $p \neq 0.54$, what is the P-value?

(A $P\left(z > \frac{0.56 - 0.54}{\sqrt{500(0.54)(1 - 0.54)}}\right)$

(B) $2P\left(z > \frac{0.56 - 0.54}{\sqrt{500(0.54)(1 - 0.54)}}\right)$

(C) $P\left(z > \frac{0.56 - 0.54}{\sqrt{\frac{(0.54)(1 - 0.54)}{500}}}\right)$

(D) $2P\left(z > \frac{0.56 - 0.54}{\sqrt{\frac{(0.54)(1 - 0.54)}{500}}}\right)$

(E) $2P\left(z > \frac{0.56 - 0.54}{\sqrt{500(0.56)(1 - 0.56)}}\right)$

407. Two students using the same data perform different significant tests. The first student performs the test H_0: $p = 0.76$ with H_a: $p \neq 0.76$, while the second student performs the test H_0: $p = 0.76$ with H_a: $p < 0.76$. Although both use the $\alpha = 0.01$ level of significance, the first student claims there is not enough evidence to reject H_0, while the second student says there is enough evidence to reject H_0. Which of the following could have been the value for the test statistic?

(A) $z = -2.6$
(B) $z = -2.4$
(C) $z = -0.01$
(D) $z = 0.76$
(E) $z = 2.6$

408. A hypothesis test is being run to see if there is significant evidence that fewer than 75 percent of elementary school children regularly eat cereal. In a random sample of 500 elementary school children, 66 percent say they regularly eat cereal. What is the test statistic for the appropriate test?

(A) $\dfrac{0.66 - 0.75}{\sqrt{\dfrac{(0.66)(0.34)}{500}}}$

(B) $\dfrac{0.66 - 0.75}{\sqrt{\dfrac{(0.75)(0.25)}{500}}}$

(C) $\dfrac{0.66 - 0.75}{\sqrt{\dfrac{(0.5)(0.5)}{500}}}$

(D) $\dfrac{0.75 - 0.66}{\sqrt{\dfrac{(0.66)(0.34)}{500}}}$

(E) $\dfrac{0.75 - 0.66}{\sqrt{\dfrac{(0.75)(0.25)}{500}}}$

409. Five years ago, a claim stated, "22 percent of American adults are functionally illiterate." For her dissertation, a graduate student tests the hypothesis that this percent has gone down, where H_0: $p = 0.22$ and H_a: $p < 0.22$. An appropriate z-test gives a P-value of 0.18. Is there sufficient evidence at the 10% significance level to say that the proportion of American adults who are functionally illiterate is now less than 22 percent?

(A) Yes, because $0.18 < 0.22$
(B) Yes, because $0.18 > 0.10$
(C) Yes, because $0.10 < 0.22$
(D) No, because $0.18 < 0.22$
(E) No, because $0.18 > 0.10$

410. **A test of the hypotheses H_0: $p = 0.4$ versus H_a: $p \neq 0.4$ was conducted using a sample of size $n = 175$. The test statistic was $z = 1.15$. What was the *P*-value of the test?**

(A) 0.125
(B) 0.25
(C) 0.4
(D) 0.5
(E) 0.75

411. **A survey of 850 Europeans reveals that 725 believe that Atlantic bluefin tuna are an endangered animal and should have protections from the fishing industry. In Japan, bluefin tuna are considered a sushi delicacy. In a survey of 720 Japanese, only 216 believe that bluefin tuna are endangered and should be protected. To test at the 5% significance level whether or not the data are significant evidence that the proportion of Japanese who believe that tuna need protection is less than the proportion of Americans with this belief, a student notes that $\frac{216}{720} = 0.3$ and sets up the following: H_0: $p = 0.3$ and H_a: $p > 0.3$, where p is the proportion of Europeans who believe that bluefin tuna need protection. Which of the following is a true statement?**

(A) The student has set up a correct hypothesis test.
(B) Given the large sample sizes, a 1% significance level would be more appropriate.
(C) A two-sided test would be more appropriate.
(D) Given that $\frac{725 + 216}{850 + 720} = 0.60$, H_a: $p > 0.60$ would be more appropriate.
(E) A two-population difference in proportions hypothesis test would be more appropriate.

412. In a random survey of 500 college STEM majors, 315 said they were more interested in being challenged than in receiving high grades. In a random survey of 400 college business majors, 220 said they were more interested in being challenged than in receiving high grades. Is there sufficient evidence to show that the proportion of STEM majors who are more interested in being challenged than in receiving high grades is greater than the proportion of business majors who are more interested in being challenged than in receiving high grades?

(A) Because 0.63 > 0.55, there is strong evidence that the proportion of STEM majors who are more interested in being challenged is greater than that of business majors.
(B) Because 0.0075 < 0.01, there is very strong evidence that the proportion of STEM majors who are more interested in being challenged is greater than that of business majors.
(C) Because 0.01 < 0.0329 < 0.05, there is strong evidence that the proportion of STEM majors who are more interested in being challenged is greater than that of business majors.
(D) There is insufficient evidence that the proportion of STEM majors who are more interested in being challenged is greater than that of business majors.
(E) There is insufficient information to determine whether the proportion of STEM majors who are more interested in being challenged is greater than that of business majors.

413. In a well-known basketball study, it was reported that Larry Bird hit a second free throw in 48 out of 53 attempts after the first free throw was missed and hit a second free throw in 251 of 285 attempts after the first free throw was made. Suppose an appropriate test is performed to determine whether there is sufficient evidence to say that the probability that Bird will make a second free throw is different depending on whether or not he made the first free throw. What is the *P*-value of the appropriate test?

(A) $P\left(z > \dfrac{\frac{48}{53} - \frac{251}{285}}{\sqrt{\left(\frac{48+251}{53+285}\right)\left(1 - \frac{48+251}{53+285}\right)\left(\frac{1}{53} + \frac{1}{285}\right)}}\right)$

(B) $2P\left(z > \dfrac{\frac{48}{53} - \frac{251}{285}}{\sqrt{\left(\frac{48+251}{53+285}\right)\left(1 - \frac{48+251}{53+285}\right)\left(\frac{1}{53} + \frac{1}{285}\right)}}\right)$

(C) $P\left(z > \dfrac{\frac{48}{53} - \frac{251}{285}}{\sqrt{\frac{\frac{48}{53}\left(1 - \frac{48}{53}\right)}{53} + \frac{\frac{251}{285}\left(1 - \frac{251}{285}\right)}{285}}}\right)$

(D) $2P\left(z > \dfrac{\frac{48}{53} - \frac{251}{285}}{\sqrt{\frac{\frac{48}{53}\left(1 - \frac{48}{53}\right)}{53} + \frac{\frac{251}{285}\left(1 - \frac{251}{285}\right)}{285}}}\right)$

(E) $P\left(z > \dfrac{\frac{48}{53} - \frac{251}{285}}{\sqrt{(0.5)(1 - 0.5)\left(\frac{1}{53} + \frac{1}{285}\right)}}\right)$

414. A poll of 800 moviegoers showed 560 liked a new action film, while a poll of 600 moviegoers showed 450 liked the sequel. In a hypothesis test on whether a higher proportion of moviegoers like the sequel than like the original, what is the test statistic?

(A) $z = \dfrac{0.7 - 0.75}{\sqrt{(0.7)(1 - 0.7)/800 + (0.75)(1 - 0.75)/600}}$

(B) $z = \dfrac{0.7 - 0.75}{\sqrt{(0.7214)(1 - 0.7214)\left(1/800 + 1/600\right)}}$

(C) $z = \dfrac{0.7 - 0.75}{\sqrt{800(0.7)(1 - 0.7) + 600(0.75)(1 - 0.75)}}$

(D) $z = \dfrac{0.7 - 0.75}{\sqrt{\dfrac{(0.725)(1 - 0.725)}{700}}}$

(E) $z = \dfrac{0.7 - 0.75}{\sqrt{700(0.725)(1 - 0.725)}}$

415. In independent random samples of 600 men and 500 women, 80 percent of the men and 76 percent of the women say they are satisfied with their physical attractiveness. Is there sufficient evidence to say that the proportion of men who are satisfied with their physical attractiveness is greater than the proportion of women who are satisfied with their physical attractiveness?

(A) Yes, because the difference in sample proportions, 0.04, is less than 0.05
(B) Yes, because the probability of observing a difference at least as large as the sample difference, if the two population proportions are the same, is less than 0.05
(C) Yes, because the probability of observing a difference at least as large as the sample difference, if the two population proportions are the same, is greater than 0.05
(D) No, because the probability of observing a difference at least as large as the sample difference, if the two population proportions are the same, is less than 0.05
(E) No, because the probability of observing a difference at least as large as the sample difference, if the two population proportions are the same, is greater than 0.05

416. Are more immigrants coming to America or to Canada? In random surveys of 3,500 Americans and 2,800 Canadians, 525 of the Americans and 474 of the Canadians say they were born in a different country. A test is conducted with $H_0: p_1 = p_2$ versus $H_a: p_1 \neq p_2$. What is the conclusion at the 5% significance level?

(A) There is not sufficient evidence that the proportion of all Americans who say they were born in another country is different from the proportion of all Canadians who say they were born in another country because $\hat{p} = \frac{525 + 474}{3{,}500 + 2{,}800} = 0.159 > 0.05$.
(B) There is not sufficient evidence that the proportion of all Americans who say they were born in another country is different from the proportion of all Canadians who say they were born in another country because the P-value is greater than 0.05.
(C) There is sufficient evidence that the proportion of all Americans who say they were born in another country is different from the proportion of all Canadians who say they were born in another country because the P-value is positive.
(D) There is sufficient evidence that the proportion of all Americans who say they were born in another country is different from the proportion of all Canadians who say they were born in another country because the P-value is less than 0.05.
(E) There is sufficient evidence that the proportion of all Americans who say they were born in another country is different from the proportion of all Canadians who say they were born in another country because the P-value is greater than 0.05.

417. In a well-known study, two booths were set up, one on each side of a college campus. One booth sold fine Swiss chocolates for 5 cents each and Hershey chocolate kisses for a penny each. The second booth sold the fine Swiss chocolates for 4 cents each and gave away the Hershey chocolate kisses for free. Both booths limited each student to picking one piece of chocolate. A higher proportion of students bought the Swiss chocolates at the first booth, while a higher proportion of students took the free Hershey chocolate kisses at the second booth. Which of the following test procedures should be used to test for statistical significance?

(A) A one-sample z-test of proportions because the student body is a single population
(B) A two-sample z-test for a difference of proportions
(C) A one-sample t-test of the mean difference of proportions
(D) A matched pair t-test comparing one booth to the other
(E) A t-test for the slope of a regression line of proportions sold at one booth versus proportions sold at the other booth

418. A random sample of 120 AP students was asked if they would prefer to have review sessions after school rather than on Saturdays. 67 students expressed a preference for after-school sessions, while 53 expressed a preference for Saturday sessions. A two-sample z-test for the difference of proportions using $\hat{p} = \frac{67}{120}$ and $\hat{p} = \frac{53}{120}$ results in a P-value of 0.035. What was the error in this analysis?

(A) A two-sided test should have been performed rather than the one-sided test that was performed.
(B) A one-sample z-test should have been performed rather than the two-sample test.
(C) A chi-square test of independence would have been the proper test.
(D) It is not clear whether or not the 10% rule was violated.
(E) There was no error.

419. A hypothesis test comparing two population proportions results in a P-value of 0.028. Which of the following is a proper conclusion?

(A) The probability that the null hypothesis is true is 0.028.
(B) The probability that the alternative hypothesis is true is 0.028.
(C) The difference in sample proportions is 0.028.
(D) The difference in population proportions is 0.028.
(E) None of the above are proper conclusions.

HYPOTHESIS TESTS FOR MEANS

420. An efficiency expert is interested in comparing the mean time taken for breaks by employees in areas with access to the Internet and those in areas that do not have this access. She interviews a simple random sample (SRS) of 10 employees with access to the Internet and an SRS of 10 without access. The efficiency expert then proceeds to run a t-test to compare the mean time taken for breaks in each group. Which of the following is a necessary assumption?

(A) The population standard deviations from each group are known.
(B) The population standard deviations from each group are unknown.
(C) The population standard deviations from each group are equal.
(D) The population of break times from each group is normally distributed.
(E) The samples must be independent samples and for each sample np and $n(1 - p)$, must both be at least 10.

421. **A high school coach claims that the average pulse rate of those trying out for sports is 62.4 beats per minute (bpm). The AP Statistics instructor suspects this is a made-up number and runs a hypothesis test on a simple random sample (SRS) of 32 students trying out for sports, calculating a mean of 65.0 bpm with a standard deviation of 10.3 bpm. What is the *P*-value?**

(A) $\left(z > \dfrac{65.0-62.4}{\left(\dfrac{10.3}{\sqrt{32}}\right)}\right)$

(B) $2P\left(z > \dfrac{65.0-62.4}{\left(\dfrac{10.3}{\sqrt{32}}\right)}\right)$

(C) $P\left(t > \dfrac{65.0-62.4}{\left(\dfrac{10.3}{\sqrt{32}}\right)}\right)$ with $df = 31$

(D) $2P\left(t > \dfrac{65.0-62.4}{\left(\dfrac{10.3}{\sqrt{32}}\right)}\right)$ with $df = 31$

(E) $P\left(t > \dfrac{65.0-62.4}{10.3}\right)$ with $df = 32$

422. **In a one-sided hypothesis test for the mean for a random sample of size 20, the *t*-score of the sample mean is 2.615 in the direction of the alternative. Is this significant at the 5% level? At the 1% level?**

(A) Significant at the 1% level but not at the 5% level
(B) Significant at the 5% level but not at the 1% level
(C) Significant at both the 1% and 5% levels
(D) Significant at neither the 1% nor 5% levels
(E) Cannot be determined from the given information

423. **A confidence interval estimate is determined from the summer earnings of a simple random sample (SRS) of *n* students. All other things being equal, which of the following results in a larger margin of error?**

(A) A lesser confidence level
(B) A smaller sample standard deviation
(C) A smaller sample size
(D) Introducing bias into the sampling
(E) Introducing blinding into an experiment

424. A spokesperson for the National Council of Teachers of Mathematics (NCTM) states that middle school math teachers spend an average of $1,250 per year of their own funds on classroom materials. A member of a Board of Education believes that the real figure is lower. So he interviews 12 randomly chosen middle school math teachers and comes up with a mean of $1,092 and a standard deviation of $308 spent out of pocket by middle school math teachers during the past academic year for classroom materials. Where is the *P*-value?

(A) Below 0.01
(B) Between 0.01 and 0.025
(C) Between 0.025 and 0.05
(D) Between 0.05 and 0.10
(E) Over 0.10

425. A critic of public education claims that during their four years of high school, students read an average of only 38 books. An English teacher plans to sample 100 high school graduates randomly as to the number of books read during their high school years and will reject the critic's claim if the sample mean exceeds 40 books. If the critic's claim is wrong and the true mean is 43 books, what is the probability that the random sample will lead to a mistaken failure to reject the critic's claim? Assume that the standard deviation in number of books read from the sample is 12 books.

(A) $P\left(t < \dfrac{43-38}{\left(\dfrac{12}{\sqrt{100}}\right)}\right)$

(B) $P\left(t > \dfrac{40-38}{\left(\dfrac{12}{\sqrt{100}}\right)}\right)$

(C) $P\left(t < \dfrac{40-43}{\left(\dfrac{12}{\sqrt{100}}\right)}\right)$

(D) $P\left(t > \dfrac{40-43}{\left(\dfrac{12}{\sqrt{100}}\right)}\right)$

(E) $P\left(t > \dfrac{43-38}{\left(\dfrac{12}{\sqrt{100}}\right)}\right)$

426. **A dietician claims that a new weight loss program will result in an average loss of 9 pounds in the first month. The program developer believes that the average weight loss in the first month will be greater than this. The program developer runs a test on a random sample of 64 overweight volunteers. What conclusion is reached if the sample mean loss is 9.55 pounds with a standard deviation of 3.00 pounds?**

(A) The *P*-value is less than 0.001, indicating very strong evidence against the 9-pound claim.
(B) The *P*-value is 0.01, indicating strong evidence against the 9-pound claim.
(C) The *P*-value is 0.07, indicating some evidence against the 9-pound claim.
(D) The *P*-value is 0.18, indicating very little evidence against the 9-pound claim.
(E) The *P*-value is 0.43, indicating no evidence against the 9-pound claim.

427. **According to Major League Baseball (MLB) official rules, the average weight of manufactured baseballs should be 5.25 ounces. An MLB inspector weighs a simple random sample (SRS) of five baseballs, obtaining weights of 5.3, 5.15, 5.15, 5.2, and 5.25 ounces. The inspector then runs an appropriate hypothesis test. Which of the following gives the *P*-value of this test?**

(A) $P\left(t < \dfrac{5.21 - 5.25}{\left(\dfrac{0.0652}{\sqrt{5}}\right)}\right)$ with $df = 4$

(B) $2P\left(t < \dfrac{5.21 - 5.25}{\left(\dfrac{0.0652}{\sqrt{5}}\right)}\right)$ with $df = 4$

(C) $P\left(t < \dfrac{5.21 - 5.25}{\left(\dfrac{0.0652}{\sqrt{5}}\right)}\right)$ with $df = 5$

(D) $2P\left(t < \dfrac{5.21 - 5.25}{\left(\dfrac{0.0652}{\sqrt{5}}\right)}\right)$ with $df = 5$

(E) $0.5P\left(t < \dfrac{5.21 - 5.25}{\left(\dfrac{0.0652}{\sqrt{5}}\right)}\right)$ with $df = 5$

428. An editor for a college magazine claims that the mean cost for 4-year tuition at private colleges is $258,000. A high school guidance counselor checks a random sample of 10 students who attended private colleges for four years and found the mean cost for tuition alone was $245,000 with a standard deviation of $38,000. What is the test statistic in performing a hypothesis test of H_0: $\mu = 258{,}000$ and H_a: $\mu \neq 258{,}000$?

(A) $t = \dfrac{245{,}000 - 258{,}000}{\sqrt{\dfrac{38{,}000}{10-1}}}$

(B) $t = 2\dfrac{245{,}000 - 258{,}000}{\sqrt{\dfrac{38{,}000}{10}}}$

(C) $t = 2\dfrac{245{,}000 - 258{,}000}{\left(\dfrac{38{,}000}{\sqrt{10-1}}\right)}$

(D) $t = \dfrac{245{,}000 - 258{,}000}{\left(\dfrac{38{,}000}{\sqrt{10}}\right)}$

(E) $t = \dfrac{245{,}000 - 258{,}000}{\left(\dfrac{38{,}000}{\sqrt{10}}\right)}$

429. In testing the null hypothesis H_0: $\mu = 51$ against the alternative hypothesis H_0: $\mu > 51$, a sample from a normal population has a mean of 52.1 with a corresponding *t*-score of 2.2 and a *P*-value of 0.0377. Which of the following is *not* a reasonable conclusion?

(A) If the null hypothesis is assumed to be true, the probability of obtaining a sample mean as extreme as or more extreme than 52.1 is only 0.0377.

(B) In all samples using the same sample size and the same sampling technique, the null hypothesis will be wrong 3.77 percent of the time.

(C) There is sufficient evidence to reject the null hypothesis at the 5% significance level.

(D) If the population standard deviation σ had been known, the *z*-score of 2.2 would have resulted in a smaller *P*-value.

(E) If the alternative hypothesis had been two-sided, the same *t*-score would have resulted in a *P*-value of 0.0754.

430. **A cellphone salesperson claims that the average teen processes 3,700 texts per month. A teacher believes this claim is high and randomly samples 30 teens. What conclusion is reached if the sample mean is 3,490 with a standard deviation of 1,120?**

(A) There is sufficient evidence to prove the salesperson's claim is true.
(B) There is sufficient evidence to prove the salesperson's claim is false.
(C) The teacher has sufficient evidence to reject the salesperson's claim.
(D) The teacher does not have sufficient evidence to reject the salesperson's claim.
(E) There are not sufficient data to reach any conclusion.

431. **The claim is made that the students at Lake Wobegon High School are smarter than other high school students. To test this, the principal gathers a simple random sample (SRS) of 30 students and finds their mean IQ is 104 with a standard deviation of 7.5. What is the test statistic in performing a hypothesis test of H_0: $\mu = 100$ and H_a: $\mu > 100$ (average IQ is 100)?**

(A) $t = \dfrac{104 - 100}{7.5}$

(B) $t = \dfrac{104 - 100}{\sqrt{\dfrac{7.5}{30}}}$

(C) $t = \dfrac{104 - 100}{\sqrt{\dfrac{7.5}{29}}}$

(D) $t = \dfrac{104 - 100}{\left(\dfrac{7.5}{\sqrt{30}}\right)}$

(E) $t = \dfrac{104 - 100}{\left(\dfrac{7.5}{\sqrt{29}}\right)}$

432. A t-test is being conducted at the 5% significance level with H_0: $\mu = 20$ and H_a: $\mu \neq 20$. If a 95 percent t-interval constructed from the same data set contains the value 20, which of the following can be concluded about the t-test?

(A) The P-value is less than 0.05, and there is sufficient evidence to reject H_0.
(B) The P-value is greater than 0.05, and there is sufficient evidence to reject H_0.
(C) The P-value is less than 0.05, and there is not sufficient evidence to reject H_0.
(D) The P-value is greater than 0.05, and there is not sufficient evidence to reject H_0.
(E) Without knowing the sample standard deviation, none of the above can be concluded.

433. A test of the hypotheses H_0: $\mu = 10$ versus H_a: $\mu \neq 10$ was conducted using a sample of size $n = 6$. The test statistic was $t = 1.641$. What was the P-value of the test?

(A) 0.0760
(B) 0.0809
(C) 0.1008
(D) 0.1519
(E) 0.1617

434. Suppose in a small town with a single Walmart, it is known that customers spend an average of \$75 per visit. A new Target is opening, and its sales manager believes that the new store is taking in a higher mean amount per customer visit than Walmart. To test the Target manager's belief, which hypotheses should be used?

(A) H_0: the mean amount spent at Target is equal to \$75
H_a: the mean amount spent at Target is less than \$75
(B) H_0: the mean amount spent at Target is equal to \$75
H_a: the mean amount spent at Target is greater than \$75
(C) H_0: the mean amount spent at Target is less than \$75
H_a: the mean amount spent at Target is greater than \$75
(D) H_0: the mean amount spent at Walmart is equal to \$75
H_a: the mean amount spent at Walmart is less than \$75
(E) H_0: the mean amount spent at Walmart is equal to \$75
H_a: the mean amount spent at Walmart is greater than \$75

435. A spokesperson for the airline industry states that the mean number of pieces of lost luggage per 1,000 passengers is 3.09. A consumer agency believes the true figure is higher and runs an appropriate hypothesis test that results in a *P*-value of 0.075. What is an appropriate conclusion?

(A) Because 0.075 > 0.05, there is sufficient evidence at the 5% significance level to conclude that the mean number of pieces of lost luggage per 1,000 passengers is greater than 3.09.
(B) Because 0.075 > 0.05, there is sufficient evidence at the 5% significance level to conclude that the mean number of pieces of lost luggage per 1,000 passengers is less than 3.09.
(C) Because 0.075 < 0.10, there is sufficient evidence at the 10% significance level to conclude that the mean number of pieces of lost luggage per 1,000 passengers is less than 3.09.
(D) Because 0.075 < 0.10, there is not sufficient evidence at the 10% significance level to conclude that the mean number of pieces of lost luggage per 1,000 passengers is greater than 3.09.
(E) Because 0.05 < 0.075 < 0.10, there is sufficient evidence at the 10% significance level but not at the 5% significance level to conclude that the mean number of pieces of lost luggage per 1,000 passengers is greater than 3.09.

436. A researcher believes that a new diet should improve weight gain in laboratory mice. The average weight gain for 18 mice on the new diet is 4.2 ounces with a standard deviation of 0.4 ounces. The average weight gain for 16 control mice on the old diet is 3.8 ounces with a standard deviation of 0.3 ounces. Where is the *P*-value?

(A) Below 0.01
(B) Between 0.01 and 0.025
(C) Between 0.025 and 0.05
(D) Between 0.05 and 0.10
(E) Over 0.10

437. A teacher is interested in comparing the mean writing speeds of teenagers using cursive versus those printing. A simple random sample (SRS) of 50 students is chosen. The students are timed copying the same paragraph, once in cursive and once in printing. For each student, a coin flip decides which method he/she must use first. Which of the following is a proper test?

(A) Test of difference in two population means
(B) Test of difference in two population proportions
(C) One-sample test on differences of paired data
(D) Chi-square goodness-of-fit test
(E) Chi-square test for homogeneity

438. An entomologist believes that a certain species of large beetles decreased in size between two time periods. He plans to sample the remains of 45 fossilized beetles randomly from each period and reject any equality claim if the mean size in the second time period sample is at least 1 centimeter less than the mean size from the first period sample. If the standard deviation in size in all time periods for this species of beetle is known to be 1.8 centimeters, what is the probability the anthropologist will commit a Type I error and mistakenly reject a correct null hypothesis of equality?

(A) $P\left(z > \dfrac{1-0}{\sqrt{\dfrac{1.8^2}{45}+\dfrac{1.8^2}{45}}}\right)$

(B) $P\left(t > \dfrac{1-0}{\sqrt{\dfrac{1.8^2}{45}+\dfrac{1.8^2}{45}}}\right)$ with $df = 44$

(C) $P\left(t > \dfrac{1-0}{\sqrt{\dfrac{1.8^2}{45}+\dfrac{1.8^2}{45}}}\right)$ with $df = 88$

(D) $P\left(z > \dfrac{1-0}{\left(\dfrac{1.8}{\sqrt{45}}\right)}\right)$

(E) $P\left(t > \dfrac{1-0}{\left(\dfrac{1.8}{\sqrt{45}}\right)}\right)$ with $df = 44$

439. College students majoring in computer science and interested in pursuing careers in computer security were randomly put into either Python or PowerShell classes. In a standard security certification exam, 230 students who took Python scored an average of 76.5 with a standard deviation of 7.4, while 185 students who took PowerShell averaged 78.8 with a standard deviation of 5.6. Is there sufficient evidence of a difference in the test results? What is the test statistic for H_0: $\mu_1 - \mu_2 = 0$ and H_a: $\mu_1 - \mu_2 \neq 0$?

(A) $t = \dfrac{76.5 - 78.8}{\sqrt{\dfrac{(7.4)^2}{230} + \dfrac{(5.6)^2}{185}}}$

(B) $t = 2\dfrac{76.5 - 78.8}{\sqrt{\dfrac{(7.4)^2}{230} + \dfrac{(5.6)^2}{185}}}$

(C) $t = \dfrac{76.5 - 78.8}{\left(\dfrac{7.4}{\sqrt{230}} + \dfrac{5.6}{\sqrt{185}}\right)}$

(D) $t = 2\dfrac{76.5 - 78.8}{\left(\dfrac{7.4}{\sqrt{230}} + \dfrac{5.6}{\sqrt{185}}\right)}$

(E) $t = \dfrac{76.5 - 78.8}{\left(\dfrac{6.5}{\sqrt{207.5}}\right)}$

440. To test whether husbands or wives have greater manual agility, a simple random sample (SRS) of 75 married couples is chosen. All 150 people are given the Illinois Agility Run Test, a commonly used fitness test of agility, and their scores are recorded. What is the proper conclusion at a 5% significance level if a two-sample hypothesis test, H_0: $\mu_1 - \mu_2 = 0$, H_a: $\mu_1 - \mu_2 \neq 0$, results in a *P*-value of 0.038?

(A) The observed difference between husbands and wives is significant.
(B) The observed difference is not significant.
(C) A conclusion is not possible without knowing the mean scores of the husbands and of the wives.
(D) A conclusion is not possible without knowing both the mean and standard deviation of the scores of the husbands and of the wives.
(E) A two-sample hypothesis test should not be used in this example.

441. A random sample of college students tries out two different strategies of card counting at blackjack. The data give the following table.

Strategy	Sample Size	Mean Winnings	Standard Deviation
A	20	131	12
B	15	140	15

Assuming all conditions of inference are met, and performing a two-sample *t*-test with H_0: $\mu_1 - \mu_2 = 0$, is there evidence of a significant difference in outcomes at the 5% significance level?

(A) $P < 0.05$, so reject H_0
(B) $P < 0.05$, so fail to reject H_0
(C) $P > 0.05$, so reject H_0
(D) $P > 0.05$, so fail to reject H_0
(E) The 5% significance level is inappropriate for gambling experiments.

442. To compare online prices between Amazon and Walmart, a consumer-oriented research organization picks 50 basic items and checks the prices of these items online at Amazon and Walmart. Which test should be used to determine if the prices are different at the two sites?

(A) χ^2-test for goodness-of-fit
(B) χ^2-test for independence
(C) Two-sample *z*-test
(D) Two-sample *t*-test
(E) Matched pairs *t*-test

443. The prescription drugs Coumadin and Plavix are both blood thinners known to increase clotting time. In one double-blind study, Coumadin outperformed Plavix. The 95% confidence interval estimate of the difference in mean clotting time (prothrombin time) increase was (2.1, 3.7) seconds. Which of the following is a reasonable conclusion?

(A) Plavix raises clotting time an average of 2.1 seconds, while Coumadin raises clotting time an average of 3.7 seconds.
(B) There is a 0.95 probability that Coumadin will outperform Plavix in raising clotting time for any given individual.
(C) There is a 0.95 probability that Coumadin will outperform Plavix by at least 2.1 seconds in raising the clotting time for any given individual.
(D) We should be 95% confident that Coumadin will outperform Plavix as the drug that raises a clotting time.
(E) None of the above.

444. Material thickness plays a vital part in the quality of soccer balls. Researchers plan to test two layers versus five layers of lining between the cover and the bladder of soccer balls. Two identical-looking soccer balls, one with two internal layers and the other with five, are each kicked and given scores ranging from 1–10 by 12 professional soccer players. A score of 10 is given for a top-quality feel to the ball. A score of 1 is given for a poor-quality feel to the ball. For each player, a coin flip decides which ball is to be kicked first. What type of study and what type of inference test are appropriate here?

(A) Observational study and *t*-test of difference of means
(B) Matched pairs design and one-sample *t*-test for a mean difference
(C) Matched pairs design and two-sample *t*-test for difference of population means
(D) Completely randomized design and one-sample *t*-test for a mean difference
(E) Completely randomized design and two-sample *t*-test for difference of population means

445. An experiment was performed to determine whether a waitress would receive larger tips if she gave her name when greeting customers. A sample of waitresses was randomly assigned to two groups. One group was instructed to give their names when greeting customers, while the other group was instructed not to give their names. After one week, the mean tip received for each group was calculated. The *P*-value of this one-sided test was 0.065. Consider the three factors: original sample size, two sample standard deviations, and magnitude of the difference of the two sample means. Which of the following would have resulted in a smaller *P*-value?

(A) The magnitude of the difference of the two sample means and the two sample standard deviations remains the same, but the original sample size is smaller.
(B) The original sample size remains the same, but the magnitude of the difference of the two sample means is larger and the two sample standard deviations are smaller.
(C) The original sample size and the magnitude of the difference of the two sample means remain the same, but the two sample standard deviations are larger.
(D) The original sample size and the two sample standard deviations remain the same, but the magnitude of the difference of the two sample means is smaller.
(E) The original sample size remains the same, but the magnitude of the difference of the two sample means is smaller and the two sample standard deviations are larger.

446. In which of the following is a matched pairs *t*-test not appropriate?

(A) Heights of twins for 100 randomly selected pairs of twins
(B) Heights of 100 randomly selected children when they are age 10 and again when they are age 11
(C) Heights of both spouses of 100 randomly selected married couples
(D) Heights of both people in pairs formed by 100 randomly selected people
(E) Heights of 100 randomly selected children from England and from France in pairs by matching family income.

447. A doctors' association gathers data on weights from a random sample of elementary school children who are allowed to watch TV all week and from an independent random sample of elementary school children who are allowed to watch TV only on weekends. The data are summarized in the table below.

	TV All Week	**Limited TV**
Number in Sample	175	160
Mean Weight	75.4	71.3
Standard Deviation	9.8	8.2

What is the *t*-statistic for an appropriate test of whether elementary school children who watch TV all week have a higher mean weight than elementary school children who watch TV only on weekends: H_0: $\mu_1 - \mu_2 = 0$ and H_a: $\mu_1 - \mu_2 > 0$?

(A) $t = \dfrac{75.4 - 71.3}{\left(\dfrac{9.8}{\sqrt{175}} + \dfrac{8.2}{\sqrt{160}}\right)}$

(B) $t = \dfrac{9.8 - 8.2}{\sqrt{\dfrac{75.4^2}{175} + \dfrac{71.3^2}{160}}}$

(C) $t = \dfrac{9.8 - 8.2}{\sqrt{\dfrac{9.8^2}{175} + \dfrac{8.2^2}{160}}}$

(D) $t = \dfrac{75.4 - 71.3}{\sqrt{\dfrac{9.8^2}{175} + \dfrac{8.2^2}{160}}}$

(E) $t = \dfrac{75.4 - 71.3}{\sqrt{\dfrac{9.8^2 + 8.2^2}{175 + 160}}}$

448. A sports equipment manufacturer is considering producing one of two new golf tees. The manufacturer chooses 60 sports stores. The manufacturer randomly assigns 30 stores to sell a prototype of one of the tees and the other 30 stores to sell a prototype of the other tee. After a period of time, the mean number of sales in each group of 30 stores is calculated. A statistician concludes that the difference in means is statistically significant. Which of the following is the appropriate conclusion?

(A) It is reasonable to conclude that the difference in sales is caused by the difference in tees because the tees were randomly assigned to stores.
(B) It is reasonable to conclude that the difference in sales is caused by the difference in tees because the sample size was ≥30.
(C) It is not reasonable to conclude that the difference in sales is caused by the difference in tees because the 60 stores were not randomly selected.
(D) It is not reasonable to conclude that the difference in sales is caused by the difference in tees because there was no control group for comparison.
(E) It is not reasonable to conclude that the difference in sales is caused by the difference in tees because this was an observational study, not an experiment.

449. A car simulator was used to compare the effect on reaction time between DWI (driving while intoxicated) and DWT (driving while texting). A total of 20 volunteers were instructed to drive at 60 mph and then hit the brakes in response to the sudden image of a child darting into the road. One day, each driver was tested for stopping distance while driving while texting. Another day, each driver was tested after consuming a quantity of alcohol. For each driver, which test was done on the first day was decided by a coin toss. Which of the following is an appropriate test and hypotheses to determine if there is a difference in stopping distances between DWI and DWT?

(A) A two-sample t-test with H_0: $\mu_{\text{DWI}} = \mu_{\text{DWT}}$ and H_a: $\mu_{\text{DWI}} \neq \mu_{\text{DWT}}$
(B) A two-sample t-test with H_0: $\mu_{\text{DWI}} \neq \mu_{\text{DWT}}$ and H_a: $\mu_{\text{DWI}} = \mu_{\text{DWT}}$
(C) A two-sample t-test with H_0: $\mu_{\text{DWI}} = \mu_{\text{DWT}}$ and H_a: $\mu_{\text{DWI}} > \mu_{\text{DWT}}$
(D) A matched pairs t-test with H_0: $\mu_{\text{Diff}} = 0$ and H_a: $\mu_{\text{Diff}} \neq 0$
(E) A matched pairs t-test with H_0: $\mu_{\text{Diff}} = 0$ and H_a: $\mu_{\text{Diff}} > 0$

450. A sports statistician is interested in whether the mean number of home runs hit in American League stadiums (with designated hitters batting instead of pitchers) is greater than the mean number hit in National League stadiums (where pitchers do bat). A random sample of games is analyzed as to the number of home runs. Computer output for a random sample of American and National League games is as follows.

```
     Two-Sample T for American vs National League Games
                  N      Mean       StDev        SE Mean
American         100    1.216       0.493        0.0493
National         100    1.093       0.486        0.0486
95% CI for mu American – National: (0.0086, 0.2374)
T-test mu American = National (vs >): T = 1.777 P = 0.0386
```

Do the confidence interval and *t*-test lead to the same conclusion, and was this expected?

(A) With $0.0386 < 0.05$ and the entire confidence interval > 0, both lead to the conclusion that there is sufficient evidence that the mean number of home runs hit in American League games is greater than the mean number hit in National League games. It should be expected that both lead to the same conclusion.

(B) With $0.0386 < 0.05$ and the entire confidence interval > 0, both lead to the conclusion that there is sufficient evidence that the mean number of home runs hit in American League games is greater than the mean number hit in National League games. It is possible that this confidence interval and *t*-test contradict each other.

(C) With $0.0086 < 0.0386 < 0.2374$, both lead to the conclusion that there is sufficient evidence that the mean number of home runs hit in American League games is greater than the mean number hit in National League games. It should be expected that both lead to the same conclusion.

(D) With $0.0386 < 0.05$ and $0.05 < 0.2374$, they lead to opposite conclusions, with the *t*-test showing a significant difference and the confidence interval not. This is not unexpected because these are two different approaches.

(E) With $0.0386 < 0.05$ and $0.05 < 0.2374$, they lead to opposite conclusions with the *t*-test showing a significant difference and the confidence interval not. This is unexpected as only one conclusion at the given significance level is possible.

451. In an often-quoted study reported in the *British Medical Journal*, researchers examined the number of people admitted to the ER for auto accidents on Friday the 6th of several months versus on Friday the 13th of several months. Is there evidence that more people are admitted, on average, on Friday the 13th? Some data are summarized below.

Number of ER Admissions from Auto Accidents

	Oct 1989	Jul 1990	Sep 1991	Dec 1991	Mar 1992	Nov 1992
Friday 6th	9	6	11	11	3	5
Friday 13th	13	12	14	10	4	12

Let μ_1 = the mean number of ER admissions from auto accidents on all Fridays that fall on the 6th

Let μ_2 = the mean number of ER admissions from auto accidents on all Fridays that fall on the 13th

Let μ_D = the mean of the differences (the 6th minus the 13th) in the number of ER admissions from auto accidents on all Fridays that fall on the 6th and 13th.

What is the appropriate test and alternative hypothesis?

(A) Two-sample t-test with H_a: $\mu_1 \neq \mu_2$
(B) Two-sample t-test with H_a: $\mu_1 < \mu_2$
(C) Two-sample t-test with H_a: $\mu_1 > \mu_2$
(D) Match paired t-test with H_a: $\mu_D < 0$
(E) Match paired t-test with H_a: $\mu_D > 0$

452. A study compared the math scores on a standardized test for home-schooled and public-schooled high school students. The scores of the eight students in each group are shown in the following table.

Home-Schooled	78	65	70	83	96	66	70	81
Public-Schooled	75	74	78	89	91	70	75	89

Assume the groups were simple random samples and all conditions for inference were met. What is the appropriate test to determine if there is a significant difference in the average test scores of the two groups?

(A) Matched pair t-test for means
(B) Two-sample t-test for means
(C) Chi-square test of independence
(D) Chi-square goodness-of-fit test
(E) Linear regression t-test

TYPE I AND II ERRORS, POWER

453. An industry spokesperson claims that the mean income of entry-level employees is $29,500 with a standard deviation of $2,500. A reporter plans to test this claim through interviews with a random sample of 40 entry-level employees. If the reporter finds a sample mean more than $500 less than the claimed $29,500, she will dispute the spokesperson's claim. What is the probability that the reporter will commit a Type I error?

(A) $P\left(t < \dfrac{29{,}000 - 29{,}500}{\left(\dfrac{2{,}500}{\sqrt{40}}\right)}\right)$ with $df = 40 - 1$

(B) $P\left(z < \dfrac{29{,}000 - 29{,}500}{\left(\dfrac{2{,}500}{\sqrt{40}}\right)}\right)$

(C) $P\left(t < \dfrac{29{,}000 - 29{,}500}{2{,}500}\right)$ with $df = 40 - 1$

(D) $P\left(z < \dfrac{29{,}000 - 29{,}500}{2{,}500}\right)$

(E) This cannot be calculated without knowing the sample standard deviation.

454. When leaving for school on an overcast morning, you make a judgment on the null hypothesis: The weather will remain dry. What would the results be of Type I and Type II errors?

(A) Type I error: carry an umbrella and it rains
Type II error: carry no umbrella, but weather remains dry
(B) Type I error: get drenched
Type II error: carry no umbrella, but weather remains dry
(C) Type I error: get drenched
Type II error: carry an umbrella, and it rains.
(D) Type I error: get drenched
Type II error: needlessly carry around an umbrella
(E) Type I error: needlessly carry around an umbrella
Type II error: get drenched

455. Consider a hypothesis test with H_0: $\mu = 127$ and H_a: $\mu < 127$. Which of the following choices of significance level and sample size results in the greatest power of the test when $\mu = 125$?

(A) $\alpha = 0.05$, $n = 30$
(B) $\alpha = 0.01$, $n = 30$
(C) $\alpha = 0.05$, $n = 45$
(D) $\alpha = 0.01$, $n = 45$
(E) There is no way of answering without knowing the strength of the given power.

456. An assembly line machine is supposed to turn out fidget spinners with spin times of 4 minutes. Each day, a simple random sample (SRS) of five fidget spinners are pulled and their spin times measured. If their mean is under 3.5 minutes, the machinery is stopped and an engineer is called to make adjustments before production is resumed. The quality control procedure may be viewed as a hypothesis test with H_0: $\mu = 4.0$ and H_a: $\mu < 4.0$. What would a Type II error result in?

(A) A warranted halt in production to adjust the machinery
(B) An unnecessary stoppage of the production process
(C) Continued production of fidget spinners with low spin times
(D) Continued production of fidget spinners with the proper spin times
(E) Continued production of fidget spinners that may or may not have the proper spin times.

457. Given an experiment with H_0: $\mu = 31$, H_a: $\mu > 31$, and a possible correct value of 32, which of the following will increase with an increase in the sample size n?

(A) The probability of a Type I error
(B) The probability of a Type II error
(C) The power of the test
(D) The significance level α
(E) 1 – power

458. A quality inspector plans to test a sample of 50 bottles of water with a claimed pH level of 6.0. If she finds the mean pH to be less than 5.5, she will have the bottling machinery stopped and the water sources inspected. From previous analysis, it is known that the standard deviation in pH level among bottles is 0.75. What is the probability that the inspector will commit a Type I error and mistakenly stop a machine when the pH level of the water really is the claimed 6.0?

(A) $P\left(z < \frac{5.5 - 6.0}{0.75}\right)$

(B) $P\left(z < \frac{5.5 - 6.0}{\left(\frac{0.75}{\sqrt{50}}\right)}\right)$

(C) $P\left(t < \frac{5.5 - 6.0}{0.75}\right)$ with $df = 50 - 1$

(D) $P\left(t < \frac{5.5 - 6.0}{\left(\frac{0.75}{\sqrt{50}}\right)}\right)$ with $df = 50 - 1$

(E) $P\left(t < \frac{5.5 - 6.0}{\left(\frac{0.75}{\sqrt{50}}\right)}\right)$ with $df = 50$

459. Which of the following is *incorrect*?

(A) The power of a test concerns its ability to detect a true alternative hypothesis.
(B) The significance level of a test is the probability of rejecting a true null hypothesis.
(C) The probability of a Type I error plus the probability of a Type II error always equals 1.
(D) Power equals 1 minus the probability of failing to reject a false null hypothesis.
(E) Anything that makes a null hypothesis harder to reject increases the probability of committing a Type II error.

460. Suppose H_0: $p = 0.4$, H_a: $p < 0.4$, and against the alternative $p = 0.3$, the power is 0.8. Which of the following is a valid conclusion?

(A) The probability of committing a Type I error is 0.1.
(B) If $p = 0.3$ is true, the probability of failing to reject H_0 is 0.2.
(C) The probability of committing a Type II error is 0.7.
(D) All of the above are valid conclusions.
(E) None of the above are valid conclusions.

461. Which of the following is a true statement?

(A) A Type I error is a conditional probability.
(B) A Type II error results if one incorrectly assumes the data are normally distributed.
(C) Types I and II errors are caused by mistakes, however small, by the person conducting the test.
(D) The probability of a Type II error does not depend on the probability of a Type I error.
(E) In conducting a hypothesis test, it is possible to make both a Type I and Type II error simultaneously.

462. Which of the following statements is *incorrect*?

(A) The significance level of a test is the probability of a Type II error.
(B) Given a particular alternative, the power of a test against that alternative is 1 minus the probability of the Type II error associated with that alternative.
(C) If the significance level remains fixed, increasing the sample size reduces the probability of a Type II error.
(D) If the significance level remains fixed, increasing the sample size raises the power.
(E) With the sample size held fixed, increasing the significance level decreases the probability of a Type II error.

463. A manufacturer of surgical gowns (to protect surgeons from contamination) periodically checks a sample of its product and performs a major shutdown and inspection if any leaks are detected. Similarly, a manufacturer of blankets periodically checks the sizes of its blankets coming off an assembly line and halts production if measurements are sufficiently off target. In both situations, we have the null hypothesis that the production equipment is performing satisfactorily. For each situation, which is the more serious concern, a Type I or Type II error?

(A) Surgical gown producer: Type I error
Blanket manufacturer: Type I error
(B) Surgical gown producer: Type I error
Blanket manufacturer: Type II error
(C) Surgical gown producer: Type II error
Blanket manufacturer: Type I error
(D) Surgical gown producer: Type II error
Blanket manufacturer: Type II error
(E) This is impossible to answer without making an expected value judgment between human life and accurate blanket sizes.

464. If all other variables remain constant, which of the following will *not* increase the power of a hypothesis test?

(A) Increasing the sample size
(B) Increasing the significance level
(C) Increasing the probability of a Type II error
(D) Decreased variability in the data
(E) An increased difference between the true and the hypothesized parameters

465. A company that produces bungee cords continually monitors the strength of the cords. If the mean strength from a sample drops below a specified level, the production process is halted and the machinery inspected. Which of the following would result from a Type I error?

(A) Halting the production process when sufficient customer complaints are received
(B) Halting the production process when the bungee cord strength is below specifications
(C) Halting the production process when the bungee cord strength is within specifications
(D) Allowing the production process to continue when the bungee cord strength is below specifications
(E) Allowing the production process to continue when the bungee cord strength is within specifications

466. Given that the power of a significance test against a particular alternative is 97 percent, which of the following is true?

(A) The probability of mistakenly rejecting a true null hypothesis is less than 3 percent.
(B) The probability of mistakenly rejecting a true null hypothesis is 3 percent.
(C) The probability of mistakenly rejecting a true null hypothesis is greater than 3 percent.
(D) The probability of mistakenly failing to reject the false null hypothesis is 3 percent.
(E) The probability of mistakenly failing to reject the false null hypothesis is different from 3 percent.

467. A company will market a new hybrid luxury smart phone only if it can sell the phone for more than $900 (otherwise there is not enough profit to make the venture worthwhile). The company does a random survey of 80 potential customers and runs a hypothesis test with H_0: $\mu = 900$ and H_a: $\mu > 900$. What would be the consequences of Type I and Type II errors?

(A) Type I error: produce a nonprofitable smartphone
Type II error: fail to produce a profitable smartphone
(B) Type I error: fail to produce a profitable smartphone
Type II error: produce a nonprofitable smartphone
(C) Type I error: fail to produce a nonprofitable smartphone
Type II error: produce a profitable smartphone
(D) Type I error: fail to produce a profitable smartphone
Type II error: produce a profitable smartphone
(E) Type I error: produce a nonprofitable smartphone
Type II error: fail to produce a nonprofitable smartphone

468. A Human Resource manager suspects that employees are abusing the company's sick day policy and plans to investigate this issue. The hypotheses under consideration are the following:

H_0: employees are not abusing the company's sick day policy
H_a: employees are abusing the policy by taking sick days when they are not actually sick

What is the power of this test?

(A) The probability of mistakenly accusing employees of abusing the sick day policy when in fact they aren't abusing the policy
(B) The probability of mistakenly not accusing employees who are abusing the sick day policy
(C) The probability of correctly not accusing employees who are not abusing the sick day policy
(D) The probability of correctly accusing employees who are abusing the sick day policy
(E) The probability that the test will come to a correct decision on whether or not employees are abusing the sick day policy

469. A hypothesis test is to be performed at a significance level of $\alpha = 0.10$. What is the effect on the probability of committing a Type I error if the sample size is increased?

(A) The probability of committing a Type I error decreases.
(B) The probability of committing a Type I error increases.
(C) The probability of committing a Type I error remains the same.
(D) This cannot be determined without knowing the relevant standard deviation.
(E) This cannot be determined without knowing if a Type II error is committed.

470. A TV manufacturer claims that 75 percent of purchasers understand the setup instructions. The sales department of a TV superstore disagrees and decides to do a large-sample *z*-test for a population proportion at the 5% significance level. The null hypothesis that the proportion of TV purchasers who understand the instructions is 75 percent is tested against the alternative hypothesis that the proportion of TV purchasers who understand the instructions is less than 75 percent. For which of the following will the power of the test be highest?

(A) Sample size n = 50 and 55 percent of purchasers understand the instructions
(B) Sample size n = 100 and 55 percent of purchasers understand the instructions
(C) Sample size n = 50 and 65 percent of purchasers understand the instructions
(D) Sample size n = 100 and 65 percent of purchasers understand the instructions
(E) Sample size n = 100 and 95 percent of purchasers understand the instructions

471. Dogs have far more sensitive noses than humans, estimated to be up to 100,000 times better than humans. Researchers believe that dogs can sniff out lung cancer from breath samples of sufferers. Using the dogs, the researchers plan to test thousands of samples and prescribe chemotherapy only if the dogs indicate the presence of cancer. Consider the following hypotheses:

H_0: The subject does not have cancer.
H_a: The subject does have cancer.

What would a Type II error be in context?

(A) A subject with cancer is correctly diagnosed with cancer.
(B) A subject with cancer is mistakenly thought to be cancer free.
(C) A cancer-free subject is correctly diagnosed to be cancer free.
(D) A cancer-free subject is mistakenly diagnosed with cancer.
(E) Correct and mistaken diagnoses of cancer are confounded.

472. What is the probability of a Type II error when a hypothesis test is being conducted at the 10% significance level (α = 0.10)?

(A) 0.05
(B) 0.10
(C) 0.90
(D) 0.95
(E) There is insufficient information to answer this question.

P-VALUES FROM SIMULATIONS

473. A marriage counselor plans a study on whether a husband or a wife is happier on a greater proportion of days when the wife has a job outside the home and the husband is a stay-at-home dad. In a random sample of married couples where the wife works and the husband is at home, the difference in proportions of happy days between husband and wife is 0.1106. The counselor, who took a required statistics class in college, is concerned that the proportions she subtracted to get 0.1106 are not from independent samples. So she runs 200 simulations of couples with the null hypothesis that there is no difference in proportions of happy days. The dotplot of resulting proportion differences is below.

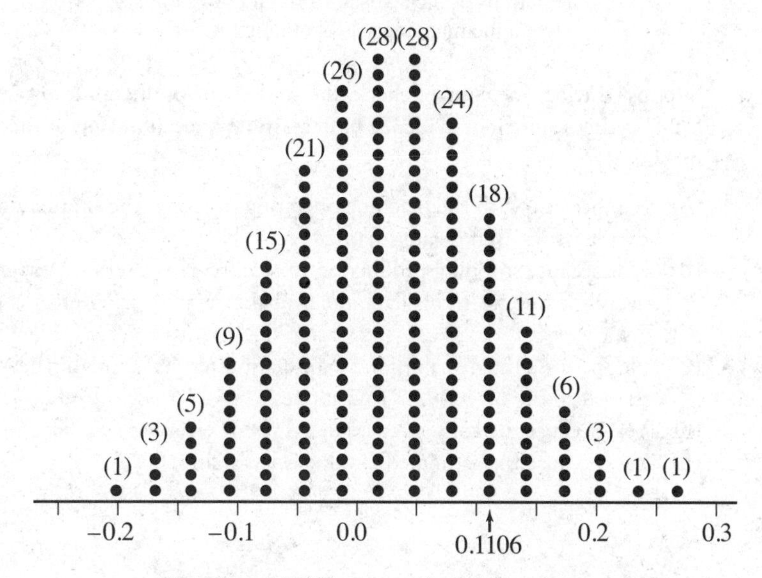

Is there sufficient evidence of a significant difference in the proportion of happy days experienced by husbands and wives in marriages where the wife works and the husband is a stay-at-home dad?

(A) Yes, because 0.1106 > 0.0
(B) Yes, because 0.1106 > 0.05
(C) Yes, because the distribution of simulated differences is approximately normal so the central limit theorem applies
(D) No, because 0.1106 > 0.05
(E) No, because the simulated *P*-value is large

474. **An industrial control check is as follows. Random samples are periodically gathered. If a particular statistic is significantly greater than what is expected during proper operation of the machinery, a recalibration is necessary. In a simulation of 100 such samples where the machinery is working properly, the resulting statistic is summarized in the following dotplot.**

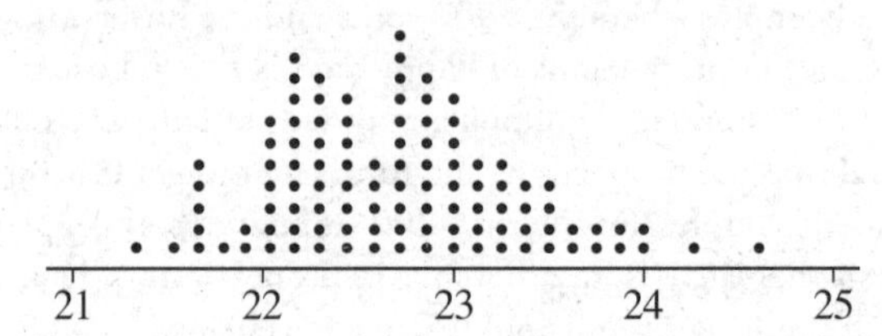

Calculations of the statistic from random samples when the machinery is operating properly

Suppose during one control check, the statistic from the random sample is 24. Is there sufficient evidence to necessitate a recalibration of the machinery?

(A) No, because with the machine operating properly, the simulation gave two statistics even greater than 24
(B) No, because stopping production to recalibrate is a serious, expensive decision, and given this data, the probability of a Type I error is too great
(C) No, because the distribution of simulation results is roughly bell-shaped so the 68-95-99.7 rule applies
(D) Yes, because the consequences of a Type II error are significant
(E) Yes, because the estimated P-value is less than 0.05

CHI-SQUARE TESTS

475. **Data from a simple random sample (SRS) of teachers is cross-classified by gender and support for a new educational initiative, resulting in the following table.**

	Male	Female
For	37	41
Against	42	62
No Opinion	41	29

Is there evidence of a relationship between gender and support for the initiative among teachers?

(A) There is strong evidence of a relationship between gender and support for the initiative among teachers, $P < 0.05$.
(B) There is weak evidence of a relationship between gender and support for the initiative among teachers, $0.05 < P < 0.10$.
(C) There is not sufficient evidence of a relationship between gender and support for the initiative among teachers, $P > 0.10$.
(D) Further information is needed to be able to perform a chi-square test of independence.
(E) The test is inconclusive.

476. A survey of 150 high school seniors is conducted to see if there is a relationship between whether or not a student is taking AP Statistics and whether students have positive or negative outlooks on their future. The data are summarized in the following table.

	Positive Outlook	Negative Outlook
Taking AP Stats	46	24
Not Taking AP Stats	37	43

What is the appropriate test statistic for a hypothesis test with H_0: Student outlook on the future is independent of taking AP Statistics?

(A) $\frac{(46-38.7)^2}{38.7}+\frac{(24-31.3)^2}{31.3}+\frac{(37-44.3)^2}{44.3}+\frac{(43-35.7)^2}{35.7}$

(B) $\frac{(46-38.7)^2}{46}+\frac{(24-31.3)^2}{24}+\frac{(37-44.3)^2}{37}+\frac{(43-35.7)^2}{43}$

(C) $\frac{(46-35)^2}{35}+\frac{(24-35)^2}{35}+\frac{(37-40)^2}{40}+\frac{(43-40)^2}{40}$

(D) $\frac{(46-41.5)^2}{41.5}+\frac{(24-33.5)^2}{33.5}+\frac{(37-41.5)^2}{41.5}+\frac{(43-33.5)^2}{33.5}$

(E) $\frac{(46-37.5)^2}{46}+\frac{(24-37.5)^2}{24}+\frac{(37-37.5)^2}{37}+\frac{(43-37.5)^2}{43}$

477. Which of the following is *not* true with regard to contingency tables for chi-square tests for independence?

(A) Categorical rather than quantitative variables are being considered.
(B) Observed frequencies should be whole numbers.
(C) Expected frequencies should be whole numbers.
(D) Expected frequencies in each cell should be at least 5, and to achieve this, one sometimes combines categories for one or the other or for both of the variables.
(E) The expected frequency for any cell can be found by multiplying the row total by the column total and then dividing by the sample size.

478. The table below shows the number of employees arriving late for work at an office and is broken down by the day of the week. An HR officer would like to know if such late arrivals are related to the day of the week.

Monday	Tuesday	Wednesday	Thursday	Friday
12	5	9	4	15

What is the value of chi-square for the appropriate test?

(A) $\frac{(12-9)^2}{12} + \frac{(5-9)^2}{5} + \frac{(9-9)^2}{9} + \frac{(4-9)^2}{4} + \frac{(15-9)^2}{15}$

(B) $\frac{(12-9)^2}{9} + \frac{(5-9)^2}{9} + \frac{(9-9)^2}{9} + \frac{(4-9)^2}{9} + \frac{(15-9)^2}{9}$

(C) $\frac{(12-9)^2}{45} + \frac{(5-9)^2}{45} + \frac{(9-9)^2}{45} + \frac{(4-9)^2}{45} + \frac{(15-9)^2}{45}$

(D) $\frac{(12-5)^2}{12} + \frac{(5-5)^2}{5} + \frac{(9-5)^2}{9} + \frac{(4-5)^2}{4} + \frac{(15-5)^2}{15}$

(E) $\frac{(12-5)^2}{5} + \frac{(5-5)^2}{5} + \frac{(9-5)^2}{5} + \frac{(4-5)^2}{5} + \frac{(15-5)^2}{5}$

479. A geneticist claims that three species of fruit flies should appear in the ratio 1:6:9. Suppose that a random sample of 320 flies contained 25, 100, and 195 flies of each species, respectively. Does a chi-square test show sufficient evidence to reject the geneticist's claim at the 5% significance level?

(A) The test proves the geneticist's claim.
(B) The test proves the geneticist's claim is false.
(C) The test does not give sufficient evidence to reject the geneticist's claim at the 5% significance level.
(D) The test gives sufficient evidence to reject the geneticist's claim at the 5% significance level.
(E) The test is inconclusive.

480. Three English professors are interviewed regarding a random sampling of 20 of their term paper grades. The following table gives the resulting counts.

	Professor *A*	Professor *B*	Professor *C*
Grades A, B	5	7	9
Grade C	13	10	7
Grades D, F	2	3	4

A statistics student runs a chi-square test of homogeneity. What is the most proper conclusion?

(A) There is no evidence that these professors give different distributions of grades.
(B) There is evidence at the 10% level, but not at the 5% level, that the professors give different grade distributions.
(C) There is evidence at the 5% level, but not at the 1% level, that the professors give different grade distributions.
(D) There is evidence at the 1% level that the professors give different grade distributions.
(E) A chi-square test of homogeneity is not appropriate.

481. Researchers wanted to determine if preference between baseball and chess is independent of age. 210 interviews yielded the following numbers.

Age	**10–19**	**20–49**	**50–89**	***Total***
Prefer Baseball	55	40	15	110
Prefer Chess	10	15	35	60
Total	65	55	50	170

What is the *P*-value for an appropriate test?

(A) $P\left(\chi^2 > \frac{(55-42.1)^2}{55} + \frac{(40-35.6)^2}{40} + \frac{(15-32.4)^2}{15} + \frac{(10-22.9)^2}{10} + \frac{(15-19.4)^2}{15} + \frac{(35-17.6)^2}{35}\right)$, $df = 2$

(B) $P\left(\chi^2 > \frac{(55-42.1)^2}{42.1} + \frac{(40-35.6)^2}{35.6} + \frac{(15-32.4)^2}{32.4} + \frac{(10-22.9)^2}{22.9} + \frac{(15-19.4)^2}{19.4} + \frac{(35-17.6)^2}{17.6}\right)$, $df = 2$

(C) $P\left(\chi^2 > \frac{(55-42.1)^2}{55} + \frac{(40-35.6)^2}{40} + \frac{(15-32.4)^2}{15} + \frac{(10-22.9)^2}{10} + \frac{(15-19.4)^2}{15} + \frac{(35-17.6)^2}{35}\right)$, $df = 3$

(D) $P\left(\chi^2 > \frac{(55-42.1)^2}{42.1} + \frac{(40-35.6)^2}{35.6} + \frac{(15-32.4)^2}{32.4} + \frac{(10-22.9)^2}{22.9} + \frac{(15-19.4)^2}{19.4} + \frac{(35-17.6)^2}{17.6}\right)$, $df = 3$

(E) $P\left(\chi^2 > \frac{(55-42.1)^2}{28.33} + \frac{(40-35.6)^2}{28.33} + \frac{(15-32.4)^2}{28.33} + \frac{(10-22.9)^2}{28.33} + \frac{(15-19.4)^2}{28.33} + \frac{(35-17.6)^2}{28.33}\right)$, $df = 6$

482. Which of the following is the proper use of a chi-square test of independence?

(A) To test whether the distribution of counts on a categorical variable matches a claimed distribution
(B) To test whether the distribution of counts on a numerical variable matches a claimed distribution
(C) To test whether the distribution of two different groups on the same categorical variable matches
(D) To test whether two categorical variables on one set of subjects are related
(E) To test whether two numerical variables on one set of subjects are related

483. Last year, college students learned about major news events from the following sources:

MSNBC: 21% CNN: 26% FOX: 12% *The Daily Show*: 41%.

In a random sample of 80 college students this year, 23 watched MSNBC, 14 watched CNN, 10 watched FOX, and 33 watched *The Daily Show* to learn about major news events. If a goodness-of-fit test is performed, what will be the *P*-value?

(A) $P\left(\chi^2 > \frac{(23-16.8)^2}{23} + \frac{(14-20.8)^2}{14} + \frac{(10-9.6)^2}{10} + \frac{(33-32.8)^2}{32.8}\right)$
with $df = 3$

(B) $P\left(\chi^2 > \frac{(23-16.8)^2}{23} + \frac{(14-20.8)^2}{14} + \frac{(10-9.6)^2}{10} + \frac{(33-32.8)^2}{32.8}\right)$
with $df = 4$

(C) $P\left(\chi^2 > \frac{(23-16.8)^2}{16.8} + \frac{(14-20.8)^2}{20.8} + \frac{(10-9.6)^2}{9.6} + \frac{(33-32.8)^2}{32.8}\right)$
with $df = 3$

(D) $P\left(\chi^2 > \frac{(23-16.8)^2}{16.8} + \frac{(14-20.8)^2}{20.8} + \frac{(10-9.6)^2}{9.6} + \frac{(33-32.8)^2}{32.8}\right)$
with $df = 4$

(E) $P\left(\chi^2 > \frac{(23-16.8)^2}{0.25} + \frac{(14-20.8)^2}{0.25} + \frac{(10-9.6)^2}{0.25} + \frac{(33-32.8)^2}{0.25}\right)$
with $df = 79$

484. Researchers wanted to determine if people know Mitt Romney's real first name. Data from random samples of 170 Democrats and 170 Republicans are summarized in the following table.

	Mittens	Willard	Something Else
Democrats	23	41	106
Republicans	55	52	63

Assuming that people from each party have the same pattern of responses when asked about Romney's real first name, what is the expected number of Republicans who answer Willard (which is actually the correct answer) from this group of 340 people?

(A) $\frac{93}{170}$

(B) $\frac{52}{93}$

(C) $\frac{52}{170}$

(D) $\frac{(170)(93)}{211}$

(E) $\frac{(170)(93)}{340}$

485. A study of hospital admissions in a small town reported the following numbers for three different days broken down by phase of the moon.

	New Moon	Full Moon	Half Moon
Accidents	35	77	53

Is there sufficient evidence to say that the number of admissions on the three days is not the same?

(A) There is sufficient evidence at the 0.001 significance level that the number of admissions on each day is not the same.
(B) There is sufficient evidence at the 0.01 level, but not at the 0.001 level, that the number of admissions on each day is not the same.
(C) There is sufficient evidence at the 0.05 level, but not at the 0.01 level, that the number of admissions on each day is not the same.
(D) There is sufficient evidence at the 0.10 level, but not at the 0.05 level, that the number of admissions on each day is not the same.
(E) There is not sufficient evidence to say that the number of admissions on each day is not the same.

486. A random sample of 100 men was cross-classified by heart disease and baldness. The table below summarizes the tallies.

	No Baldness	Some Baldness	Extreme Baldness
Heart Disease	120	85	38
Healthy	256	51	5

Which of the procedures below is most appropriate to investigate whether there is an association between heart disease and baldness?

(A) Two-sample *t*-test of the difference between population means
(B) Two-sample *z*-test of the difference between population proportions
(C) Matched pair *t*-test
(D) Chi-square goodness-of-fit test
(E) Chi-square test of independence

487. An electronics store sales representative believes men and women have different preferences with regard to TV sizes. To test his belief, he shows models of four different TVs with different sizes to a random sample of 80 men and a random sample of 70 women. He plans to use a chi-square test of homogeneity. Assuming that conditions for inference are met, which of the following is a true statement?

(A) The test is not valid because the sample sizes are different.
(B) The test would be more appropriate if 75 married couples had been used.
(C) Given the number of choices, four, the sample sizes are too small for a test of homogeneity of proportions.
(D) The null hypothesis is that the proportion of each gender who prefer each TV is $\frac{1}{4}$.
(E) The more that men and women differ in their TV preferences, the larger the chi-square statistic will be.

488. A highway engineer claims that the four highways leading into a city are used in the ratio 3:2:3:4 during the morning rush hour. A study involving a simple random sample (SRS) of 3,600 cars counted 920, 570, 700, and 1,410 cars using the four highways, respectively. What is the *P*-value for the appropriate test?

(A) $P\left(\frac{(920-900)^2}{920}+\frac{(570-600)^2}{570}+\frac{(700-900)^2}{700}+\frac{(1{,}410-1{,}200)^2}{1{,}410}\right)$ with $df = 3$

(B) $P\left(\frac{(920-900)^2}{920}+\frac{(570-600)^2}{570}+\frac{(700-900)^2}{700}+\frac{(1{,}410-1{,}200)^2}{1{,}410}\right)$ with $df = 4$

(C) $P\left(\frac{(920-900)^2}{900}+\frac{(570-600)^2}{600}+\frac{(700-900)^2}{900}+\frac{(1{,}410-1{,}200)^2}{1{,}200}\right)$ with $df = 3$

(D) $P\left(\frac{(920-900)^2}{900}+\frac{(570-600)^2}{600}+\frac{(700-900)^2}{900}+\frac{(1{,}410-1{,}200)^2}{1{,}200}\right)$ with $df = 4$

(E) $P\left(\frac{(920-900)^2}{900}+\frac{(570-600)^2}{900}+\frac{(700-900)^2}{900}+\frac{(1{,}410-1{,}200)^2}{900}\right)$ with $df = 3$

489. In a random sample of 920 college students, all were surveyed as to whether they had parlor tattoos, tattoos from elsewhere, or no tattoos. Then all were tested for hepatitis C. The data are displayed in the table below.

	Parlor Tattoo	**Nonparlor Tattoo**	**No Tattoo**
Hepatitis C	24	11	32
Healthy	37	52	764

In a test of independence, which of the following is used as an expected cell count for healthy students with no tattoo?

(A) 764

(B) $\frac{1}{6}(920)$

(C) $\frac{1}{2}(32 + 764)$

(D) $\frac{1}{3}(37 + 52 + 764)$

(E) $\frac{(37 + 52 + 764)(32 + 764)}{920}$

***490. It is hypothesized that scores on a certain intelligence test are normally distributed with a mean of 100 and a standard deviation of 15. A psychologist runs a goodness-of-fit test on a simple random sample (SRS) of 200 scores, which results in the table below.**

Score	Below 85	85–100	100–115	Above 115
Number of People	21	84	67	28

What is the χ^2 statistic for this test?

(A) $\frac{(21-32)^2}{21}+\frac{(84-68)^2}{84}+\frac{(67-68)^2}{67}+\frac{(28-32)^2}{28}$

(B) $\frac{(21-32)^2}{32}+\frac{(84-68)^2}{68}+\frac{(67-68)^2}{68}+\frac{(28-32)^2}{32}$

(C) $\frac{(21-50)^2}{50}+\frac{(84-50)^2}{50}+\frac{(67-50)^2}{50}+\frac{(28-50)^2}{50}$

(D) $\frac{(21-50)^2}{21}+\frac{(84-50)^2}{84}+\frac{(67-50)^2}{67}+\frac{(28-50)^2}{28}$

(E) $\frac{(21-50)^2}{32}+\frac{(84-50)^2}{68}+\frac{(67-50)^2}{68}+\frac{(28-50)^2}{32}$

491. A small study was made to investigate whether police searches of stopped vehicles are independent of the driver's race. The data counts for one day are summarized in the table below.

	Race			
Search	**White**	**Black**	**Other**	***Total***
Yes	4	8	11	23
No	26	15	4	45
Total	30	23	15	68

A chi-square test of independence gives $\chi^2 = 16.1$ and a *P*-value of 0.00032. Which of the following is a correct statement?

(A) The degrees of freedom are (3 – 1)(4 – 1).
(B) The chi-square test should not have been used because two of the cells are less than 5.
(C) The null hypothesis states that there is an association between whether police search a stopped vehicle and the driver's race.
(D) The very small *P*-value suggests that there is an association between whether police search a stopped vehicle and the driver's race.
(E) The chi-square test shows that police treat different races differently.

492. An ice cream distributor wants to know if adults and children prefer the same flavors of ice cream. A random sample of 100 adults and an independent random sample of 100 children are surveyed as to ice cream preferences. The data are shown in the table below.

	Ice Cream Flavor				
	Vanilla	**Chocolate**	**Strawberry**	**Coffee**	**Cookie Dough**
Adults	25	20	15	35	5
Children	20	25	25	0	30

Which of the following procedures would be most appropriate to investigate whether there is a relationship between adults/children and ice cream flavor preference?

(A) Chi-square test of independence
(B) Chi-square test of homogeneity
(C) Chi-square test of goodness-of-fit
(D) Two-sample t-test
(E) Matched pairs t-test

*493. A restaurant receives two food deliveries per day, and timeliness is critical. The restaurant manager tabulates the number of on-time deliveries for a random sample of 200 days, as shown in the table.

Number of On-Time Deliveries	0	1	2
Observed Number of Days	12	75	113

What is the χ^2 statistic for a goodness-of-fit test that the distribution is binomial with the probability equal to 0.8 that a delivery is on time?

(A) $\frac{(12-8)^2}{8}+\frac{(75-64)^2}{64}+\frac{(113-128)^2}{128}$

(B) $\frac{(12-8)^2}{12}+\frac{(75-64)^2}{75}+\frac{(113-128)^2}{113}$

(C) $\frac{(12-10)^2}{10}+\frac{(75-30)^2}{30}+\frac{(113-160)^2}{160}$

(D) $\frac{(12-10)^2}{12}+\frac{(75-30)^2}{75}+\frac{(113-160)^2}{113}$

(E) $\frac{(12-66)^2}{12}+\frac{(75-67)^2}{75}+\frac{(113-67)^2}{113}$

494. One survey of what middle school and high school students enjoy reading listed the top four choices as fantasy, science fiction, horror, and fairy tales. To test whether middle school and high school students make similar choices, independent random samples are taken from each group of students. Each student is asked what he/she enjoys reading most among the choices: fantasy, science fiction, horror, fairy tales, or other. A chi-square test of homogeneity is performed, and the resulting *P*-value is below 0.05. Which of the following is a proper conclusion?

(A) There is sufficient evidence that for all five choices, the proportion of middle school students who prefer each choice is equal to the corresponding proportion of high school students.
(B) There is sufficient evidence that the proportion of middle school students who prefer science fiction is different from the proportion of high school students who prefer science fiction.
(C) There is sufficient evidence that for all five choices, the proportion of middle school students who prefer each choice is different from the corresponding proportion of high school students.
(D) There is sufficient evidence that for at least one of the five choices, the proportion of middle school students who prefer that choice is equal to the corresponding proportion of high school students.
(E) There is sufficient evidence that for at least one of the five choices, the proportion of middle school students who prefer that choice is different from the corresponding proportion of high school students.

HYPOTHESIS TESTS FOR SLOPES

495. A retailer believes that sales of some luxury items go up when the price is higher. Below is the computer printout for the regression analysis of sales versus price (in $1,000s) for a random sample of these luxury items.

```
Dependent variable is: Sales
Source        SSq         df     MSq         F
Regression    105.226     1      105.226     6.15
Residual      102.649     6      17.1082

Variable      Coeff       s.e.        t      prob
Constant      18.2961     5.182       3.53   0.0124
Price         0.573437    0.2312      2.48   0.0478
R-sq = 50.6%   R-sq(adj) = 42.4%
s = 4.136       df = 8-2 = 6
```

What is the *P*-value for a *t*-test with H_0: $\beta = 0$ and H_a: $\beta > 0$?

(A) 0.0124
(B) 0.0239
(C) 0.0478
(D) 0.0956
(E) 0.5060

496. A linear regression analysis is performed on two variables. Which of the following tells you that another model probably gives a better fit?

(A) The correlation *r* is low.
(B) The mean of the residuals is 0.
(C) The *P*-value for H_0: $\beta = 0$ and H_a: $\beta > 0$ is low.
(D) The coefficient of determination is high.
(E) The residual plot has a pattern.

497. The space shuttle *Challenger* took off when the temperature was 31°F. Before this launch, 23 earlier launches experienced from zero to three O-ring failures. After the *Challenger* disaster, there was some speculation that the number of O-ring failures was related to temperature at liftoff. A computer printout, performed too late, is shown below.

```
Dependent variable: Failures
s = 0.6673  R-sq = 31.5%  R-sq(adj) = 28.2%
Variable      Coef          s.e. Coeff    t-ratio     prob
Constant      4.79365       1.409         3.4         0.0027
Temperature  −0.0626587     0.02016       -3.11       0.0053
Source        df    SS          MS          F
Regression    1     4.30166     4.30166     9.66
Residual      21    9.35052     0.445263
```

Is there evidence of a relationship between O-ring failures and temperature at liftoff?

(A) There is no evidence of a relationship between the number of O-ring failures and the temperature at liftoff.
(B) There is evidence of a relationship at the 0.10 level but not at the 0.05 level.
(C) There is evidence of a relationship at the 0.05 level but not at the 0.01 level.
(D) There is evidence of a relationship at the 0.01 level but not at the 0.001 level.
(E) There is evidence of a relationship at the 0.001 level.

498. In a random sample of 20 college students, each was interviewed as to GPA and weekly work-study hours on campus. What is the critical t-value in calculating a 95% confidence interval estimate for the slope of the resulting least squares regression line?

(A) 1.96
(B) 2.074
(C) 2.086
(D) 2.093
(E) 2.101

499. A scatterplot of the ages of 30 couples applying for marriage licenses shows a strong, positive, linear relationship between the men's and women's ages. A significance test on the slope with H_0: $\beta = 0$ and H_a: $\beta > 0$ results in a P-value of 0.035. Which of the following statements is an appropriate conclusion?

(A) With this small of a P-value, $0.035 < 0.05$, there is not sufficient evidence of a linear relationship between the men's and women's ages of couples applying for marriage licenses.
(B) With this small of a P-value, $0.035 < 0.05$, there is proof of a linear relationship between the men's and women's ages of couples applying for marriage licenses.
(C) With this small of a P-value, $0.035 < 0.05$, there is sufficient evidence that among couples applying for marriage licenses, a one-year difference in ages among men corresponds to a constant positive difference in ages among women on average.
(D) Among couples applying for marriage licenses, 3.5 percent of the difference among the women's ages can be explained by the linear model of women's ages versus men's ages.
(E) Among couples applying for marriage licenses, 12.25 percent of the difference among the women's ages can be explained by the linear model of women's ages versus men's ages.

500. An auto magazine report analyzes the association between the weight of a car (in thousands of pounds) and the fuel efficiency (in miles per gallon). Some of the regression analysis is shown below.

Variable	**Sample size**	**Mean**	**SE Mean**	**St Dev**
MPG	30	24.1	0.867	4.75
Predictor	Coef	SE Coef		
Constant	45.7	2.01		
Weight	−8.73	0.74		
s = 2.39	R-sq = 73.7%			

Assuming that all conditions for inference are met, what is the appropriate test statistic for testing the null hypothesis, H_0: $\beta = 0$, that the slope of the regression line is 0?

(A) $\dfrac{45.7}{2.01}$

(B) $\dfrac{24.1}{0.867}$

(C) $\dfrac{-8.73}{2.39}$

(D) $\dfrac{-8.73}{0.74}$

(E) $\dfrac{-8.73}{\left(\dfrac{0.74}{\sqrt{30}}\right)}$

FREE-RESPONSE QUESTIONS

FR Exploratory Analysis

CHAPTER 5

Answers for Chapter 5 are on pages 301–310.

ONE-VARIABLE DATA ANALYSIS

501. In a random sample of 20 high school students, the number of text messages sent by each during the past day were

{0, 4, 8, 12, 13, 15, 15, 18, 23, 24, 40, 43, 47, 50, 53, 57, 59, 61, 61, 65}.

(a) Create a stemplot of the number of text messages sent by students.
(b) Describe the distribution.
(c) Would a boxplot give more, less, or basically the same information? Explain.

502. The histogram below shows the distribution of weights of the 21 bags of candy coming off an assembly line during a one-minute production interval.

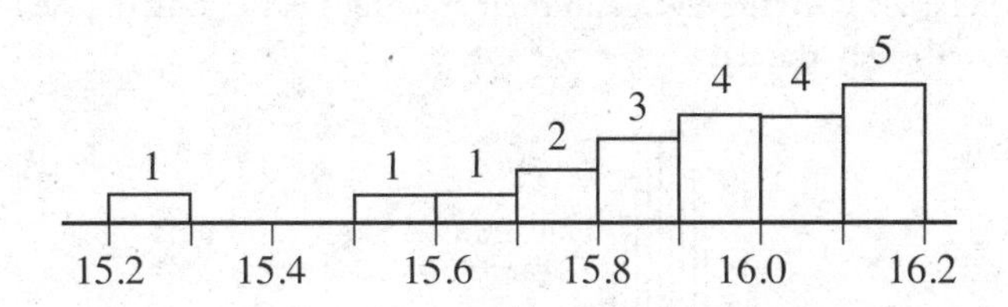

Bag Weights (ounces)

(a) Write a few sentences to describe the distribution of bag weights for the one-minute interval.
(b) One of the bag weights is 15.85 ounces. If this bag weighed 15.75 ounces instead of 15.85 ounces, what effect would the decrease have on the mean and on the median?

503. A bank cybersecurity engineer notes the number of malware intrusions intercepted each hour during a random sample of 109 hour-long periods.

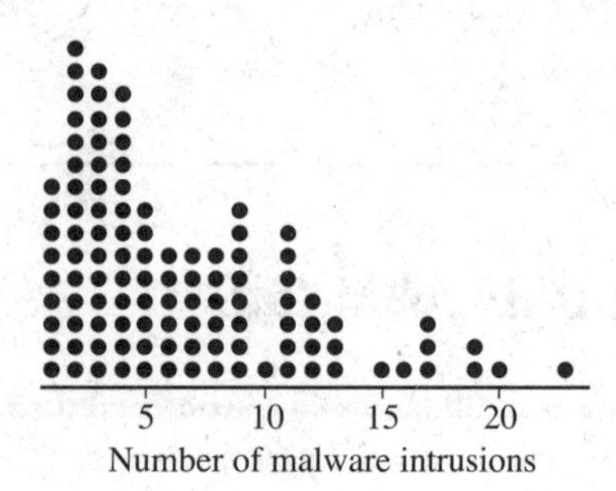

(a) Describe the distribution.

(b) Is the quotient, $\dfrac{\text{mean intrusions}}{\text{median intrusions}}$, greater than 1, less than 1, or approximately 1? Explain.

(c) Are conditions for inference (constructing a confidence interval for the mean number of intrusions per hour) met? Explain.

504. Do graduates from different universities obtain different starting salaries? In one study of starting salaries, random samples of graduates from two universities were surveyed and their starting salaries noted. The histograms below display the data.

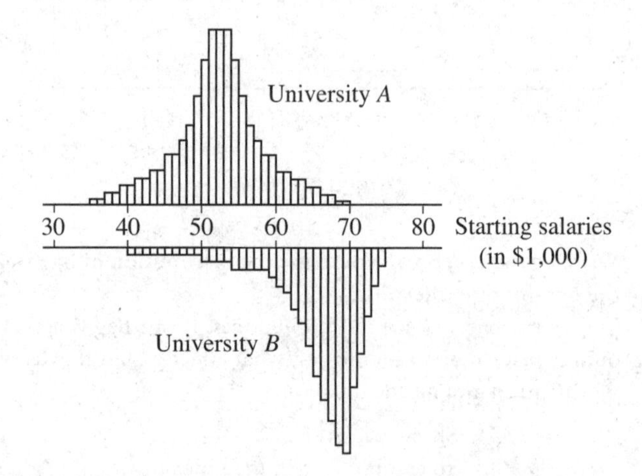

(a) Compare the distributions of starting salaries of graduates from the two universities.

(b) Which is greater, the difference between the two university mean starting salaries or the difference between the two university median starting salaries? Explain.

505. In independent random samples of 20 high school students and 20 college students, each student gave a rating between 0 and 100 for a new music website.

HS students: 27, 32, 82, 36, 43, 75, 45, 16, 23, 48, 51, 57, 60, 64, 39, 40, 69, 72, 54, 57

College students: 49, 50, 35, 69, 75, 35, 49, 54, 98, 58, 22, 34, 60, 38, 47, 65, 79, 38, 42, 87

Using back-to-back stemplots, compare the distributions.

506. The A1C test is used to diagnose diabetes. Cumulative graphs of A1C level readings in the populations of three countries (*A*, *B*, and *C*) are given below.

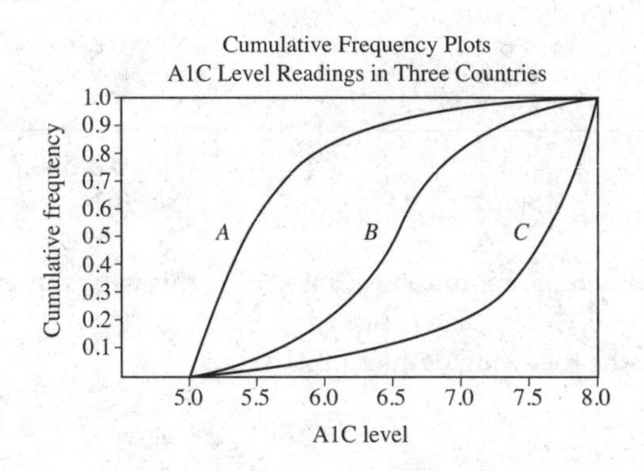

Write a few sentences comparing the distributions of A1C level readings in the three countries.

507. Ms. D. has 12 students in her AP Statistics class. They all did very well on the final exam with scores of {81, 87, 88, 88, 88, 89, 89, 90, 90, 90, 90, 95}.

(a) The trimean is defined as $\frac{Q_1 + 2\text{Med} + Q_3}{4}$. Compute the trimean for the above data. Show your work.

(b) The mean trimmed 50% is computed by trimming the upper 25% of the scores and the lower 25% of the scores and then computing the mean of the remaining scores. Compute the mean trimmed 50% for the above data. Show your work.

(c) More generally, to compute a trimmed mean, you remove some of the higher and lower scores and then compute the mean of the remaining scores. In what way can the median be considered a trimmed mean?

(d) In general, what advantage might both the trimean and a trimmed mean have over using the standard mean, $\bar{x} = \frac{\Sigma x}{n}$?

508. Suppose we calculate the melting points of 10 nuggets claimed to be pure silver and find the mean to be 1,760°F with a standard deviation of 27°F.

(a) What would be the mean and standard deviation of the melting points of the 10 nuggets if the 10 measurements were converted to Celsius? (°F = (1.8)(°C) + 32)

(b) How would the conversion above have changed if the sample size had been 20 nuggets instead of 10 nuggets?

509. The mean absolute deviation (MAD) of a set $\{x_1, \ldots, x_n\}$ is defined to be

$$\text{MAD} = \frac{1}{n}\sum_{i=1}^{n}\left|x_i - \bar{x}\right|.$$

(a) Compare the MADs of the sets with the two dotplots.

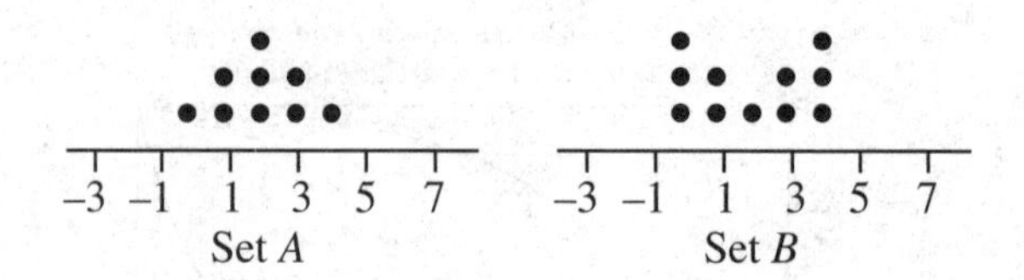

(b) After comparing the above answers, is this what was expected?

510. Consider the following dotplot of 21 values.

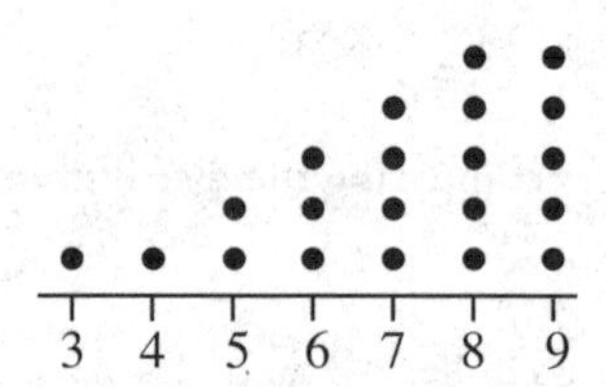

(a) Calculate and compare the mean and median.

(b) Comment on whether your answer was as expected.

511. The selling prices of a simple random sample of 20 Florida condos is summarized below.

Price ($1,000)	0–100	100–200	200–300	300–400	400–500
Frequency	1	0	10	5	4

(a) Draw a cumulative frequency plot of this data.

(b) If given the raw data, would the mean or the median best describe the typical selling price for Florida condos? Explain.

512. Fifty subjects with advanced heart disease participated in an experiment to compare two new blood pressure medications, *A* and *B*. The subjects were randomly assigned into two groups, one group to take medication *A* and the other medication *B*. Every person's blood pressure was taken prior to assignment and again after three months on the medication. The calculated differences, old pressure minus new pressure (+ represents improvement), are shown in the following table.

Medication	Values < Q_1	Q_1	Median	Q_3	Values > Q_3
A	−2, −1	0	3	10	12, 15
B	−2, 2	3	4	5	6, 10

(a) Draw parallel boxplots of the above data showing outliers, if any.
(b) Which medication should be chosen if the goal is to have the greatest percentage of subjects lower their blood pressures? Explain.
(c) Which medication should be chosen if the goal is to have the greatest mean improvement (lowering) in subjects' blood pressures? Explain.

513. The salaries of a random sample of 91 faculty members at a large university are summarized in the following histogram. (Assume salaries are at the center of each bar.)

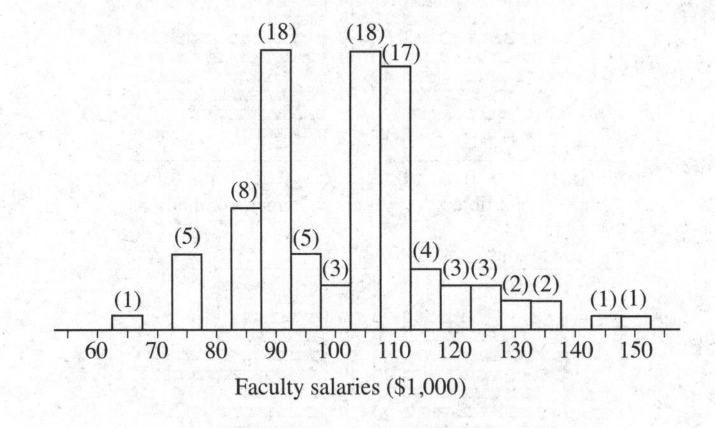

(a) Construct a boxplot of the above data.
(b) What feature does the histogram show that is missed by the boxplot?
(c) What feature is more clearly distinguished in the boxplot than in the histogram?

514. It is difficult to know when a minute has passed without looking at a clock. A random sample of 18 students consisted of 11 boys and 7 girls. With eyes closed, each student raised his/her hand when he/she guessed that a minute had passed. Their times were recorded and are displayed below.

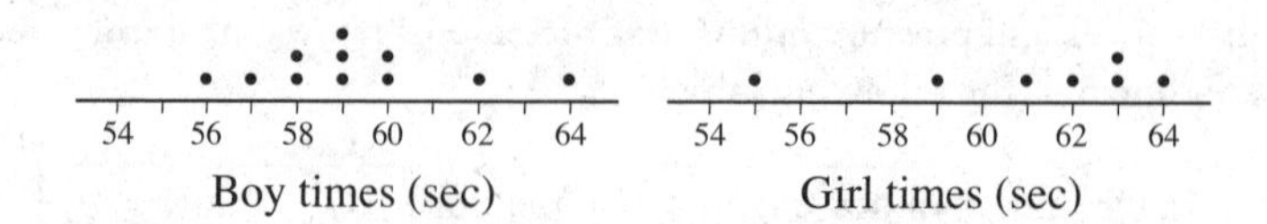

(a) In either the boy or girl distributions, are any of the times considered to be outliers? Explain.
(b) If all the times are combined, does the new set of times have any outliers? Explain.
(c) Explain the reason for the answer to (b) in relation to the answer to (a).

515. Two species of insects were continually genetically modified over the course of a 90-day period of time (their normal life expectancy). The resulting age (in days) and gender distributions are given in the population pyramids below.

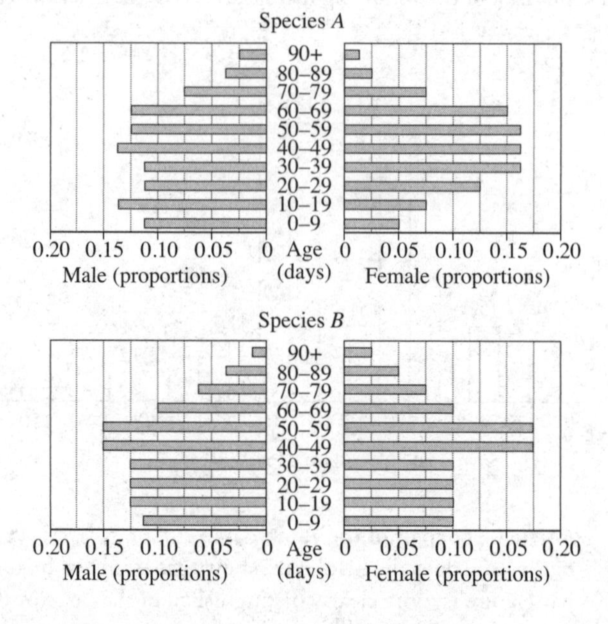

(a) In which interval is the median age of the Species *B* insects? Explain.
(b) In either or both species, have the modifications resulted in longer life expectancies for males or females? Explain.
(c) Have the most recent modifications resulted in favoring more births of a particular gender in either or both species?

516. A furniture store sells low-end and high-end outdoor patio furniture. The selling prices ($) for their lounge chairs (low- and high-end) are summarized in the boxplots below.

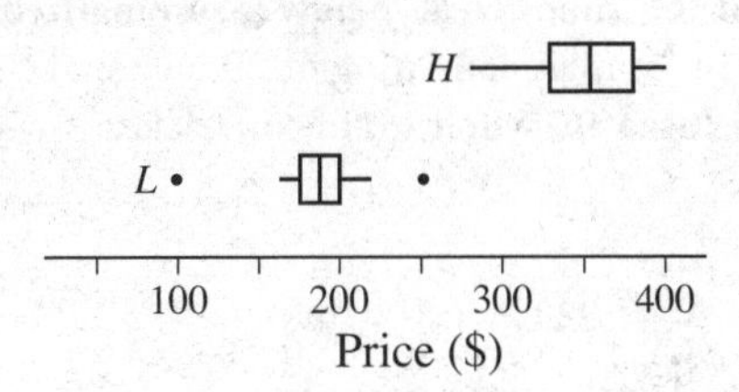

(a) Compare the distributions of the prices of the low- and high-end lounge chairs.
(b) If a single histogram was made of all the lounge chair prices, describe its shape.

517. A cumulative relative frequency plot for scores (maximum possible 120) on an arcade video game during one 24-hour period is as follows.

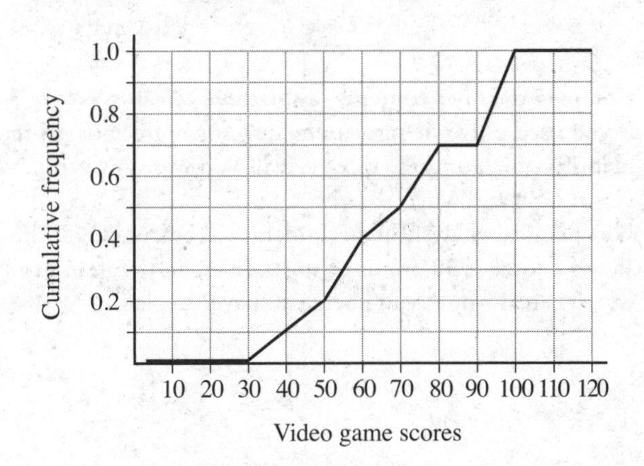

(a) Interpret in context the meaning of the point (50, 0.2).
(b) Can you make an estimate for the number of scores below 60?
(c) What is the median score?
(d) What proportion of the scores are over 80?
(e) What does the flat portion between 80 and 90 mean?

518. Cereal is the breakfast of choice for many children. Nutritional content varies widely, with dieticians saying that the best cereals are low in sugar, high in fiber, and full of protein. Nutritional content also varies widely among boxes of the same cereal. Below is summarized the nutritional content found in samples consisting of 20 boxes taken of each of three cereals, all advertised as "whole grain corn flakes."

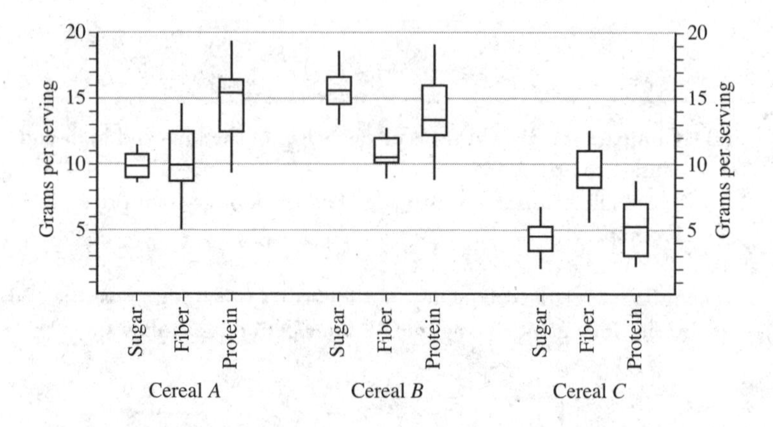

(a) Compare the fiber contents of the three cereals.

(b) If you used only one measurement—sugar, fiber, or protein—to identify which of these three cereals you were analyzing, what would you choose and why?

(c) If a chemical analysis of an unidentified box of one of these cereals shows a total of 30 grams of sugar, fiber, and protein per serving, which cereal would you guess you have? Explain.

TWO-VARIABLE DATA ANALYSIS

519. A simple random sample (SRS) of beachfront condo listings in Myrtle Beach comparing weekly rent ($) versus size (ft^2) yields the following computer output.

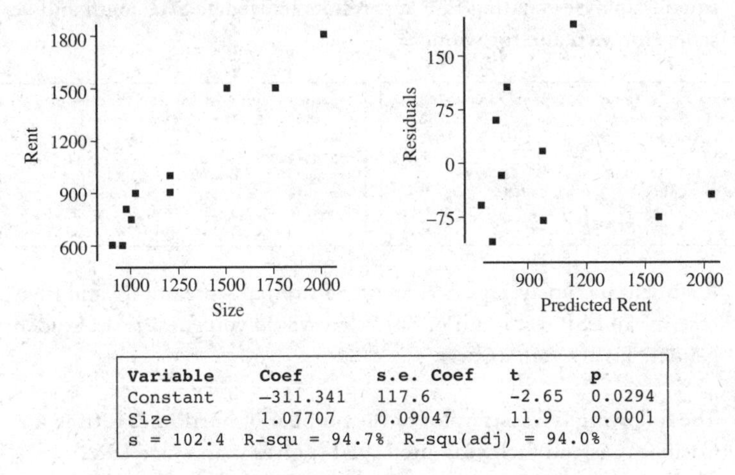

Variable	Coef	s.e. Coef	t	p
Constant	–311.341	117.6	-2.65	0.0294
Size	1.07707	0.09047	11.9	0.0001
s = 102.4	R-squ = 94.7%	R-squ(adj) = 94.0%		

(a) Is a linear model appropriate for these data? Explain.
(b) Interpret the slope of the regression line.
(c) Interpret r^2 in context.

520. The calories and fat content per serving size of 10 brands of potato chips are fitted with a least squares regression line with computer output.

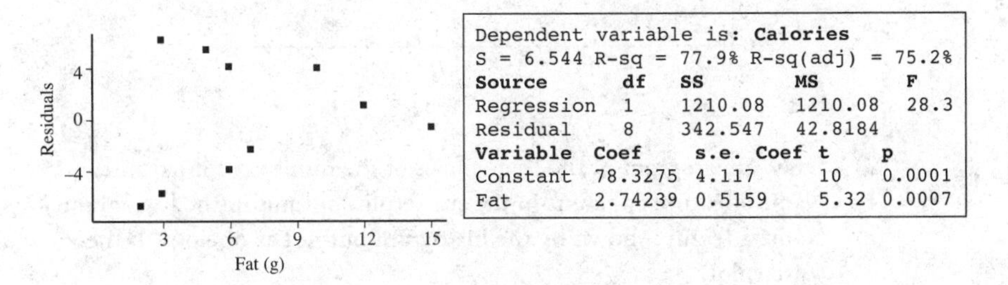

Dependent variable is: **Calories**
S = 6.544 R-sq = 77.9% R-sq(adj) = 75.2%

Source	df	SS	MS	F
Regression	1	1210.08	1210.08	28.3
Residual	8	342.547	42.8184	

Variable	Coef	s.e. Coef	t	p
Constant	78.3275	4.117	10	0.0001
Fat	2.74239	0.5159	5.32	0.0007

(a) Is a line an appropriate model? Explain.
(b) Interpret the slope of the regression line in context.
(c) Interpret the y-intercept of the regression line in context.
(d) What are the predicted calories for a brand with 10 g of fat per serving?
(e) What are the actual calories for the brand with 10 g of fat per serving?

521. A research group believes it can predict a subject's ESP (telepathy and clairvoyance) testing result based on the average SAT math and verbal scores of those given ESP testing. The research group suspects that gender also makes a difference. A stratified random sample of high school students who have taken the SATs are given ESP testing. Below are two least squares analyses relating ESP test results to average SAT math and verbal scores for men and for women.

Dependent variable is: ESP test result (male)
s = 5.018 R-sq = 37.7% R-sq(adj) = 29.9%

Source	SS	df	MS	F
Regression	121.787	1	121.787	4.84
Residual	201.438	8	25.1797	

Variable	Coef	s.e. Coef	t	p
Constant	50.89	13.58	3.75	0.0056
SAT	0.0476848	0.02168	2.2	0.0591

Dependent variable is: ESP test result (female)
s = 3.981 R-sq = 79.6% R-sq(adj) = 77.1%

Source	SS	df	MS	F
Regression	495.192	1	495.192	31.2
Residual	126.808	8	15.851	

Variable	Coef	s.e. Coef	t	p
Constant	22.1923	10.77	2.06	0.0734
SAT	0.0961538	0.0172	5.59	0.0005

A student has an average SAT math and verbal score of 689 and then receives an ESP test result of 84. What would you guess is the student's gender? Justify your answer.

522. The following is a scatterplot of the number of hurricanes hitting a Caribbean island each year after 1982 (t gives years since 1982).

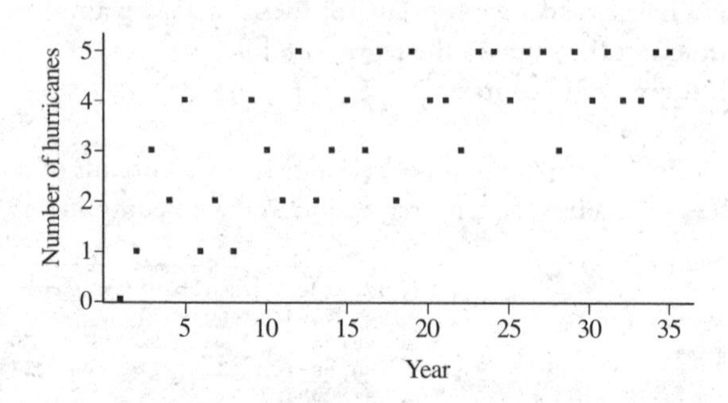

(a) Draw a histogram of the frequencies of the number of hurricanes.
(b) Name a feature apparent in the scatterplot but not in the histogram.
(c) Name a feature shown by the histogram but not as obvious in the scatterplot.

523. Data are collected on the distance (in feet) that students sit from the front of the class and their exam averages for the class (on a 0–100 scale). A scatterplot of exam average y versus distance x from the front is described as linear, negative, and strong.

(a) In the above context, what is meant by linear, by negative, and by strong? The regression equation is $\hat{y} = 97.5 - 1.15x$.
(b) Interpret the slope in context.
(c) One student who sat 12 feet from the front had a residual of –4.7. What was that student's exam average?

524. It is believed that customers will hesitate buying a luxury item if it seems to be priced too high or too low. A new-model boat is offered at different prices in eight different showrooms. The number sold versus price ($1,000) is fitted with a least squares regression line, yielding the following summary statistics and residual plot.

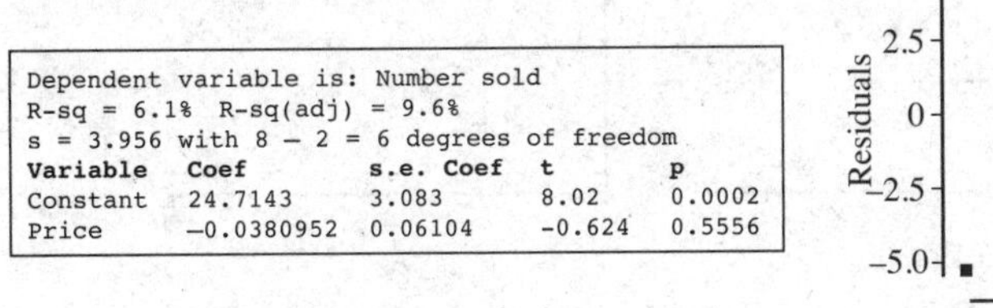

```
Dependent variable is: Number sold
R-sq = 6.1%  R-sq(adj) = 9.6%
s = 3.956 with 8 - 2 = 6 degrees of freedom
Variable   Coef         s.e. Coef   t        p
Constant   24.7143      3.083       8.02     0.0002
Price      -0.0380952   0.06104     -0.624   0.5556
```

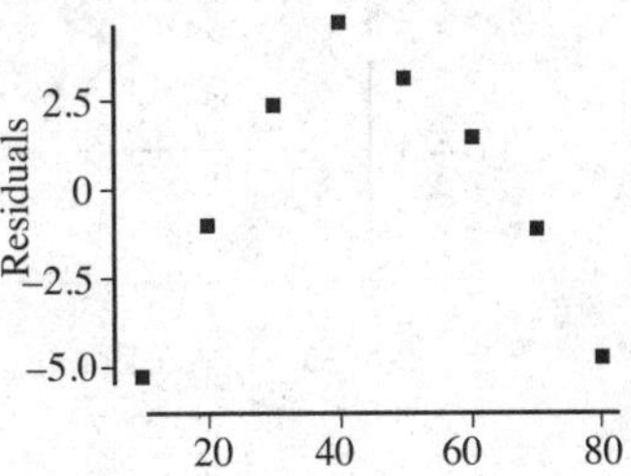

(a) Comment on the use of a linear model for this data.
(b) How can this computer output be used to recommend the best cost to achieve the most sales? Explain.

525. The table below gives weight loss (in pounds) during the first month for six overweight patients on varying dosages of an experimental drug.

Dosage (grams), x	0.5	1.0	1.5	2.0	2.5	3.0
Weight Loss (pounds), y	7	10	12	14	16	17

Linear regression lines on (x, y), on the transformed data $(x, \log y)$, and on the transformed data $(\sqrt{x}, y)$ result in the following computer output, respectively.

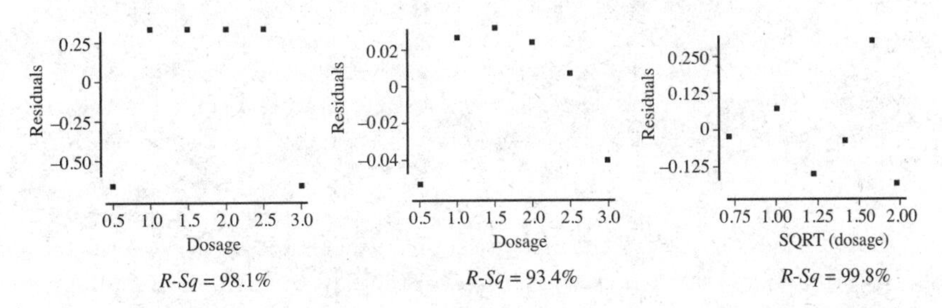

R-Sq = 98.1%   *R-Sq* = 93.4%   *R-Sq* = 99.8%

(a) Interpret the coefficient of determination for the transformed data $(\sqrt{x}, y)$.
(b) Compare the three regression lines as to goodness-of-fit for a linear model.

526. In a study of salaries of young engineers at a software company, two regression analyses are run, the first on salary versus years of college education and the second on salary versus age. The graphs of the residuals follow.

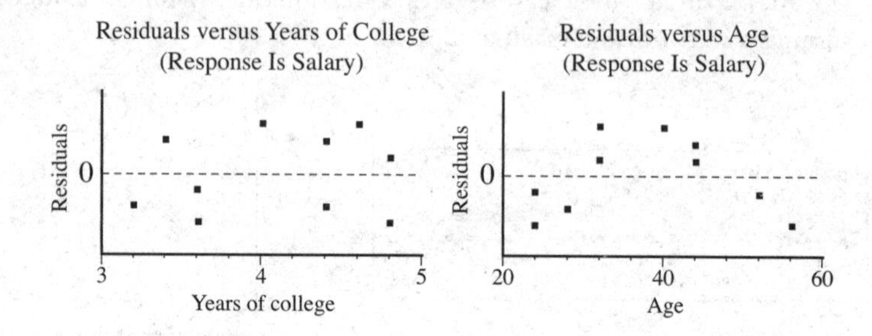

(a) Which of the regression lines indicates a better linear fit? Explain.

(b) The two regression lines are used to find estimates for the salary of a 40-year-old software engineer with 4 years of college education. One of the engineers in the sample is 40 years old and has 4 years of college education. Which of the estimates are underestimates and which are overestimates of the salary of this engineer? Explain.

FR Collecting and Producing Data

Answers for Chapter 6 are on pages 310–314.

527. A researcher wants to determine if watching TV from five feet or closer damages a person's eyes. The researcher wants to know if this is truth or a myth.
   (a) How can an observational study be performed?
   (b) How can an experiment be performed?
   (c) Which approach is more appropriate here? Explain.
   (d) If an experimental design was implemented, give an ethical consideration that would cause the researchers to halt the experiment early.

528. High A1C levels (a test for diabetes) can be reduced by either a low-fat diet or a medication such as metformin. Researchers would like to test the effectiveness of metformin and to note whether the effectiveness, if any, is enhanced by diet. A random sample of adults with high A1C levels, on no special diets, and not on medication, are recruited for a study.
   (a) Conclusions will apply to what population?
   (b) Explain how you would design a completely randomized experiment.
   (c) How might you incorporate blocking and for what purpose?
   (d) How might blinding be incorporated in this study and for what purpose?

529. A company efficiency expert believes that employees who eat at least 1,000 calories at breakfast have higher productivity levels at work. She interviews a simple random sample (SRS) of 30 employees who claim to eat under 1,000 calories at breakfast and an SRS of 25 employees who claim to eat over 1,000 calories at breakfast. In each group, she looks up productivity levels on the job.
   (a) Explain why this is an observational study and not an experiment.
   (b) Give an example of a possible confounding variable with an explanation in the context of this study.
   (c) If the employees who eat over 1,000 calories have higher productivity records, is it reasonable to encourage all employees to eat larger breakfasts? Explain.
   (d) How could the efficiency expert design an experiment to study caloric intake at breakfast with productivity in the workplace?

530. A pastor expects his parishioners to spend some time each week in Bible study. He randomly selects 30 parishioners from among those coming to church one Sunday and asks each the number of minutes spent reading the Bible during the previous week. Using this data, the pastor's secretary, who once took a college statistics class, calculates the following.

$$n = 30 \quad \bar{x} = 31.2 \quad s = 13.2 \quad \text{Min} = 13.5$$

$$Q_1 = 18.7 \quad \text{Med} = 27 \quad Q_3 = 41.2 \quad \text{Max} = 57$$

The pastor asks the secretary to compute a confidence interval estimate of the mean number of minutes parishioners read the Bible weekly. Comment on this study, including necessary assumptions and how well they are met. (You are not asked to calculate the confidence interval.)

531. A manufacturer of circuit boards wishes to test which of two soldering irons leads to fewer soldering errors. The workroom consists of eight large tables, and the manufacturer plans to have everyone at a given table use the same soldering iron. On one side of the room are windows with a pretty view, and on one side is the floor manager's desk.

(a) Suppose you decide to block using the Scheme *A* below (one block is white, one gray). How would you use randomization, and what is the purpose of the randomization? (You are not asked to give an actual randomization procedure.)

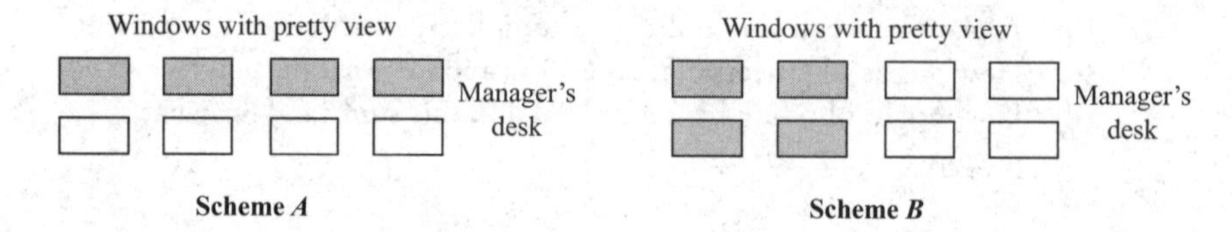

(b) Comment on the strength and weakness of Scheme *A* as compared to blocking Scheme *B* (one block is white, one gray).

532. A high school math department conducts a study to determine whether a classroom with windows leads to higher exam scores than a classroom without any windows. Two algebra classes are scheduled, each with 25 students. It is randomly decided which class will use which classroom. During the year, each teacher administers the same exams. At the end of the academic year, overall exam grades are compared.

(a) Identify the response variable, the treatments, and the experimental units.
(b) Was randomization properly used? Explain.
(c) Was replication properly used? Explain.
(d) Teacher is a confounding variable. Explain.

533. How does the American public now feel about leaving Afghanistan? Two surveys are conducted. The survey in *VFW* (Veterans of Foreign Wars) magazine asks readers to respond concerning whether or not they support the "retreat from Afghanistan." The survey in the newspaper the *Washington Post* asks readers to respond concerning whether or not they support the "phased troop redeployment from Afghanistan."

   (a) Give two examples of bias that may have been introduced given the ways in which the samples were selected.
   (b) Explain how bias may have been introduced given the ways in which the surveys were worded. Suggest different wording to avoid that bias.

534. A popular coffee chain has 11,500 stores in the U.S. and 13,500 stores outside the U.S. Each store has approximately 10 full-time employees. Suppose the company would like to survey 250 of its employees about new coffee drinks under consideration. Under discussion are three sampling methods.

   Method 1: A simple random sample
   Method 2: Randomly pick 125 of their 115,000 U.S. full-time employees and randomly pick 125 of their 135,000 non-U.S. full-time employees
   Method 3: Randomly pick 25 of their 25,000 stores and pick all full-time employees at these 25 stores

   (a) Give a design for carrying out Method 1, and give a disadvantage of using Method 1.
   (b) What is Method 2 called? What is an advantage of Method 2 in this context?
   (c) What is Method 3 called? What is an advantage of Method 3 in this context?

535. A medical insurance adjuster wants to interview a random sample of 50 doctors in a city. Suppose there are 25 hospitals, each with 50 doctors.

   (a) Describe a procedure for using stratified sampling using the hospitals as strata.
   (b) Describe a procedure for using cluster sampling using the hospitals as clusters.
   (c) In this context, give an advantage of using stratified sampling over cluster sampling.
   (d) In this context, give an advantage of using cluster sampling over stratified sampling.

536. A reading specialist plans a study to determine if high school students read faster from a physical book or on an iPad. The reading specialist randomly selects 60 high school students for the study.

   (a) Describe a randomization process and an inference procedure for the study to be conducted with a completely randomized design.
   (b) Describe a randomization process and an inference procedure for the study to be conducted with a matched pairs design.

FR Probability

CHAPTER 7

Answers for Chapter 7 are on pages 314–320.

537. A chess match is played with a series of games. The winner of a game receives 1 point, while a drawn game gives each player $\frac{1}{2}$ point. A player who scores 2 points before his or her opponent wins the match. Suppose Player A is the stronger player and when playing Player B will win with a probability of 0.6 and will draw with a probability 0.3. The outcome of any game is independent of the outcome of any other game. Player A and Player B begin a match.

    (a) What is the probability Player A wins the match in 2 games?
    (b) What is the probability Player A wins the match after exactly 3 games?
    (c) What is the probability that the first game was a draw given that Player A wins the match in exactly 3 games?
    (d) What is the probability Player A wins the second game given that he or she draws the first game?

538. You are asked to choose between two envelopes, one of which has twice as much money as the other. You arbitrarily pick one, open it, and find \$2. You are then given the chance to switch envelopes. You reason that the other envelope has either \$1 or \$4, each with a probability of 0.5. Applying your understanding of expected value, you calculate 0.5(\$1) + 0.5(\$4) = \$2.50 and conclude that you should switch envelopes. Comment on this reasoning.

539. The probability that Anthony Rizzo of the Chicago Cubs hits a homer on any given at bat is 0.06, and we assume each at bat is independent.

    (a) What is the probability the next homer will be on his fifth at bat?
    (b) What is the probability he hits exactly one homer in five at bats?
    (c) What is the expected number of homers in every 10 at bats?
    (d) What is the expected number of at bats until the next homer?

540. The weights of babies born to nonsmokers have a normal distribution with a mean of 7.0 pounds and a standard deviation of 0.8 pounds. Babies are considered low birth weight (LBW) if they weigh less than 5.5 pounds. Note that 5 percent of babies born to smokers are LBW and that 16 percent of pregnant women are smokers.

(a) What is the probability that a nonsmoker will have an LBW baby?
(b) What is the probability a baby is LBW?
(c) Given that a baby is LBW, what is the probability the baby was born to a nonsmoker?

541. An online business charges shipping and handling (S&H) fees based on the sizes of the sales. 30 percent of the company's sales are under $10, 40 percent are between $10 and $25, 20 percent are between $25 and $50, 5 percent are between $50 and $100, and 5 percent are over $100. The charges are as follows, and note that S&H is free for orders over $100!

Sale ($)	<10	10–25	25–50	50–100	>100
S&H Charge ($)	2.50	5.00	7.50	10.00	0.00

(a) What is the mean S&H charge? Show your work.
(b) What is the median S&H charge? In other words, what charge C has the property that $P(x \geq C) \geq 0.5$ and $P(x \leq C) \geq 0.5$? Explain.

542. Die *A* has three 5s, two 3s, and one 1 on its six faces. Die *B* has two 2s and four 4s on its six faces. Both dice are fair. Each player simultaneously rolls one of the dice, and the winner is the player with the higher number showing.

(a) If you want to win, would you rather roll die *A* or die *B*? Explain.
(b) If the winner receives whatever shows on his/her winning die, what is the expected value for one roll to each player? Explain.

543. The probability distribution for the number of days per week that college students get a good night's sleep is as follows.

Number of Days	0	1	2	3	4
Relative Frequency	0.12	0.41	0.25	0.15	0.07

(a) Calculate and give a brief interpretation of the mean of this probability distribution.
(b) In a random sample of 10 college students, there are a total of 20 good nights of sleep. A new random sample of 50 students is planned. How do you expect the average number of good nights of sleep for this new sample to compare to that of the first sample? Explain.
(c) Find the median of the above distribution, where the median M is defined to be a value such that $P(x \geq M) \geq 0.5$ and $P(x \leq M) \geq 0.5$.

544. A school must choose between two snow removal services. Service *A* charges $2,500 yearly plus $2,000 for every month with over three snowfalls necessitating their services. Service *B* charges $450 per snow removal. Relevant probabilities are shown in the following tables.

Month	Oct	Nov	Dec	Jan	Feb	Mar	Apr
***P*(> 3 snowfall services)**	0.01	0.02	0.18	0.28	0.22	0.13	0.02

Annual Number of Snow Services	6	7	8	9	10	11	12
Probability	0.05	0.10	0.20	0.25	0.20	0.15	0.05

Which service should the school use to minimize expected cost? Justify your answer.

545. Suppose a pregnancy test correctly tests positive for 98 percent of pregnant women but also gives a false positive reading for 3 percent of women who are not pregnant. Suppose 5 percent of women who purchase this over-the-counter test are actually pregnant.

(a) What is the probability a woman tests positive?
(b) If a woman tests positive, what is the probability she is pregnant?
(c) If two women purchase the test and both test positive, what is the probability that exactly one of the two is pregnant?

546. One model of a popular tablet computer has a life expectancy that is roughly normally distributed with a mean of 24 months and a standard deviation of 3 months.

(a) The warranty will repair any computer failing within 18 months free of charge. The company's contracted cost to repair computers is $350. What is the expected value of the cost to the company per computer?

(b) Suppose the company decides it is willing to extend the warranty to an additional 5 percent of the computers. What warranty should the company offer?

*547. Simple random samples of younger adults and older adults in a large city were surveyed as to the number of apps on their smart phones. The data are summarized in the following table.

	n	Mean	Median	Min	Max	Q_1	Q_3	SD
Younger Adults	75	25	25	13	37	19	31	10
Older Adults	80	15	15	3	27	12	18	4

Is there evidence that either of these samples were drawn from populations with roughly normal distributions? Explain.

548. Simple random samples of younger adults and older adults in a large city were surveyed as to the amount of credit card debt, and the data are summarized in the following table.

	n	Mean	Min	Max	Q_1	Q_3	SD
Younger Adults	100	6,500	2,900	10,100	5,700	7,300	1,200
Older Adults	100	5,550	0	10,950	4,350	6,750	1,800

Assume that both samples are drawn from roughly normally distributed populations. Would a greater percentage of younger or of older adults more likely be able to pay off their credit debt with $8,000? Explain.

*549. Medical scientists applying to the CDC for grants are awarded scores for their grant applications. These scores follow a roughly normal distribution. All applicants with scores more than 1.5 standard deviations above the mean are awarded grants, and all applicants with scores that are outliers on the high end are awarded additional lab equipment in addition to the grants.

(a) What is the probability a randomly selected scientist receives the grant?
(b) What is the probability that at least one out of three randomly chosen scientists receives the grant?
(c) What is the probability a randomly selected scientist receives the grant and the additional lab equipment?
(d) What is the probability a randomly selected scientist receives the grant but not the additional lab equipment?

550. New Caledonian crows are among the smartest animals on this planet. An experiment testing the length of time it takes these crows to solve a particular puzzle (and receive a food reward) finds that 5 percent of the crows fail. Of those that do solve the puzzle, their times are roughly normally distributed with a mean of 1.8 minutes and a standard deviation of 0.4 minutes.

(a) What is the probability a randomly selected crow solves the puzzle in under 1.5 minutes?
(b) What is the probability a randomly selected crow has not solved the puzzle in 2.0 minutes?

551. MLB signing bonuses for different position players with respective probabilities are given in the table below.

	Position		
Bonus	**Pitcher**	**Infielder**	**Outfielder**
$2,000,000	0.30	0.30	0.20
$5,000,000	0.10	0.05	0.05

(a) What is the probability a given bonus was for a pitcher?
(b) What is the expected value for a bonus?
(c) Are position and bonus independent? Explain.

*552. A cereal box is advertised to hold 16 ounces of corn flakes. Suppose the amounts per box are roughly normally distributed with a standard deviation of 0.05 ounces and a mean *m* that the producer can set. An inspector randomly samples two boxes, and the company is fined $10,000 for each box found to be under 16 ounces.

(a) If the company is willing to accept an expected value of $125 for the fine, what probability *p* should it be willing to accept where *p* is the probability that a box is under 16 ounces?
(b) What should *m* be set at if the company is willing to accept an expected value of $125 for the fine?

553. The weights of individual apples are approximately normally distributed with a mean of 8 ounces and a standard deviation of 0.5 ounces. The weights of individual oranges are approximately normally distributed with a mean of 6 ounces and a standard deviation of 0.4 ounces. The weights of individual pieces of fruit are independent.

(a) What is the distribution of the total fruit weight of fruit gift boxes containing 6 randomly selected apples and 6 randomly selected oranges?
(b) The gift boxes are advertised as containing 5 pounds of fruit. What is the probability that a gift box contains at least 5 pounds of fruit?
(c) An empty gift box weighs exactly 12 ounces. What is the distribution of total weights (box plus fruit) of this gift offering?

554. The standard triathlon is a 1.5 km swim, 40 km bike ride, and 10 km run. In a random sampling of recent competitions, the mean and SD of the participants' times for each event were as shown in the table.

	Swim	**Bike Ride**	**Run**
Mean (minutes)	30	90	70
SD (minutes)	5	10	10

Assume the times for the three legs of the race are each roughly normally distributed and independent.

(a) What is the distribution of total times to complete the triathlon?
(b) What is the probability that a participant will complete the triathlon in less than 180 minutes?

555. An adult amusement park ride has a carrying capacity of 1,650 lb. Suppose the men at the amusement park have a mean weight of 175 lb with a standard deviation of 15 lb. Suppose the women have a mean weight of 130 lb with a standard deviation of 10 lb. Assume both weight distributions are roughly normal.

 (a) What are the mean μ_{SUM} and standard deviation μ_{SUM} of the combined weight of 6 men and 4 women, assuming all weights are independent?
 (b) What is the probability that the 6 men and 4 women will overload the ride?

556. Suppose that women's times for the 200-meter sprint have a roughly normal distribution with a mean of 25.2 seconds and a standard deviation of 1.2 seconds. Suppose that men's times have a roughly normal distribution with a mean of 22.8 seconds and a standard deviation of 0.9 seconds. A male and a female sprinter are picked at random. Assume their times are independent.

 (a) What is the probability the sum of their sprints is over 50 seconds?
 (b) What is the probability that the man sprinted faster than the woman?

557. Eddie Feigner, the world's best softball pitcher, routinely threw an underhand softball at over 100 mph. Suppose his fastball speeds were roughly normally distributed with a mean of 98.3 mph and a standard deviation of 3.2 mph.

 (a) Would it have been very unusual for him to throw a 102 mph fastball? Explain.
 (b) Would it have been very unusual for him to average 102 mph over 35 random fastballs? Explain.

558. Suppose the number of minutes per day high school students spend eating lunch in the school cafeteria is roughly normally distributed with a mean of 19 minutes.

 (a) Which is more likely: a simple random sample (SRS) of 25 students eating lunch an average of less than 18 minutes per day or an SRS of 75 students eating lunch an average of less than 18 minutes per day? Explain.
 (b) Suppose the sampling distribution of $\bar{x}$ for samples of size 100 has a standard deviation of 0.8 minutes. What is the probability of an SRS of 100 students eating lunch an average of more than 21 minutes?
 (c) Suppose the original population is not roughly normal but, rather, is skewed right (toward the higher values). How would your calculation in (b) change?

FR Statistical Inference

CHAPTER 8

Answers for Chapter 8 are on pages 320–339.

559. In a random sample of 150 adults over the age of 45, 30 say they have played in a band at least one time in their lives.

  (a) Construct a 99% confidence interval for the proportion of all adults over the age of 45 who have played in a band at least one time in their lives.
  (b) Suppose all adults over the age of 45 who have played in a band at least one time in their lives each send in a donation of $10 to the organization Musicians without Borders, which aims to empower musicians as social activists. Assuming there are 125,000,000 adults over the age of 45 in the U.S., what is a 99% confidence interval for what these donations would total for this worthwhile charity?

560. There are 12,500 high school students in a large city school district. Administrators and teachers want to determine the extent to which parents do homework for their children. In an anonymous survey, 150 of the 500 students say that someone else has done their homework at least once. Of those 150 students, 90 or $\frac{90}{150} = 0.6$ or 60% say that a parent has done their homework for them at least once.

  (a) Explain how to pick a simple random sample of 500 students for the anonymous survey.
  (b) What is wrong with using 0.6 to calculate a confidence interval of the proportion of all high school students in the city for whom a parent has done their homework for them at least once?
  (c) Given the above sample, what is an estimate of the number of high school students in the city for whom a parent has done their homework for them at least once?

561. In a random sample of 915 adults, 366 say they believe in ghosts.

  (a) With what margin of error can we find a 95% confidence interval of the proportion of adults who believe in ghosts?
  (b) With what confidence can we report a margin of error of ±2 percent in giving a confidence interval of the proportion of adults who believe in ghosts?

562. Among all the banking firms in a large city, five are randomly selected. From each of these, 10 employees are randomly chosen.

   (a) Is this a simple random sample (SRS)? Why or why not? Explain why your answer would or would not change if you knew that all banking firms in the city had the same number of employees.
   (b) Suppose that 38 of the 50 selected employees are banking assistants. Calculate a 95% confidence interval estimate for the proportion of this city's banking firm employees who are banking assistants.

563. A survey is to be conducted to determine the proportion of young adults who earn more than $7.25 per hour (minimum wage). Two sampling methods are being considered. Method *A* involves standing on a downtown street corner and randomly picking several young adults to interview every 10 minutes during a 12-hour period. Method *B* involves posting the question on several popular young adult websites; the viewer simply has to click on one of two possible answers to participate.

   (a) State a possible source of bias for Method *A*, and describe how it may affect the results.
   (b) State a possible source of bias for Method *B*, and describe how it may affect the results.
   (c) How many young adults should be interviewed to estimate the proportion of the young adult population who earn more than $7.25 to within ± 0.05 with 95% confidence?
   (d) If separate estimates are desired for males and females, identify and describe a proper sampling method.

564. A senator's approval rating stood at 65 percent before she took a crucial vote.

   (a) Her staff believes the rating is still around 65 percent. To confirm this, how large of a simple random sample (SRS) should the staff sample to obtain a 94% confidence interval estimate with a margin of error $\leq 3.5\%$?
   (b) The senator's staff randomly samples 700 people and finds 432 people approve of the senator's job performance. Is there evidence that the rating has changed from 65 percent? Perform an appropriate statistical test.
   (c) In part (b) above, suppose the staff suspects the rating has gone down. Is there evidence that the approval rating has slipped down from 65 percent? Perform an appropriate statistical test.
   (d) Are the answers in (b) and (c) contradictory? Explain.

565. Below are all the scores of a school's AP Statistics students on a practice 40 question multiple-choice exam.

33 31 37 39 27 31 40 36 27 27
27 30 34 38 27 29 27 38 37 40
33 36 29 26 34 32 39 32 39 36
32 32 25 31 26 40 33 37 29 26
35 26 37 33 27 28 32 37 33 32

(a) Using the following line from a random number table, explain and carry out a procedure to select a simple random sample of size 10 from the population above.

77219 48190 20235 26836 23590
44492 14607 09431 75299 42662

(b) Using this sample, and assuming all conditions for inference are met, construct a 90 percent confidence interval for the population mean μ.

(c) The true population mean is 32.44. Is it in your interval, and is this unexpected?

566. The goals scored per game during a 21-game span for each of two hockey teams is shown below.

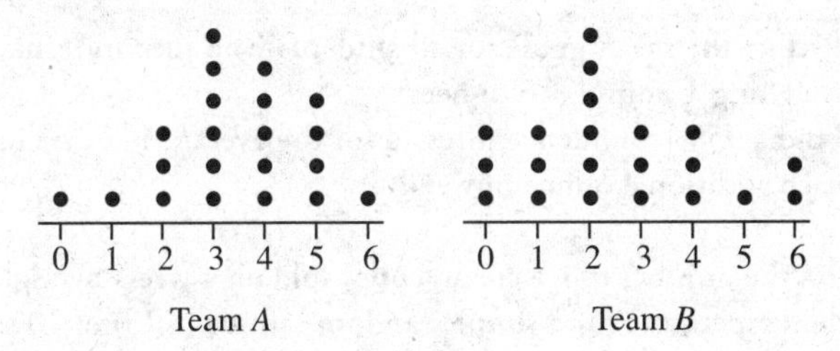

(a) The standard deviation of goals scored for Team *A* is 1.5. Explain what this says about variability.

(b) Looking at the dotplots, what can be said about the difference between goals scored by the two teams? Explain.

(c) A 95% confidence interval estimate for the difference in mean goals scored per game is (–0.17, 1.88). From this calculation, is there evidence of a difference? Explain.

567. In 2015, processed meat was classified as a carcinogen by the World Health Organization (WHO). In a representative sample of eight months, a student counts how many times a month he eats processed meat at school. In an independent representative sample of eight months, he counts how many times he eats processed meat at home. The data are shown in the following table.

School	4	5	5	3	6	4	5	7
Home	4	3	4	4	3	2	5	4

Calculate a 90% confidence interval estimate for the difference between the mean number of times a month this student eats processed meat at school and at home.

568. In a random sample of eight students taking a college writing class, the weights in ounces of their term papers and their resulting grades are tabulated below. Assume all conditions for inference are satisfied.

Weight (oz)	12.1	11.1	11.1	6.5	4.7	10.7	5.9	14.4
Grade	78	79	76	65	67	79	77	94

(a) Predict the mean grade for all students who turn in term papers weighing 1 pound (16 ounces).
(b) Find a 95% confidence interval for the average increase in grade for each additional ounce in weight.

569. Data on the number of ice cream cones sold on a weekday night and outside temperature for a simple random sample (SRS) of 10 summer days are gathered. A scatterplot and regression output follow.

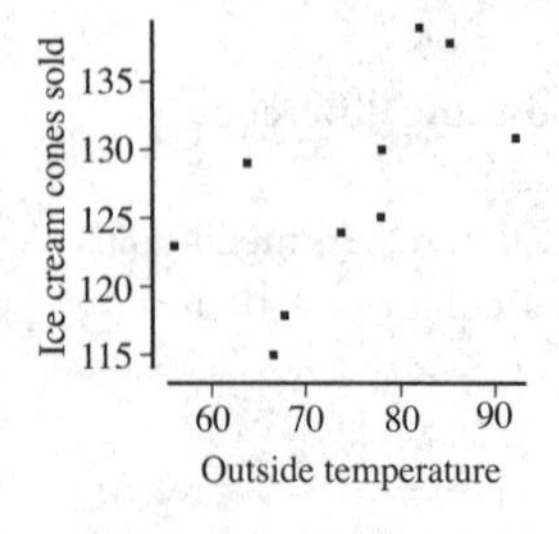


```
Regression Analysis: Cones vs Temperature
Variable   Coeff      s.e. Coeff    t      p
Constant   92.9885    14.8          6.28   0.0002
Temp       0.459939   0.1971        2.33   0.0479
s = 6.382 R-sq= 40.5% R-sq(adj)= 33.1%
Source       df   SS        MS        F
Regression   1    221.783   221.783   5.45
Residual     8    325.817   40.7272
```

(a) From the above output, what is the equation of the regression line?
(b) Interpret the slope of the regression line in context.
(c) Assuming all conditions for inference are satisfied, compute the margin of error for a 95% confidence interval for the slope.

570. A dispatcher for a trucking firm believes the greater the number of hours slept before a trucker begins a long haul, the higher the trucker will score on an alertness test required at the halfway distance. The dispatcher looks up the records of a random sample of 18 long-haul truckers and has them fill out a questionnaire about hours of sleep. Some computer graphical output is shown below.

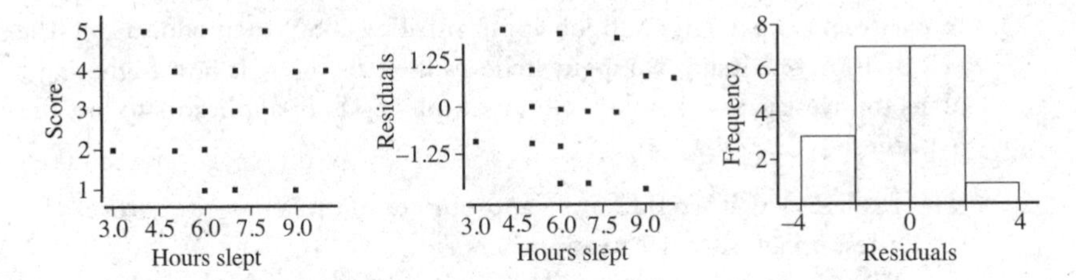

(a) Is this an observational study or an experiment? Explain.
(b) Comment on the design of the study.
(c) The dispatcher took a statistics course long ago and plans to do a test of significance for evidence of an association between score and hours sleep. Comment on whether the necessary assumptions are met for inference on slope for a least squares regression line.

*571. A random sample of adults with various THC levels (a urine test for cannabis) were given a concentration test consisting of a series of small objects, each of which could be fit into an appropriate shaped hole. Each person had five minutes to find the proper holes for as many objects as possible. Computer regression output of number of objects successfully placed plotted against urine THC level (100 ng/mL) is as follows.

```
Dependent variable is: Objects
Variable    Coefficient   s.e.      t      P
Constant    20.0499       0.969     20.7   0.0001
THC         −0.7809       0.0713    −11    0.0001
R-sq = 93.8%    R-sq(adj) = 93.0%
s = 1.398 with 10 − 2 = 8 degrees of freedom
```

(a) What is the equation of the regression line?
(b) Find a 95% confidence interval for the slope, and interpret in context. Assume all conditions for inference are satisfied.
(c) Find a 95% confidence interval for the y-intercept, and interpret in context. Assume all conditions for inference are satisfied.

572. A D1 university recruiter claims that 10 percent of its baseball players go on to play professionally after graduation. A reporter contacts a simple random sample (SRS) of baseball players who graduated during the past 20 years and finds that only 32 out of 450 went on to play professionally. Is there sufficient evidence to write an article disputing the university's claim? Give statistical justification for your conclusion.

573. In past years, 3 percent of all job applicants lied about their education. The HR division of a major company believes the true figure is now higher and plans to investigate a simple random sample (SRS) of applicants to test the hypothesis.

 (a) Explain why it would not be appropriate to run a one-proportion z-test on an SRS of 150 applicants.
 (b) What is the minimum sample size necessary to run this hypothesis test?
 (c) Suppose the HR division uses your result from part (b) and finds that 16 of the applicants lied about their salary. Is this sufficient evidence to say that the percentage of applicants lying about their salary is now over 3 percent?

574. (a) In a random survey of 500 men who asked a woman's father for permission before asking the woman for her hand in marriage, 143 said the marriage ended in divorce. In a random survey of 500 men who did not first ask a woman's father for permission, 169 of the marriages ended in divorce. Is there convincing statistical evidence that men who ask a woman's father for permission before asking the woman have marriages with lower divorce rates?
 (b) Based on the above study, a marriage counselor tells young men that if they want to lower the probability of divorce, they should ask a father's permission before asking a woman for her hand in marriage. Is the counselor's advice justified based on this study? Explain.

575. It is difficult to distinguish between marshmallows and mushrooms by taste alone if one is not allowed to see or smell. A person claims he can distinguish between these, and the following test is designed. He will be given a sample of each in random order to taste while blindfolded with his nose pinched. This will be repeated 16 times. Let p be the proportion of times the person answers correctly.

 (a) What are the null and alternative hypotheses?
 (b) Suppose he correctly answers in 12 out of the 16 trials. What is the probability of answering exactly 12 of 16 if he is simply guessing?
 (c) What is the P-value if he answers 12 of 16, and interpret this in context.
 (d) Is there sufficient evidence to reject the null hypothesis? Give an answer in context.

576. A company advertises it has a process that can extract a mean of 35 grams of dissolved salts from 1 liter of seawater. A geologist believes the true figure is lower. Using this process, a sample of fifteen 1-liter containers of seawater from 15 random locations yields a mean of 34.82 grams of dissolved salts with a standard deviation of 0.65 grams. Assume the sample distribution is symmetric and unimodal with no outliers.

(a) Is there sufficient evidence for the geologist to dispute the advertisement? Justify your answer.
(b) A large-scale test of a second company's process shows yields of dissolved salts that are roughly normally distributed with a mean of 34.75 grams and a standard deviation of 0.83 grams. What is the probability that using this second process, a 1-liter container of seawater will yield at least 35 grams of dissolved salts?
(c) What is the probability that using this second process on 10 randomly selected 1-liter containers of seawater, at least 2 of them yield at least 35 grams of dissolved salts?

*577. The cash register totals are noted at a random sampling of 100 sales at each of 2 competing hardware stores.

Store	$5	$10	$25	$50	Mean	SD
A	10	35	48	7	$19.50	$11.56
B	40	30	15	15	$16.25	$15.72

(a) Is the mean sale in Store *A* significantly greater than the mean sale in Store *B*? Explain.
(b) Is the proportion of $50 sales in Store *A* significantly less than the proportion of $50 sales in Store *B*? Explain.
(c) Using the above results, what should each hardware store emphasize to its stockholders?

578. To test whether the company's secretaries type letters quicker from listening to someone speak live or from recorded dictation, an office manager randomly selects five secretaries. He instructs them to time themselves (in seconds) typing a long letter while someone reads the letter to them and then time themselves again typing the same letter from a recording of someone reading the letter. The results are summarized in the following table.

Secretary	1	2	3	4	5
Live Dictation Time (in seconds)	241	247	217	255	255
Recorded Dictation Time (in seconds)	239	248	217	257	259

(a) Comment on the design of this experiment.
(b) Is the office manager justified in claiming that recorded dictation leads to slower typing speeds? Support your answer with appropriate statistical evidence.

579. In a random sample of 20 junior boxers, their weights are noted and a histogram and summary statistics are as follows.

$n = 20, \quad \bar{x} = 164.75, \quad s = 5.02$

155 165 175

Their trainer wants all of those over 167 pounds to lose weight.

(a) Is 167 pounds in the 95% confidence interval for the mean weight?
(b) Is there evidence at the 5% significance level that the mean weight is less than 167 pounds? Perform an appropriate hypothesis test.
(c) Is there a contradiction between the answers in (a) and (b) above? Explain.

580. Each of 300 students was randomly (coin toss) placed into either Professor *A*'s or Professor *B*'s section of a required college writing course. At the end of the semester, students gave a score from 1 to 5 with "poor" = 1 and "great" = 5 for their professor. The results are shown in the following table.

Score	1	2	3	4	5
Professor *A* Frequency	16	41	53	38	13
Professor *B* Frequency	45	50	29	11	4

(a) Use a graphical display to compare the scores received. Write a few sentences, based on your graphs, to compare Professor *A* and Professor *B*.
(b) Does there appear to be significantly greater satisfaction with Professor *A* as compared to Professor *B*? Give a statistical justification for your answer.

581. A handyman wishes to test which radial arm saw can cut through lumber more quickly. He picks out a random sampling of five pieces of lumber of various species and thicknesses. The cutting times (in seconds) are summarized in the following table.

Board	***A***	***B***	***C***	***D***	***E***
Saw *S*	3.1	6.8	1.5	4.2	2.8
Saw *T*	3.3	6.7	1.8	4.5	3.0

(a) What is the mean cutting time for each saw?
(b) The handyman performs a two-sample t-test with H_0: $\mu_S = \mu_T$ and H_a: $\mu_S \neq \mu_T$, which results in a *P*-value of 0.886. He concludes that there is not significant evidence of a difference in cutting times. Comment on his choice of test.
(c) Perform the proper test to decide if there is significant evidence of a difference in cutting times. Use a significance level of 0.10.

582. At schools using an innovative math program, a simple random sample (SRS) of 100 students results in an average score of 178 with a standard deviation of 27 on a state test. At schools using a traditional approach, an SRS of 150 students results in an average score of 171 with a standard deviation of 31 on the same state test.

(a) Is there evidence that students using the innovative approach have a higher average score than students using the traditional approach? Give statistical justification for your answer.
(b) Suppose a study using this design resulted in a *P*-value less than 0.01. Would it be reasonable for all school boards to push for adoption of the innovative approach? Explain.
(c) Assuming standard deviations of 27 and 31 as listed above, how large a sample (same number for both) should be used to be 95 percent sure of knowing the difference in scores to within 5 points?

*583. An athletic trainer wishes to determine if a newly designed gym shoe enhances athletic performance. Ten professional high jumpers competitively jump on two successive days. For each jumper, a coin toss determines whether he uses his old gym shoes or the new pair on the first day. Each jumper then switches shoes on the next day. Their jump heights (feet) are as follows.

Runner	1	2	3	4	5	6	7	8	9	10
Jump Heights With New Shoes (ft)	7.7	7.3	7.4	7.9	7.7	7.3	7.6	8.2	7.2	7.3
Jump Heights With Old Shoes (ft)	7.7	7.1	7.3	7.6	7.3	7.5	7.7	7.9	7.1	7.3

(a) To determine if there is sufficient statistical evidence on whether professional high jumpers improve their jumps when wearing the new shoes, what statistical test should be used, what are the hypotheses, and are the conditions for inference met? (You are not asked to perform the test.)
(b) Does knowing a jumper's jump height with the old shoes help predict his jump height using the new shoes? Perform an appropriate statistical test.

*584. In a baseball game, the slugging average is the mean number of bases per hit. It can be calculated using the following formula.

$$\text{Slugging average} = \frac{(\text{Singles}) + (\text{Doubles} \times 2) + (\text{Triples} \times 3) + (\text{Home runs} \times 4)}{\text{Total at bats}}$$

In a random sampling of 75 hits from each of two major league teams, the distributions of singles (1B), doubles (2B), triples (3B), and homers (HR) are shown in the following table.

	1B	2B	3B	HR
Team *A*	44	17	10	4
Team *B*	34	22	12	7

Is there evidence that the two teams have a different slugging average? Give statistical justification for your answer.

585. Can a particular video game improve a batter's reaction time? Batters' reaction times (fraction of a second between the ball leaving a pitcher's hand and the start of a swing) are measured before and after playing the video game for 25 hours.

(a) What is the appropriate hypothesis test, the hypotheses, and the conditions to check?

(b) Suppose the test is run and no statistically significant improvement is detected in batter reaction times after the video game training. If the researcher plans a second test, name two specific changes that can be made to increase the power of the test. Explain your choices.

586. A study is performed on 20 different models of luxury SUVs to see if a new gas additive is effective in reducing carbon dioxide (CO_2) emissions. Data on CO_2 emissions (pounds per gallon of gas) are collected before and after input of the additive. A partial computer output follows.

	N	Mean	StDev	SE Mean
Before	20	20.60	7.35	1.64
After	20	20.05	6.84	1.53
Difference	20	0.55	1.90	0.42

(a) Assuming that graphical displays indicate that assumptions of normality are reasonable, construct a 95% confidence interval for the mean difference in CO_2 emissions before and after the additive. Interpret your interval in context.
(b) Does this confidence interval give sufficient evidence that this additive reduces mean CO_2 emissions for luxury SUVs?

587. A World Health Organization (WHO) report gives average life expectancies for 10 randomly chosen countries in each of two regions of the world:

Region 1: 63, 72, 48, 67, 68, 70, 59, 51, 63, 52
Region 2: 78, 73, 57, 69, 74, 86, 70, 66, 91, 77

(a) Create a back-to-back stemplot to display this data.
(b) Compare the two distributions.
(c) A two-sample *t*-test was performed on the above data with H_0: $\mu_1 - \mu_2 = 0$ and H_a: $\mu_1 - \mu_2 < 0$. Conditions for inference were met, the test statistic was $t = -3.134$, and the *P*-value was 0.0029. Write a proper conclusion.

*588. A random sample of five men were measured for waist sizes (in inches) and percent body fat, giving the following data.

Waist (inches)	32	35	36	39	44
Body Fat (%)	12	23	24	30	40

(a) Is there evidence of a positive linear relationship? You may refer to the following computer regression printouts.

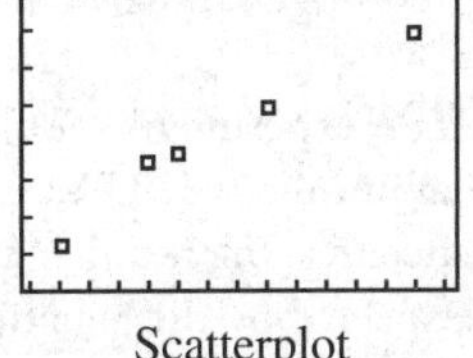

Scatterplot

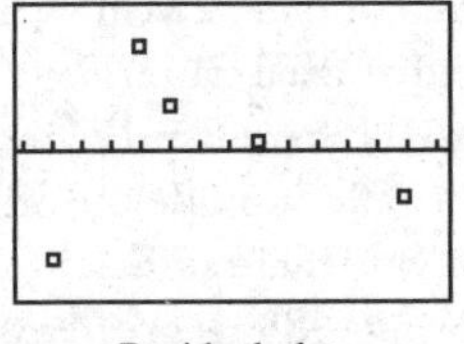

Residual plot

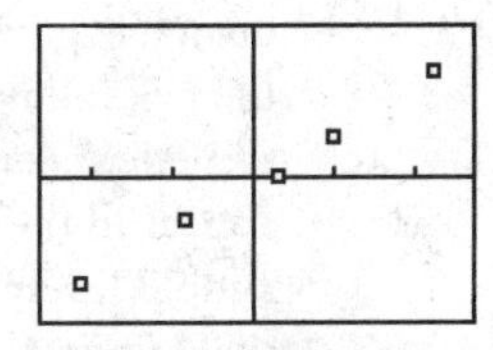

Normal probability plot of residuals

(b) Predict the percent body fat for a man with a 37-inch waist.
(c) Give an approximate range for percent body fat for a man with a 37-inch waist.

589. Doctors are planning a clinical trial of a new "molecularly targeted therapy" for the treatment of leukemia. A random sample of children with leukemia will be treated. After a suitable time period, their disease remission rate will be compared with children being treated with standard chemotherapy for leukemia.

(a) What are the null and alternative hypotheses for this significance test?
(b) What is a Type I error in this context, and what is the consequence of committing this error?
(c) What is a Type II error in this context, and what is the consequence of committing this error?

590. Executives at a company that manufactures a cooler are thinking of moving to the manufacture of a better cooler but only if a test shows that the new cooler will keep drinks cold for over 24 hours. The company's research department proposes a study with H_0: $\mu = 24$ and H_a: $\mu > 24$.

(a) Describe a Type II error in context of this study, and describe a possible consequence.
(b) Which significance level, $\alpha = 0.05$ or $\alpha = 0.10$, would result in a smaller probability of a Type II error? Explain.
(c) The statistician in the research department estimates the power of the proposed test to be 0.85 if the true mean is $\mu = 25$ hours. What does this mean in the context of this study?

591. In each of the following examples, explain why it is or is not proper to run a chi-square goodness-of-fit test.

(a) The ingredient list on a jar of fancy mixed nuts indicates 10% Brazil nuts, 10% cashews, 20% almonds, and the rest peanuts. Suppose you separate and weigh the contents of a 24-ounce jar and find 3 ounces of Brazil nuts, 2 ounces of cashews, 4 ounces of almonds, and 15 ounces of peanuts. Does the jar follow the advertised pattern?
(b) A company website claims that in a bag of its large candies, 10% are red, 20% are blue, 40% are yellow, and 30% are orange. Suppose you open a bag with 3 red, 4 blue, 2 yellow, and 10 orange candies. Does the bag follow the advertised pattern?
(c) Genetic theory predicts that offspring in a certain experiment should appear in the ratio 1:3:9. Suppose the offspring number 66, 175, and 980, respectively. Is there sufficient evidence that the number of the dominant offspring (980) is greater than expected by the genetic model?

592. In a survey of all 880 students at a college, the number of times during the academic semester in which a student pulled an all-nighter is tabulated and shown in the following table.

All-Nighters Pulled	0	1	2	3	4	5
Number of Students	280	210	175	150	50	15

(a) What are μ and σ for all-nighters pulled during a semester for this population?
(b) If a simple random sample (SRS) of size $n = 50$ is taken from this population, describe the sampling distribution of $\bar{x}$.
(c) A college counselor believes that over the years, the number of all-nighters pulled by students during a semester are 0, 1, 2, 3, 4, 5 in the ratio 15:10:10:10:4:1. Perform an appropriate hypothesis test.

593. A random survey of 250 high school teachers classified job satisfaction versus subject taught with the following results.

	Happiness Level		
Subject Taught	**Unhappy**	**Mildly Happy**	**Very Happy**
Math	10	25	30
English	20	55	45
History	20	25	20

(a) What is the probability that a randomly selected teacher from this sample teaches history and is very happy?
(b) What is the probability that a person selected at random from this sample is mildly happy given that he/she teaches math?
(c) What is the probability that a person selected at random from this sample teaches English given that he/she is unhappy?
(d) Is there evidence of a relationship between happiness level and subject taught? Explain.

594. Random samples of high school students in 2014, 2016, and 2018 were anonymously surveyed about whether or not they had ever cheated on a quiz or test. The resulting counts are as shown in the table.

	2014	**2016**	**2018**
Admitted to Cheating	78	67	51
Claimed to Never Having Cheated	44	40	59

What does a chi-square test of homogeneity indicate?

595. During the past year, the proportions of teens eating at Starbucks, Chick-fil-A, and Chipotle have been 0.3, 0.5, and 0.2, respectively. A random survey of 250 teens this past month indicates 71, 136, and 43 visits to Starbucks, Chick-fil-A, and Chipotle, respectively.

(a) Is there evidence that the overall pattern of restaurant visits has changed? Give statistical justification for your answer.
(b) A sales manager at Starbucks would like to know her company's proportion of this past month's restaurant visits with 95% confidence. What is the margin of error?
(c) A sales manager at Chick-fil-A would like to know his company's proportion of last month's restaurant visits within a margin of error of ±0.03. What would be the underlying level of confidence?

596. A social studies teacher believes that high school students taking social studies classes are more likely than students taking other classes to read newspapers. She randomly picks 40 of her social studies students and 50 students from her other classes, questions them, and gathers the following data.

	Read Newspapers	**Don't Read Newspapers**
Her Social Studies Students	37	13
Her Other Students	15	35

(a) Are the intended and sampled populations the same or different, and how does this affect the study?
(b) Discuss the appropriateness of using a chi-square test of homogeneity.
(c) Discuss the appropriateness of using a *z*-test for differences of proportions.
(d) Which would be most appropriate for a quick impression of the data: segmented bar charts, back-to-back stemplots, or parallel boxplots? Draw this visual display.

597. A random sample of 500 medical records of people who recently had major heart attacks is analyzed for numbers of died versus survived and for education level of the subjects. Some computer output is as follows.

Observed (and Expected) Counts

	No High School Degree	**High School Degree**	**College Degree**
Died	18 (9.9)	38 (42.1)	24 (28)
Survived	44 (52.1)	225 (220.9)	151 (147)

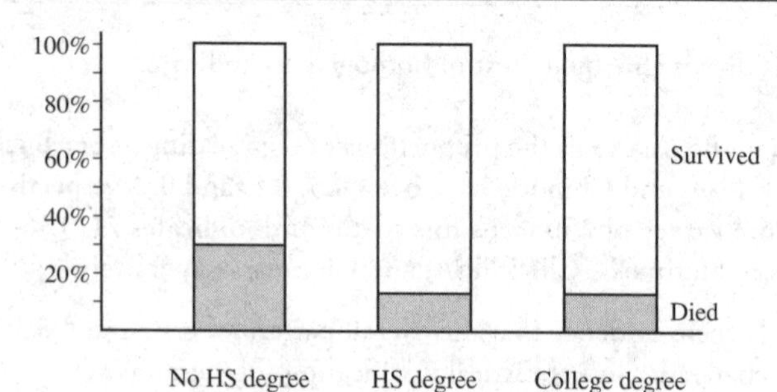

(a) Is there an association between whether or not people survive major heart attacks and their education levels? Do an appropriate statistical test.
(b) Do the segmented bar charts give any additional insights into this study? Explain.

598. A random sampling of a professor's grades (A or B, C or D, F) in both intro and advanced courses yields the following data.

	A, B	C, D	F
Intro	35	55	10
Advanced	22	15	3

(a) A chi-square test of independence gives $\chi^2 = 4.749$. What is a proper conclusion?

(b) Is there evidence that this professor gives a higher proportion of A, B grades in his advanced courses than in his intro courses? Give statistical justification for your answer.

*599. 500 patients took part in a clinical study using T-cell therapy to treat a particular lymphatic cancer. The following data show the number of these patients who survived for each given number of years.

Survival Years	0	1	2	3	4 or more
Observed Number of Patients	69	106	153	98	74

It is hypothesized that the distribution follows the Poisson distribution:

$P(x) = \frac{2^x}{x!}e^{-2}$, where $e = 2.71828$ and x is the number of survival years.

(a) Using this Poisson distribution, calculate the probabilities of 0, 1, 2, 3, and 4 or more survival years.

(b) Using this Poisson distribution, calculate the expected number of patients out of 500 who will survive 0, 1, 2, 3, and 4 or more years.

(c) Perform a goodness-of-fit test for this data and the given Poisson distribution.

*600. In an educational study, there is interest in whether or not there is more variability in the mean verbal scores of 8-year-olds or of 3-year-olds. Summary statistics are given below.

	Sample Size	Mean	StDev
Age 8	60	93.5	5.4
Age 3	60	78.6	2.7

To test H_0: $\sigma_8 = \sigma_3$ versus H_a: $\sigma_8 > \sigma_3$, the researchers will use the ratio of the sample standard deviations, $\frac{s_8}{s_3}$, as a test statistic.

(a) What is the observed test statistic from the data of this study?

(b) A simulation was conducted to study the sampling distribution of this test statistic. For each repetition of the simulation, two random samples were generated using appropriate population means with identical standard deviations, and the ratio of the sample standard deviations was noted. A histogram of the simulated ratios is given below.

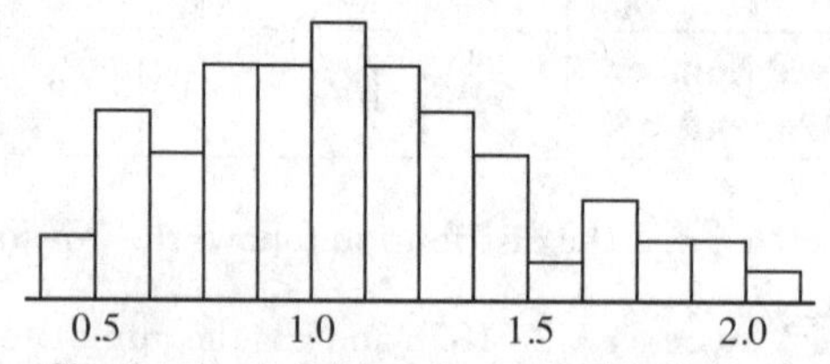

Is there statistically significant evidence that there is more variability in the mean verbal scores of 8-year-olds than of 3-year-olds?

ANSWERS

MULTIPLE-CHOICE

MC EXPLORATORY ANALYSIS

One-Variable Data Analysis (pages 3–27)

1. **(A)** Since the dotplot is symmetric, *X* must be right at the mean. So *X* has a *z*-score of 0. Seven values are smaller than *X*. Boxplots, if they show any isolated points, show only outliers that are far from the mean. There is only one set of numbers, so using a back-to-back stemplot makes no sense here.

2. **(D)** The value 50 roughly seems to split the area under the histogram in two, so the median is about 50. The area between 0 and 50 is split in two somewhere between 30 and 40, so Q_1 is between 30 and 40. Finally, boxplot (D) seems to pick up the strong left skew the best.

3. **(C)** These segmented bar charts only show relative frequencies, not actual numbers. So although relative numbers can be compared within a grade level, only percents—not actual numbers—can be compared across grade levels.

4. **(C)** Stemplots are too unwieldy to be used for very large data sets and are not used for categorical data sets. Stems should never be skipped, even if there is no data value for a particular stem. A key explaining what the stem and leaves represent should always be provided.

5. **(A)** In general, histograms give information about relative frequencies, not actual frequencies. Histograms may have vertical axes labeled with frequencies. In those situations, the height, not the relative height, gives the frequencies.

6. **(A)** The median is the middle value when the values are arranged in ascending or descending order. Note there a total of 19 values.

7. **(D)** Increasing each score by 50 increases the mean to 1,050 and leaves the standard deviation unchanged at 200. Then increasing each result by 5 percent increases both the mean and standard deviation by 5 percent to $1{,}050 + 0.05(1{,}050) = 1{,}102.5$ and $200 + 0.05(200) = 210$.

8. **(E)** Histograms refer to numerical (quantitative) data, not categorical (qualitative) data.

9. **(D)** Because of the squaring operation in the definition of standard deviation, the standard deviation (and also the variance) can be zero only if all the values in the set are equal.

10. **(B)** From the boxplot, we see that the median is 10 and $Q_3 = 50$. The only histogram that looks to have 50 percent of the area on each side of 10 and also have 25 percent of the area between 50 and 60 is (B).

11. **(A)** Fifty percent of the data are on either side of the median. The interquartile range gives the distance between Q_1 and Q_3 but doesn't say where the median falls between Q_1 and Q_3. For example, it is possible that $Q_1 = 345$ and $Q_3 = 395$. Depending on the shape of the distribution, the mean could be anything.

12. **(C)** The sum of the 25 salaries is 15(90,000) + 10(60,000) = 1,950,000. So the mean is 1,950,000 ÷ 25 = 78,000.

13. **(A)** Choice of interval width, and therefore number of bins, can dramatically change the appearance of a histogram. Using histograms, stemplots, and boxplots makes no sense with categorical variables. All graph axes should be labeled. Displaying outliers is more problematic with histograms depending on bin width. Histograms do not show individual observations.

14. **(E)** Both sets are symmetric around 10, and so both have means of 10. Set *A* is more spread out than set *B*, and so set *A* has the greater variance. For bell-shaped data, about 68% of the values fall within one standard deviation of the mean, 95% within two standard deviations, and 99.7% within three standard deviations. Thus, the standard deviation of set *B* appears to be approximately 2 and is clearly less than 4.

15. **(E)** The minimum of the combined set of scores must be the minimum of the men's scores since it is lower. The maximum of the combined set of scores must be the maximum of the women's scores since it is higher. The first quartile must be the same as the identical first quartiles of the two original distributions. There are no outliers, which are scores more than $1.5 \times IQR$ from the first and third quartiles.

16. **(B)** The distribution of heights in League *A* has a mean of 72 and a range of 4, while the distribution of heights in League *B* has a mean of 74 and a range of 4.

17. **(C)** The ranges are 20 – 5 = 15 and 40 – 25 = 15. The interquartile ranges are 15 – 10 = 5 and 35 – 30 = 5. There is no such thing as being skewed to both sides. The boxplot shows intervals in which 25 percent of the data fall, but many different distribution shapes may have the same boxplot. Boxplots say nothing about actual numbers of values.

18. **(E)** The *z*-score gives the number of standard deviations from the mean. The two exam results were 80 – 1.4(5) = 73 and 80 + 2.6(5) = 93. The difference is 93 – 73 = 20. You can also note that the two results are 2.6 – (–1.4) = 4 standard deviations apart, which is 4(5) = 20.

19. **(B)** The "Male" distribution is roughly symmetric, while the "Female" distribution is skewed left, not right. The medians are 5 and 2, respectively. The distribution with skew indicates the mean is less than the median. In the roughly symmetric distribution, though, the mean and median are close. Thus, the means are closer. Both have ranges of 5, not 6. The "Male" distribution is clustered more tightly around its mean, so it has a smaller standard deviation. Combining the male and female numbers into one set will result in a new set with min 0, max 6, and a range of 6 that is not equal to 5 + 5.

20. **(D)** A scatterplot is used to study an *association* between two quantitative variables. The variables are not associated.

21. **(A)** Note that adding 10 to every value in set *A* results in set *B*. Therefore, the means differ by 10 but the measures of variability remain the same.

22. **(D)** The range is a single number, the largest value minus the smallest value. So the range is the same no matter how the data points in the set are arranged. Outliers are extreme values. Although they do affect the range, they do not affect the interquartile range. (We say that the range is *sensitive* to extreme values while the IQR is *resistant* to extreme values.) Remember that $IQR = Q_3 - Q_1$, so the middle half of the data is between the quartiles Q_1 and Q_3. The sample comes from the population. So the smallest value in the sample cannot be smaller than the smallest value in the population. Similarly, the largest value in the sample cannot be larger than the largest value in the population.

23. **(E)** There is no comparison between the midterm and the final exam. The *z*-score gives the number of standard deviations from the mean, which is the number of standard deviations from the class average for the final exam.

24. **(A)** The empirical rule applies to bell-shaped data, which set *B* clearly is not. Both sets have ranges equal to or close to 40. For bell-shaped data, roughly 95 percent of the values fall within two standard deviations of the mean. However, although distribution *A* is roughly bell-shaped, less than 95 percent of the data are between 80 and 100 (two standard deviations from the mean if the standard deviation were only 5). So the standard deviation in set *A* is greater than 5. In set *B*, the data generally are even further from the mean than in set *A*. So the standard deviation in set *B* is even greater.

25. **(D)** Going over from 0.5 on the *y*-axis, we see that the median is about 70. The cumulative frequency plot rises very slowly at first, and thus the distribution is skewed to the left (toward the lower values). With this skew, the mean is less than the median.

26. **(D)** The interquartile range is $IQR = Q_3 - Q_1 = 139 - 126 = 13$. Outliers are any values less than $Q_1 - 1.5(IQR) = 106.5$ or greater than $Q_3 + 1.5(IQR) = 158.5$. Look at the minimum and maximum values. We

see that there are no outliers on the lower end and at least one outlier on the upper end.

27. **(A)** The empirical rule (which applies to bell-shaped data) gives that 99.7% of the data should fall within three standard deviations of the mean.

28. **(E)** For bell-shaped distributions, roughly 16 percent of the data falls more than one standard deviation above the mean. So a *z*-score of 1 has a percentile rank of about 84 percent. The third quartile has a percentile rank of 75 percent.

29. **(B)** Increasing every value by 5 percent will also increase Q_1, Q_3, and IQR as well as both $Q_1 - 1.5(\text{IQR})$ and $Q_3 + 1.5(\text{IQR})$ by the same 5 percent.

30. **(D)** Mean tip = 5 + 0.10(55.30) = 10.53 and standard deviation = 0.10(4.00) = 0.40.

31. **(A)** The *z*-score of the female is $\frac{500 - 515}{20} = -0.75$. The weight of a male with the same *z*-score is 720 – 0.75(32) = 696.

32. **(E)** The median and interquartile range are not affected by exactly how large the larger values are or by exactly how small the smaller values are. For example, suppose four of the mice succeed in under 15 minutes while the fifth mouse takes hours. In that situation, the mean is not very representative of a typical time and the standard deviation and variance are not very helpful in describing the spread.

33. **(C)** Symmetric histograms can have any number of peaks. All normal curves are bell-shaped and symmetric, but not all symmetric bell-shaped curves are normal. Even if the mean and median are equal, the shapes of the lower and upper halves can be very different.

34. **(C)** With 343 values in the set, the median is the 172nd value from either end when the values are in order. With 100 ages over 18 and 193 ages over 17, it is seen that the 172nd age must be between 17 and 18.

35. **(A)** The interquartile range IQR = $Q_3 - Q_1$ = 50 – 40 = 10. Outliers are values greater than $Q_3 + 1.5(\text{IQR})$ = 50 + 15 = 65 or less than $Q_1 - 1.5(\text{IQR})$ = 40 – 15 = 25. In this case, there are no values above 50, but there are values below 25.

36. **(B)** Although uniform and bell-shaped distributions are symmetric, symmetric distributions do not have to be either of these. Symmetric distributions cannot be skewed. Both the mean and median are at the center of symmetric distributions. However, there is nothing special about the measures of spread, such as the IQR, range, or standard deviation, of symmetric distributions.

37. **(B)** $Q_1 = 30$, $Q_3 = 40$, and IQR = 40 – 30 = 10. Outliers are values less than Q_1 – 1.5(IQR) = 15 or greater than Q_3 + 1.5(IQR) = 55.

38. **(D)** With this categorical data, the only reasonable display is a segmented bar chart.

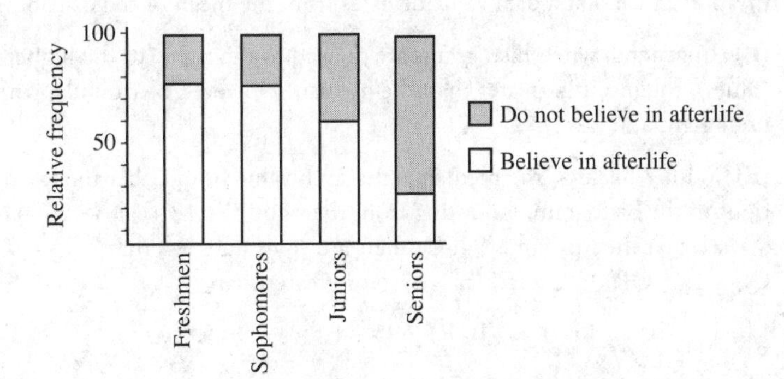

39. **(E)** The median is somewhere between 25 and 45 but not necessarily at 35. Even a single very large value can result in a mean over 45 and a standard deviation over 20. If the lower quarter of the values are all 25 and the upper quarter of values are all 45, the range would equal the interquartile range.

40. **(A)** When the two extreme values are removed, one from each side, the standard deviation will decrease while the median, which is the middle value, will remain the same. It is *possible* that the mean will remain unchanged.

41. **(E)** The median score splits the area in half, so it is not 75. The area between 50 and 60 is more than the area between 80 and 100 but less than the area between 60 and 100. The same percentage of the data is above and below the median. With data skewed to the right, the mean is usually greater than the median.

42. **(E)** Boxplots give percentages of data in specific intervals, not actual numbers.

43. **(B)** Outliers are values less than Q_1 – 1.5(IQR) and greater than Q_3 + 1.5(IQR). First calculate the IQR = 21.5 – 15.5 = 6. The outliers are values below 15.5 – 1.5(6) = 15.5 – 9 = 6.5 and greater than 21.5 + 1.5(6) = 21.5 + 9 = 30.5.

44. **(D)** If the variance of a set is zero, all the values in the set are equal. All symmetric distributions have equal mean and median, but the variance could be anything.

45. **(D)** Stemplots and histograms can show gaps and clusters that are hidden when one simply looks at calculations such as mean, median, standard deviation, quartiles, and extremes.

46. **(B)** The standard deviation can roughly be thought of as the typical or mean distance that a data value deviates from the mean of the distribution.

47. **(B)** In general, when histograms are skewed to the right (to the higher values), the mean is greater than the median. However, be careful as this is not a hard and fast rule.

48. **(E)** With 77 values, the median is the 39th value from either the left or the right in the histogram. Counting from the right, $7 + 12 + 10 + 12 = 41$. We see that the median is 94. Outliers are any values less than $Q_1 - 1.5(IQR) = 92 - 1.5(3) = 87.5$ or greater than

$$Q_3 + 1.5(IQR) = 95 + 1.5(3) = 99.5.$$

The exam scores of 85 and 87 are outliers.

49. **(A)** The median from School *A* looks to be 5; the median from School *B* looks to be 0. The skewness of each histogram results in a mean for School *A* less than 5 and a mean for School *B* greater than 0. However given the scales, the mean for School *A* is still greater than the mean for School *B*.

50. **(A)** The *z*-score itself has no units and does not change with a change in units for the data.

51. **(C)** When the values are spread far and thinly toward the higher values, the distribution is called skewed right. There is no such thing as being skewed both left and right.

52. **(E)** The empirical rule applies only to bell-shaped distributions as in set *B*, not to skewed distributions as in set *A*. The median is not the average of the minimum and maximum values. The range of set *A* is $553 - 502 = 51$, while the range of set *B* is $554 - 502 = 52$. The variance is a measure of squared deviations from the mean. Set *B* is more concentrated around its mean, so it has a smaller variance. Set *B* appears roughly symmetric, indicating the mean is about the same as the median. In contrast, set *A* is skewed to the higher values, indicating that the mean is greater than the median.

53. **(E)** The minimum top speed is somewhere between 30 and 40 mph, and the maximum top speed is somewhere between 90 and 100 mph. The median is the point that cuts the area under the histogram in half. Clearly there is less area to the left of 60.5 and more to the right. This histogram shows only relative frequencies, not actual frequencies, so there is no indication of actual numbers of coasters, such as 100. To compare the numbers of coasters between different speed intervals, one looks at areas under the his-

togram. There is less area between 70 and 80 than between 80 and 100, and there is the same area between 50 and 60 as between 80 and 90.

54. **(A)** Adding a constant to every value will increase the minimum, Q_1, median, Q_3, and maximum by the same constant but will leave measures of variability, like the interquartile range and the standard deviation, unchanged.

55. **(E)** Many data sets can have the same histogram depending on widths of the bins and exactly where values fall within each bin. Although the medians of both sets are in the interval from 25 to 30, there is no way of knowing exactly where in that interval the medians are. Similarly, the minimums of both sets are somewhere between 5 and 10, and the maximums are somewhere between 45 and 50. The values in data set *S* are clearly spread out further from the mean than is true for data set *T*, so data set *S* has the greater standard deviation.

56. **(E)** The given segmented bar charts show percentages, not actual numbers. Although *percentages* can be compared, there is no way to compare *numbers* among the different years without further information.

57. **(A)** Standardizing scores, that is, changing to *z*-scores, always results in a distribution with mean 0 and standard deviation 1.

58. **(D)** For bell-shaped distributions, almost all values are within three deviations of the mean. The mean here appears to be around 1.16, and all the data are between 0.92 and 1.40. An estimate of three standard deviations is $1.40 - 1.16 = 0.24$ or $1.16 - 0.92 = 0.24$. So an estimate for one standard deviation is one-third of 0.24, which is 0.08.

59. **(E)** The class with more students has a greater impact on the overall average than the class with fewer students. So the overall average is closer to that of the second (larger) class. This can be shown mathematically. Let n be the number of students in the first (smaller) class. So the number of students in the larger class is $n + p$ for some positive integer p. The sum of all the scores in the first class is $3.2n$, the sum of all the scores in the second class is $3.8(n + p)$, and the sum of all the scores is $3.2n + 3.8(n + p) = 7n + 3.8p$. The total number of students is $n + (n + p) = 2n + p$. So the mean score of all the students is $\dfrac{7n + 3.8p}{2n + p} > \dfrac{7n + 3.8p}{2n} = 3.5 + 1.9\dfrac{p}{n} > 3.5$.

60. **(B)** Distribution *L* is roughly symmetric, so its mean is approximately equal to its median. Therefore, roughly 50 percent of the data is below the mean. Distribution *M* is skewed right, so its mean is greater than its median. Therefore, more than 50 percent of the data is below the mean. Distribution *N* is skewed left, so its mean is less than its median. Therefore, less than 50 percent of the data is below the mean.

61. **(E)** The median, not the mean, is at the 50 percent division. 25 percent of the mpg values are below 21, which is the first quartile. Although it is difficult to determine shape from such limited statistics, these seem to indicate right skew if anything. IQR = 30 – 21 = 9. The maximum value of 48.5 is an outlier because it is greater than $Q_3 + 1.5(IQR) = 30 + 1.5(9) = 43.5$.

62. **(D)** For two independent variables *X* and *Y*,

$$\mu_{X+Y} = \mu_X + \mu_Y = 1{,}100 + 950 = 2{,}050 \text{ and}$$

$$\sigma_{X+Y} = \sqrt{\sigma_X^2 + \sigma_Y^2} = \sqrt{290^2 + 250^2} = 383.$$

63. **(C)** All three data sets have the same range: 22 – 2 =20. The interquartile ranges (IRQs) of the first and third sets are 6, while the IQR of the middle set is 4. The medians are 12, 14, and 16, respectively. There is no way of concluding "normal" or even "roughly normal" from a boxplot. Although it is not unreasonable to assume that the middle sample came from a roughly normal population, the outliers in the other two data sets make this assumption unreasonable.

64. **(E)** There will be clusters around 8 (finished middle school), 12 (finished high school), and 16 (finished college) years of schooling. There are very few normal distributions of individual observations and no reason to think that this is a roughly normal distribution. (Normal distributions in statistical inference mainly arise in sampling distributions.) There are small segments of the population with very few years of schooling, and so the distribution could be skewed to the left.

65. **(E)** The mean, standard deviation, variance, and range are all affected by outliers; the median and interquartile range are not.

66. **(D)** Many data sets can have exactly the same boxplot, so there is no way it can be said that a distribution is roughly normal or symmetric from looking at a boxplot. The number of earthquakes with magnitudes below the first quartile, 6, is the same as that between 6 and the median, 6.75. Both intervals have 25 percent of the data. Exactly half the data is between the first and third quartiles, 6 and 7.5. There were the same number of earthquakes between the median and the third quartile magnitudes as there were between the third quartile and the maximum magnitudes, both containing 25 percent of the data.

67. **(E)** There is only a very small overlap (71 to 72 inches) between the male and female heights. So the combined data will roughly be two clusters that barely intersect in the middle, with each cluster looking roughly normal.

68. **(D)** Whether or not males have an overall higher mean salary than females depends on the numbers! For example, if there were 10 male senior executives all making \$93,000, a single female senior executive making \$95,000,

a senior male junior executive making \$45,000, and 10 female junior executives all making \$48,000, the overall mean salary of the 11 male executives would be much greater than the overall mean salary of the 11 female executives, \$88,636 > \$52,273.

Two-Variable Data Analysis (pages 27–48)

69. **(A)** The equation is $\hat{y} = a + bx$, where the slope is

$$b = r\frac{s_y}{s_x} = 0.6\left(\frac{10}{40}\right) = 0.15.$$

One point on the line is $(\bar{x}, \bar{y}) = (120, 30)$, so $a = 30 - 0.15(120) = 12$.

70. **(E)** On a scatterplot, all the points would lie perfectly on a straight line sloping up to the right. So $r = 1$.

71. **(C)** The coefficient of determination, r^2, gives the proportion of the y-variance that is predictable from the linear regression model. In this case, $r^2 = (0.18)^2 = 0.0324$.

72. **(B)** The predicted winning percentage is $39 + 0.00025(75{,}000) = 57.75$. The residual is actual minus predicted, which is $55 - 57.75 = -2.75$.

73. **(D)** The slope and correlation are related by $b = r\frac{s_y}{s_x}$, so b and r have the same sign. The correlation is always between -1 and $+1$, but the slope can be any value. Choice (C) is the definition of the least square regression line. Correlation is the same no matter which variable is called x and which is called y.

74. **(D)** Least squares lines should be interpreted in the range of given values of the explanatory variable; 70 years' experience is an unreasonable extrapolation. 37.15 represents a mean value for the y-variable given $x = 0$, so starting salaries *average* \$37,150. The slope gives an average change in the y-variable for a 1 unit change in the x-variable. Regression shows relationships, not causation. There is no way to infer the correlation from the equation of the line.

75. **(A)** The y-intercept of the regression line is about 5, ruling out choice (B). The slope is negative ruling out choices (C) and (E). The correlation looks to be much closer to $-\sqrt{0.765}$ rather than $-\sqrt{0.365}$, making choice (A) the answer.

76. **(E)** A positive correlation shows a tendency for higher values on one variable to be associated with higher values of the other. However, given any two points, anything is possible.

77. **(D)** The correlation is always between –1 and +1 no matter if the association is linear or nonlinear.

78. **(C)** The regression equation gives predicted values for averages; –4.78869 is the slope. The slope gives the predicted change, on average, of the dependent variable (value in \$1,000 in this problem) for a one-unit change in the independent variable (age in this problem).

79. **(B)** The only scatterplot in which the residuals (actual minus predicted) go from positive to negative and back to positive is choice (B).

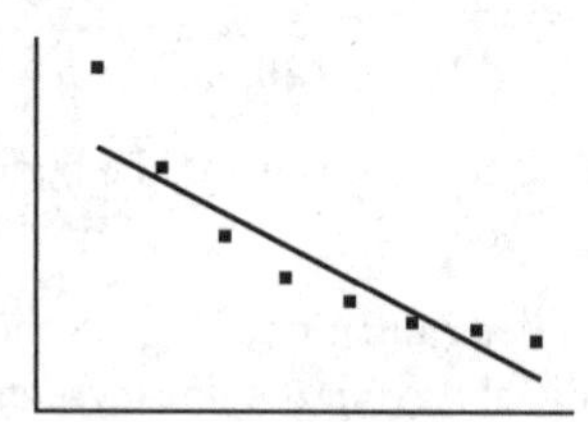

80. **(D)** There is a moderate positive slope to the data showing a moderate positive correlation. Nine executives started with salaries under \$50,000, but only six were making under \$50,000 after 5 years. Both initially and after 5 years, there were two executives who made \$100,000. Of the nine executives who initially made under \$50,000, only four were still making under \$50,000 after five years. There were at least five executives who clearly made less after 5 years than they did initially.

81. **(D)** The correlation is not changed by adding the same number to each value of one of the variables or by multiplying each value of one of the variables by the same positive number.

82. **(B)** Influential points are points whose presence or absence sharply affect the regression line. An influential point may have a small residual but still have a greater effect on the regression line than points with possibly larger residuals but average *x*-values.

83. **(E)** Reversing the dependent and independent variables leaves the correlation unchanged. We are given $r^2 = 0.465$ and a negative of –2.71, so $r = -\sqrt{0.465} = -0.682$.

84. **(D)** Since (1, 7) is on the line $\hat{y} = a + 5x$, we have $7 = a + 5(1)$ and $a = 2$. Thus, the regression line is $\hat{y} = 2 + 5x$. The point $(\bar{x}, \bar{y})$ is always on the regression line. So we have $\bar{y} = 2 + 5\bar{x}$.

85. **(E)** The regression line is $\widehat{\text{Students}} = 11 + 13.9286(\text{Year})$, where Year is measured in years since 2010. So the predicted number of students in 2013 is $11 + 13.9286(3) = 52.8$. The residual for 2013 ($x = 3$) from the residual

plot looks to be ≈ 5.2. So actual – predicted ≈ 5.2. Thus the actual number of students must have been 5.2 + 52.8 = 58.

86. **(B)** Slope $= r\dfrac{s_y}{s_x} = 0.45\left(\dfrac{4.2}{0.5}\right) = 3.78$ and

$$y\text{-intercept} = 86.5 - 3.78(1.8) = 79.7.$$

87. **(E)** The slope has the same sign, but generally not the same value, as the correlation. The coefficient of determination, which gives the proportion of the y-variance that is predictable from a knowledge of x, is equal to r^2, not r. A positive correlation indicates that higher values of x tend to be associated with higher values of y.

88. **(A)** The slope is 1.52 ppm per year. So on average, atmospheric CO_2 concentration has been rising by 1.52 ppm per year since 1960. Answer (C) may be a true statement, but it doesn't answer the question.

89. **(E)** A scatterplot, or calculations of slope, readily shows that the four points do not lie on a straight line. Thus no matter what the fifth point is, all the points cannot lie on a straight line. So r cannot be 1.

90. **(E)** Even if the regression lines are identical, the points may be more closely or less closely scattered about the lines, resulting in different sums of squares of the residuals and different correlations. It may seem that a greater sum of the squares of the residuals leads to a lesser correlation, but this depends upon the number of points. For example, {(1, 1), (2, 3), (3, 2)} has a lesser sum of squares of the residuals, 1.5 < 3.5, and a lesser correlation, 0.5 < 0.57, than {(1, 1), (2, 3), (3, 2), (4, 2), (4, 4)}.

91. **(E)** It is reasonable to assume that the diagonal line is $y = x$. There was 1 week in which the Uber driver drove more miles (evidenced by the point below the line) and 2 weeks in which they drove identical distances (the 2 points on the line). In all but these 3 weeks, the Lyft driver drove more miles than the Uber driver (the 14 points above the line).

92. **(D)** Correlation r measures association, not causation.

93. **(C)** The slope should be negative, ruling out choice (E). The y-intercept should be 10.3, ruling out choices (A) and (B). Finally, choice (C) looks much closer to a correlation of $-\sqrt{0.906} = -0.95$ than does choice (D).

94. **(B)** The slope and the correlation have the same sign. Multiplying every y-value by a negative number changes this sign. Doubling each value leaves the correlation unchanged.

95. **(A)** The least squares regression line passes through $(\bar{x}, \bar{y})$, and the slope b satisfies $b = r\dfrac{s_y}{s_x} = r\dfrac{4}{2} = 2r$. Since $-1 \le r \le 1$, we have $-2 \le b \le 2$.

96. **(C)** The mean of the residuals is 0, thus ruling out plot I. The regression line for a residual plot is a horizontal line, thus ruling our plot II. A residual plot as in plot III indicates that a nonlinear fit would be better. However, plot III is still a possible residual plot.

97. **(E)** Correlation is just a number; it has no unit of measurement. Multiplying each y-value by -1 changes the sign of both the slope and the correlation. In the case of perfect correlation, $r = \pm 1$. It can be shown that r^2, called the coefficient of determination, is the ratio of the variance of the predicted y-values to the variance of the observed y-values. Alternatively, we can say that there is a partition of the y-variance and that r^2 is the proportion of this variance that is predictable from having a knowledge of the linear regression model of y on x.

98. **(E)** The slope and the correlation are related by the formula $b = r\frac{s_y}{s_x}$.

    The standard deviation is always positive, so b and r have the same sign. Correlation r is a measurement of only *linear* association. Correlation is sensitive (nonresistant) to outliers. Positive and negative correlations with the same absolute value indicate data having the same degree of clustering around their respective regression lines, one of which slopes up to the right and the other of which slopes down to the right. Even though $r = 0.48$ indicates a better fit with a linear model than $r = 0.16$ does, we cannot say that the linearity is threefold.

99. **(B)** The log(population in 1,000s), not population, goes up an average of 0.03 per year. Calculating log(population in 1,000s) = 2.4 + 0.03(10) = 2.7 gives population = $10^{2.7}$ = 501 thousand. The coefficient of determination, $r^2 = (0.72)^2 = 0.5184$, gives that 51.84% of the variation in log(population in 1,000s), not population, can be explained by variation in time.

100. **(B)** When transforming the variables leads to a linear relationship, the original variables have a nonlinear relationship, their correlation (which measures linearity) is not close to ± 1, and the residuals do not show a random pattern. The residual plot of the variables log X and log Y may or may not show a strong nonlinear pattern; the residuals could indicate that an even stronger nonlinear model is out there.

101. **(E)** The point X swings the regression line counterclockwise toward it. If X is removed, the line would swing away. The slope of the regression line would be less positive. Without X, the remaining points form a tighter linear pattern, so the correlation is greater.

102. **(C)** The data points all fall on the straight line $\hat{y} = x - 7$, where x and y are the weights before the diet and weights after the diet, respectively.

103. **(E)** Removing X will change the regression line very little if at all, and so it is not an influential point. A point is a regression outlier if it has a large

residual. However, X is probably very close to the least squares regression line and so has a small residual. There will be a strong curved pattern in the residual plot.

104. **(E)** All three scatterplots show very strong nonlinear patterns; however, the correlation r measures the strength of only a linear association. Thus r is close to +1 in the first scatterplot, and r is approximately 0 in the other two.

105. **(D)** The regression equation is Predicted calories = 211 + 13.9286(Fat), so the predicted calories for this hamburger is 211 + 13.9286(15) = 420. The residual equals the actual value minus the predicted value. With Fat = 15, the residual plot gives a residual of approximately 5. The equation 5 = Actual – 420 gives an actual value of 425 calories.

106. **(E)** The linear regression line can be called a "line of averages" because it predicts an average y-value for a given x-value. Similarly, the slope gives an average increase in the y-direction for a 1-unit increase in the x-direction.

107. **(C)** $r_3 \approx 0$, $r_2 < 0$, $r_1 > 0$ and $|r_2|$ is closer to 1 than is r_1.

108. **(D)** Using your calculator, find the regression line to be $\hat{y} = 106 - 18x$. The regression line, also called the least squares regression line, minimizes the sum of the squares of the vertical distances between the points and the line. In this case, (3, 52), (4, 34), and (5, 16) are on the line. So the minimum sum is $(50 - 52)^2 + (38 - 34)^2 + (14 - 16)^2 = 24$. [Alternatively, note that (–2 4 –2) is in the Residual list after finding the regression line, and $(-2)^2 + 4^2 + (-2)^2 = 24$.]

109. **(C)** $\bar{x} = \dfrac{17 + 29 + 35 + 43}{4} = 31$. Since $(\bar{x}, \bar{y})$ is a point on the regression line, $\bar{y} = 8 + 2(31) = 70$

110. **(A)** The correlation is not changed by adding the same number to every value of one of the variables, by multiplying every value of one of the variables by the same positive number, or by interchanging the x- and y-variables.

111. **(E)** These are all misconceptions about correlation. Correlation measures only linearity. So when $r \approx 0$, there may be a nonlinear relationship. Correlation shows association, not cause and effect. Curved data can have a correlation near 1.

112. **(E)** The sum and thus the mean of the residuals is always zero. The correlation coefficient for the residuals and x is zero, and thus the regression line for a residual plot is a horizontal line. The standard deviation of the residuals is a "typical value" of the residuals and thus gives a measure of how the points in the scatterplot are spread around the regression line. Residual = $y - \hat{y}$, not $\hat{y} - y$.

113. **(A)** We are given $r^2 = 0.5625$, and we note that the slope of the regression line is negative. The correlation has the same sign as the slope, so we have $r = -\sqrt{0.5625} = -0.75$.

114. **(D)** All of the statements are true, but only choice (D) defines "strong." For example, choice (B) defines "linear," and choice (C) defines "positive."

115. **(E)** The correlation r cannot take a value greater than 1 or less than –1.

116. **(C)** Larger salaries are generally associated with lower happiness levels, so the relationship is *negative*. The values are close to the values predicted by a curve drawn roughly through the points, so the relationship is *strong*. A unit increase in salary does not correspond to constant changes in happiness level, so the relationship is *nonlinear*.

117. **(E)** The y-intercept has no reasonable interpretation if $x = 0$ is outside a reasonable domain. In this example, $w = 0$ is meaningless as no car weighs anywhere near 0 pounds.

118. **(D)** One equation involves x and $\hat{y}$, while the other equation involves y and $\hat{x}$. This is an important concept illustrating a difference between equations of lines in algebra and in statistics. In algebra, if $y = mx + b$ is solved for x, we find $x = \frac{1}{m}y - \frac{b}{m}$. We see that the slopes in the two equations are reciprocals. However, in statistics, we have $\hat{y} = a_1 + b_1x$, where $b_1 = r\frac{s_y}{s_x}$, and we have $\hat{x} = a_2 + b_2y$, where $b_2 = r\frac{s_x}{s_y}$. So we see that b_1 and b_2 are not reciprocals.

119. **(A)** For each of the represented distances, the distribution of the airfares for that distance has different spread and spacing from any other. The regression line for a residual plot is always a horizontal line, so it cannot show a positive linear relationship between the residuals and x-variables. The sum of the residuals is always 0. The y-values (airfares), not the residuals, appear to go from 100 to 400.

120. **(C)** The proportion of people in the sample who brush their teeth from side to side rather than up and down is $\frac{50 + 40}{300 + 200} = \frac{90}{500} = 0.18$. If the proportions are the same for men and women, it would be expected that 18 percent of the 300 men, or 54, and that 18 percent of the 200 women, or 36, brush their teeth from side to side rather than up and down.

121. **(C)** $\frac{120}{200} = 0.6$ so 60% of the sample were teenagers. With independence, we expect 60% of the 150 "yes" replies to be teenagers: 60% of 150 = 90. After 90 is placed into the upper left cell, all the other cells are forced because of the fixed row and column totals.

122. **(D)** Relative frequencies must be equal. Looking at rows gives $\frac{40}{100} = \frac{50}{50+n}$, and looking at columns gives $\frac{40}{90} = \frac{60}{60+n}$. We could also set up a proportion: $\frac{n}{50} = \frac{60}{40}$ or $\frac{n}{60} = \frac{50}{40}$. Solving any of these equations gives $n = 75$.

123. **(E)** There were 70 + 45 movies classified as PG or PG-13. Of these, 20 + 15 were action and 15 + 17 were drama.

124. **(C)** There must have been 125 – 50 = 75 foreign cars. Because $\frac{20}{50} = 0.4 = 40\%$ of the American cars have staff tags, 40% of the foreign cars also have staff tags because of independence. Calculate the number of foreign cars with staff tags: 40% of 75 = 30. If 30 of the foreign cars have staff tags, then 75 – 30 = 45 of the foreign cars must have student tags. This is shown in the following table.

	Student Tag	**Staff Tag**	***Total***
American car	30	20	50
Foreign car	45	30	75
Total	75	50	125

125. **(B)** 95 women feel "always rushed," 97 women feel either "sometimes rushed" or "almost never rushed," and $95 < 97$. The other choices are correct: 79 men feel "sometimes rushed" while 66 men feel either "always rushed" or "almost never rushed." The proportion of men who say they are "almost never rushed" is $\frac{22}{145} = 0.152$, which is greater than the proportion of women who say they are "almost never rushed," which is $\frac{28}{192} = 0.146$. The proportion of people who are either "sometimes rushed" or "almost never rushed" is 1 minus the proportion who feel "always rushed."

MC COLLECTING AND PRODUCING DATA

Sampling Strategies (pages 49–56)

126. **(D)** In stratified sampling, the population is divided into homogeneous groups called *strata* and random samples of persons from all strata are chosen. In this example, the retailer stratified by type of payment into three strata.

127. **(E)** In cluster sampling, the heterogeneous clusters all should look pretty much like the population as a whole. So we only have to pick one or more clusters randomly to find a representative sample. If a census is poorly run, it will provide less information and be less accurate than a well-designed survey. For example, having the principal ask every single student in the school whether or not he or she regularly cheats on exams produces less useful data than a carefully worded anonymous questionnaire filled out by a randomly selected sample of the student body. The sampling frame is the list from which the sample is actually drawn; this may or may not be the whole population of interest. Sampling error is a natural variation among samples; it is not an error committed by anyone. To determine if a sample is random, one must analyze the *procedure* by which the sample was obtained.

128. **(D)** Different samples give different sample statistics, all of which are estimates for the same population parameter. So error, which is called sampling error, is naturally present.

129. **(C)** In stratified sampling, the population is divided into homogeneous groups called *strata* and random samples of persons from all strata are chosen. In this case, it might well be important to be able to consider the responses from each of the three groups separately—urban, suburban, and rural.

130. **(B)** Convenience samples are based on choosing individuals who are easy to reach. A typical example is sampling based on interviews at a shopping mall. Data obtained from convenience samples tend to be highly unrepresentative of the entire population. In this example, while using the company employee database is convenient, the resulting data tell nothing about what people outside the company's employees think about the new product.

131. **(E)** In a simple random sample, every possible group of the given size has to be equally likely to be selected, and this is not true here. For example, with this procedure, it will be impossible for all players on the Black Hawks to be together in the final sample. This procedure is an example of stratified sampling, and stratified sampling does not result in simple random samples.

132. **(E)** Systematic sampling involves listing the population in some order (ranking in this example), choosing a random point to start, and then picking every *k*th person (every 20th person in this example) from the list.

133. **(C)** Method 1 is stratified sampling. The population is divided into homogeneous units called *strata* (first class and coach in this context), and we randomly select from each strata to make up the sample. Method 2 is cluster sampling. The population is divided into heterogeneous units called clusters (the individual railcars in this context), and we make up a sample by choosing everyone in one or more randomly selected clusters.

134. **(E)** The standard deviation of the sample proportions, $\sigma_{\hat{p}} = \sqrt{\frac{p(1-p)}{n}}$, is reduced if the sample size n is increased. This reduces the margin of error but has no effect on possible bias. Sampling error, which is naturally present when sample statistics are used to estimate population parameters, can generally be reduced with larger sample sizes but cannot be eliminated.

135. **(D)** Choice (A) is a systematic sample. Choice (B) is a simple random sample. Choice (C) is a cluster sample. Choice (E) is a convenience sample. A stratified sample is when the population is divided into homogeneous units called *strata* (by gender in this example), and then individuals are randomly selected from each strata.

136. **(A)** In cluster sampling, the population is divided into heterogeneous groups called *clusters* (in this example, libraries). A random sample of clusters is taken from among all the clusters.

137. **(A)** This will not result in a simple random sample because each possible set of 260 companies does not have the same chance of being selected. For example, a group of 260 companies whose names all start with *A* will not be chosen. Although the IRS agent does use chance, each company would have the same chance of being selected only if the same number of companies have names starting with each letter of the alphabet. There are probably many more companies whose name begins with *A* than with *X*. With only 10 companies chosen per letter, Amazon has less chance of being selected than Xerox. Systematic sampling involves a random starting point followed by picking every *n*th subject from an ordered list.

138. **(D)** Stratification does not reduce or remove bias. Rather, it helps reduce sampling variability when different groups within the population may contribute different statistics to a final sample.

139. **(E)** Different samples give different sample statistics, all of which are estimates of a population parameter. Sampling error relates to natural variation among samples, can never be eliminated, can be described using probability, and is generally smaller if the sample size is larger. To have higher confidence, one must accept more variability. Bias, which distorts our view of the population, is related to bad sampling design; sampling error is not related to bad design.

140. **(C)** A simple random sample (SRS) is one in which every possible sample of the desired size has an equal chance of being selected. In this case, every possible sample of 5 companies has an equal chance of being selected. Note that although it is also true that each company has an equal chance of being selected, this by itself would not insure that the analyst is using an SRS.

141. **(B)** Samples taken closer to the waste dump could contain more contaminants than samples taken farther away. Stratified sampling with vertical strips involves randomly picking some samples from every vertical strip, and these vertical strips represent all the different distances from the waste dump. Cluster sampling with horizontal strips involves randomly picking one or more horizontal strips to sample in their entirety, and each horizontal strip in its entirety represents almost all the different distances from the waste dump.

142. **(E)** It may well be that the brightest, most motivated students are the same ones who both purchase the review book and do well on the AP Exam. If students could be randomly assigned to purchase or not purchase the book, the results would be more meaningful.

143. **(E)** Response bias occurs when respondents are untruthful, usually due to people not wanting to be perceived as having unpopular views, not wanting to admit to illegal activities, or not wanting to reveal certain intimate facts. Often this can be overcome by very careful wording of the questions.

144. **(B)** A stratified sample where the six bars/stations are the strata would involve sampling a little bit of each of the different categories of food being offered. A cluster sample with the six bars/stations as clusters would result in only sampling one category of the foods being offered. A simple random sample, numbering all 235 items and using a random number generator to pick items for a meal, would be very time-consuming and would not guarantee any kind of balanced meal. A systematic sample, again involving numbering all the items, makes no sense in this context. A convenience sample is never recommended.

145. **(A)** In cluster sampling, the population is divided into heterogeneous groups called *clusters*. A random sample is then taken of one or more clusters from among all the clusters. In this example, the bottles are the clusters and are assumed to be representative of the day's production run.

146. **(E)** It may well be that very bright students are the same ones who both take multiple AP courses and eventually graduate in four years from college. If students could be randomly assigned to take or not take multiple AP courses, the results would be more meaningful. Of course, ethical considerations make it impossible to do this. Using only a sample from the observations gives less information.

147. **(C)** Given the exact same number of people surveyed in each age group, stratified sampling was probably the strategy used. In stratified sampling, the population is divided into homogeneous groups called *strata* (for example, by age), and random samples of persons from all strata are chosen. We could further do proportional sampling where the sizes of the random samples from each stratum depend on the proportion of the total population represented by the stratum (not done in this case because equal-size samples were picked from each age group). In cluster sampling, the population is divided into heterogeneous groups called *clusters*, and a random sample of clusters is taken from among all the clusters.

148. **(E)** In a simple random sample (SRS), every possible group of the given size has to be equally likely to be selected, and this is not true here. For example, with this procedure, it will be impossible for a family of four, all leaving together, to all be included in the sample. This is an example of systematic sampling, which may be easier to execute and may give a reasonable sample as long as the order in which people leave the theater is not in any way related to the variables in the survey. Still, it does not result in a simple random sample.

149. **(C)** A systematic sample involves picking every nth name on the list, not n in a row. Although the procedure does use some element of chance, all possible groups of size 80 do not have the same chance of being picked. So the result is not a simple random sample. There is a very real chance of selection bias. For example, a number of relatives with the same name and similar holiday shopping patterns might be selected. Undercoverage bias is present because those with unlisted land phones or with cell phones are not in the phone book and so are not part of the sampling frame.

150. **(E)** Teachers who attend the conference may well be the better, more enthusiastic, more motivational teachers. So the students in their classes may be more motivated to do well than students in classes with less-inspiring teachers. AP Statistics, more than any other AP course, probably teaches a set of skills that will lead to success in all college classes.

151. **(E)** Different samples give different sample statistics, all of which are estimates for the same population parameter. So error, called sampling error, is naturally present.

Bias in Sampling (pages 56–62)

152. **(C)** As long as phone numbers are randomly dialed, unlisted numbers are not a concern. Caller IDs are a major problem as many people now will not pick up a phone if they don't recognize the number. If a child answers, the procedure does not call for asking to speak to an adult. So many households with children will not be included in the survey even if their number is dialed. Persons with multiple phones have a better chance of being included

in the survey as the probability of one of their numbers being randomly chosen is higher than for those with a single phone.

153. **(B)** Low response rates lead to *nonresponsive bias*, where those who do respond may not be representative of the intended population. *Response bias*, where the responses are suspect, often occurs when people don't want to be perceived as having unpopular or unsavory views or don't want to admit to illegal activities.

154. **(E)** Sampling variability, also called sampling error, is the natural variation among samples and can be described using probability.

155. **(E)** Although choice (A) would lead to *nonresponse bias*, choice (E) is the principal reason for people chosen for a telephone survey not responding. Choice (B) refers to *wording bias* where nonneutral or poorly worded questions may lead to answers that are not representative of the population. Choice (C) refers to *response bias*, where people often don't want to be perceived as having unpopular views and so may respond untruthfully. Choice (D) refers to *undercoverage* or *selection bias*, where certain groups are inadequately represented in a survey.

156. **(D)** Method 1 is a cluster sample. However, one school may not be representative of all schools and all parents, thus resulting in *selection bias.* Method 2 is an attempt at a census. It will likely suffer from *nonresponse bias* where the response rate could be very low. The parents who do respond might have very different opinions from those parents who do not respond. Method 4 will have *voluntary response bias* where only people with strong opinions will probably bother to respond. Method 3 is a reasonable stratified sample with follow-ups.

157. **(B)** The systematic sample will first encounter 30 nondefective boards and then 20 defective boards, thus noting the exact fraction of defective boards produced that day.

158. **(D)** The variance of the estimator is inversely proportional to the sample size. If bias is present, increasing the sample size doesn't help!

159. **(C)** The *Wall Street Journal* survey has strong *selection bias.* In other words, people who read the *Journal* are not very representative of the general population. The Internet survey results in a *voluntary response sample*, which typically gives too much emphasis to people with strong opinions. The teacher's survey has strong *response bias* in that students may be uncomfortable giving truthful responses to a teacher about the teacher's class.

160. **(A)** Surveying people coming out of a church on Sunday results in a very unrepresentative sample of the adult population, especially given the question under consideration. Obtaining a high response rate, neutrally worded

questions, use of chance, and sample size will not change the selection bias and turn this into a well-designed survey.

161. **(D)** Voluntary response samples, like radio call-in surveys, are based on individuals who offer to participate. These samples typically overrepresent persons with strong opinions. Convenience samples, like shopping mall surveys, are based on choosing individuals who are easy to reach. These samples typically miss a large segment of the population. Always check carefully for bias before collecting data, because there is no recovery from a biased sampling method after the sample is collected. Nonneutral wording can readily lead to response bias. If surveyors want a particular result, they deliberately use certain wording. When a large fraction of those sampled fail to respond, the concern is whether those who do respond have different views than those who do not respond.

162. **(E)** The wording of the questions can lead to response bias. The word *sovereignty* is more positive, while the word *separate* can come across as a negative word.

163. **(E)** Poorly designed sampling techniques result in *bias*, that is, in a tendency to favor the selection of certain members of a population. For example, door-to-door surveys ignore the homeless, radio call-in programs give too much emphasis to persons with strong opinions, and interviews at shopping malls typically give the opinions of a very select sample of the population.

164. **(C)** The procedure misses all or nearly all childless families.

165. **(E)** The actual sample size is more important than the fraction of the population sampled because the standard deviation of the sampling distribution decreases as the sample size increases. If there is enough bias, the sample can be worthless. Even when the subjects are chosen randomly, there can be bias due, for example, to nonresponse or to the wording of the questions. Allowing surveyors to choose participants usually fails to result in representative samples because there are too many unknowns. If there is bias, taking a larger sample just magnifies the bias on a larger scale.

166. **(D)** Care must be taken with regard to the wording of questions so as to avoid *response bias*. Wording of questions must be neutral as subjects will give different answers depending on phrasing. Sampling error, also called sampling variation, is the natural variation among samples. Allowing researchers to choose people they think are representative is a prescription for disaster; random sampling of the population is best. A census is not worth the effort if, for example, wording bias is present.

167. **(A)** There is no reason to suspect any more bias in this stratified sample than would be present in a simple random sample. Stratified samples are often easier and more cost-effective to obtain and also make comparative data more available. In this case, responses can be compared in and among

various districts. Rural and urban districts are less varied within each district than the population as a whole, and stratified sampling recognizes this.

168. **(A)** This is an example of *voluntary response bias,* which often overrepresents negative opinions. The students who chose to respond were most likely those who were unhappy with their college choice. So there is very little chance that the 12,000 respondents were representative of the population. Knowing more about the subscribers, or taking a sample of the sample, would not have helped.

169. **(C)** There can be significant *nonresponse bias* as a large fraction of those who see a pop-up online survey do not bother responding. There can also be significant *undercoverage bias* as not everyone has access to the Internet. There are also significant demographic differences between those who do have Internet access and those who do not.

Experiments and Observational Studies (pages 62–72)

170. **(A)** Observational studies are generally cheaper and quicker to conduct than experiments. However such studies are subject to bias, and it is very difficult to conclude cause and effect from observational studies. Experiments also use randomization. (Subjects are randomly assigned to treatments in order to minimize the effect of confounding variables. Randomization in surveys is to enable generalization to populations.) Blocking in experimental design roughly corresponds to stratification in sampling design.

171. **(C)** In the first study, the families were already receiving assistance. In the second study, one of two treatments was applied to each family.

172. **(E)** Regression lines show association, not causation. Observational studies suggest relationships, which controlled experiments can help show to be cause and effect.

173. **(A)** Actual experiments would have been completely unethical, so these were probably observational studies. No conclusion about drinking providing protection again dementia is proper as there well may be confounding variables. For example, most drinkers have shorter life expectancies and so may die before dementia has a chance to develop.

174. **(E)** Although observational studies are generally more cost-effective and time-effective than experiments, they cannot be used to show direct cause-and-effect relationships like experiments can. Random sampling is important in observational studies (to be able to generalize). Randomization in assignment to treatment groups is important in experiments (to minimize the effect of confounding variables). Inference can be performed on data from either experiments or observational studies.

175. **(A)** Random sampling is the use of chance in selecting a sample from a population. It is necessary in order to generalize findings to the population. Random assignment in experiments is when subjects are randomly assigned to treatments in order to minimize the effects of possible confounding variables. In good observational studies, the responses are not influenced during the collection of data. In good experiments, treatments are compared as to differences in responses. In an experiment, there can be many factors, levels, and treatments. Randomized block design refers to randomization occurring only within groups of similar experimental units called blocks.

176. **(C)** The first two and last two studies are observational studies. (We can't ethically force drivers to speed, college students to use illegal drugs, people to eat more or less, or parents to smoke and drink.) Thus cause-and-effect conclusions are not reasonable. The third study, choice (C), sounds like an experiment. Assuming that proper techniques are used, like replication and randomization, cause-and-effect conclusions are reasonable.

177. **(A)** You should simply ask each and every teacher his or her age, because it is a small population in which data are easy to gather.

178. **(A)** No one is given apples to eat, the subjects are simply observed as they self-select whether or not to eat apples. Since no treatments are applied, this is an observational study. To make this an experiment, subjects would have to be randomly assigned to two treatment groups, one instructed to eat apples regularly and the other instructed not to eat apples regularly.

179. **(A)** No treatments were imposed, so this is an observational study. The explanatory variable is vitamin D blood levels. To perform an experiment, the researchers would randomly assign the subjects to two groups, with one group taking vitamin D capsules and the other taking a placebo.

180. **(B)** Patients who believe they have received the new drug may somehow be able to alleviate their own symptoms. Such a medical effect based on the power of suggestion is called a placebo effect.

181. **(C)** Lack of blinding is not a design fault here, because blinding is clearly impossible in this experiment. There is no indication that randomization is being used to decide which patients have the treatment of classical music and darkness and which patients are in the control group. If the treatment of classical music and darkness helps, there is no way to tell whether classical music only, darkness only, or the combination of both is responsible. So the variables will be confounded. The instructions are unclear as to factor levels, that is, loudness/softness of the music and length of time to sit in the dark. It is unclear how the response variable, "relief," is to be measured and thus compared between treatments.

182. **(E)** Use of a control group and blinding as to which subjects are in the control group are the best tools to minimize the possibility of confounding due

to the placebo effect. Replication and randomization are important marks of good experimental design. However, they do not impact the placebo effect as do the use of a control and blinding.

183. **(E)** The paired comparison design is a special case of blocking in which each pair can be considered a block. Blocking can control certain variables by bringing them directly into the picture, and thus the conclusions made are more specific. Blocking in experiment design first divides the subjects into representative groups called blocks, just as stratification in sampling design first divides the population into representative groups called strata.

184. **(A)** In a randomized block design, researchers assign subjects to treatments at random within each of the homogeneous blocks. In effect, to reduce variability, researchers run parallel experiments on the blocks before analyzing all results together at the end.

185. **(D)** There are 2 factors: hours of rest (3 levels) and energy drink (2 levels). This means there are $3 \times 2 = 6$ treatments.

186. **(A)** When there is uncertainty with regard to which variable is causing an effect, we say the variables are *confounded*. In this experiment, it might be difficult to determine if the difference in invasive procedure or if the difference in blood thinner is the real cause of observed differences in long-term incidence of stroke. However, with both males and females taking each treatment, gender and either invasive procedure or blood thinner are not confounded.

187. **(D)** Brand is the only explanatory variable, and it is being tested at four levels. Feet per second is the single response variable.

188. **(D)** In a typical K–6 elementary school, there will be a wide variety of student ages. Thus in this example, there is a confounding variable—age—that drives both the other variables. That is, older students tend to weigh more than younger students, and older students also tend to have higher math levels. Eating more will not improve math skills!

189. **(B)** The team members would like to be able to generalize to all adults, but they can generalize only to the collection of individuals from which the sample was drawn.

190. **(D)** When there is uncertainty with regard to which variable is causing an effect, we say the variables are *confounded*. Randomly putting subjects into treatment and control groups can help reduce the problems posed by confounding. Bias is a tendency to favor the selection of certain members of a population.

191. **(A)** Using volunteers instead of a random sample of people with acrophobia creates a confounding variable. It is impossible to know if the positive results were caused by taking the people to the top of the Empire State Building

or simply by the subjects' desire to be cured (the desire that made them volunteer).

192. **(D)** This is an experiment. Two treatments are applied—being given 12-ounce bowls and being given 18-ounce bowls. It is a randomized design because each attendee was randomly put in one or the other treatment group through a randomization technique (the use of random digit table). This is not a census but only a sample consisting of the attendees at the event. It would have been matched pair if, for example, each attendee was given one of each bowl and told to put some ice cream into each. It would have been a blocked design if, for example, the nutritionist suspected that men and women would respond differently and so separated the attendees by gender, handed out the two bowls randomly within each gender group, and looked at outcomes by gender.

193. **(D)** Random assignment eliminates or aims to eliminate confounding.

194. **(D)** Although all the choices are reasonable criticisms showing the weakness of the study, the strongest point is that there was one treatment and nothing to compare it to.

195. **(C)** Blocking is used to isolate the variability due to differences among the blocks so that the effects of the treatments become clearer. In this case, all subjects taking the same medication should be in the same block. Therefore, the effect of alcohol will not be confounded with the differences in medications.

196. **(B)** There are two treatments, aluminum and wood bats, that are randomly assigned to the players. A randomized block design could be followed with every player as a block if each player alternates between using aluminum and wood bats and if the difference in distance hit between the two bats is noted for each player.

197. **(D)** Blocking in experimental design first divides the subjects into groups, such as men and women, to control certain variables, like gender, by bringing these variables directly into the picture. Thus conclusions are more specific and variation is reduced.

198. **(D)** The new fertilizer is the only explanatory variable, and it is being tested at three levels. Growth is the single response variable.

199. **(C)** Each subject might receive both treatments. An example is a taste test where each subject tastes both foods being compared, in random order. Randomization is always important in sampling. In matched pair experiments, it can be used to decide which member of a pair gets which treatment or which treatment is given first if each subject is to receive both. Matched pair experiments are an example of blocking, not vice versa. Stratification refers to a sampling method, not an experimental design. The

point is to give each subject in a matched pair a different treatment and note any difference in responses.

200. **(A)** This study is an experiment because a treatment (a glass of red wine every day) is applied. There is a single factor—drinking or not drinking a glass of red wine. There is no blinding because the subjects clearly know whether or not they are drinking a glass of red wine. There is no blocking because the subjects are not divided into blocks before random assignment.

MC PROBABILITY

Basic Probability Rules (pages 73–84)

201. **(C)** This is a conditional probability:

$$P(\text{survive to 60} \mid \text{survive to 20}) = \frac{8{,}600}{9{,}800}$$

202. **(D)** $P(\text{common}) = 1 - 0.002 = 0.998$

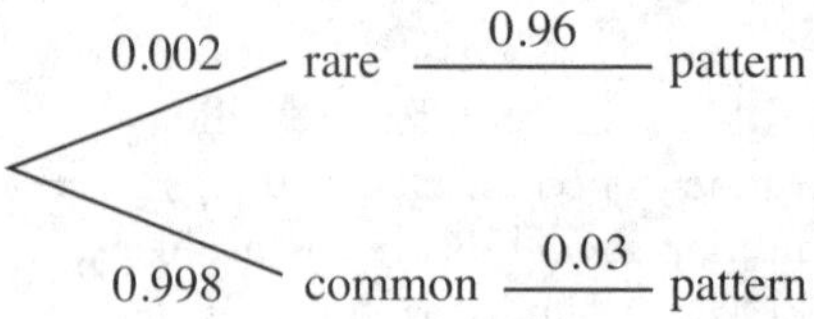

$P(\text{rare} \cap \text{pattern}) = (0.002)(0.96)$

$P(\text{common} \cap \text{pattern}) = (0.998)(0.03)$

$$P(\text{pattern}) = (0.002)(0.96) + (0.998)(0.03)$$

$$P(\text{rare} \mid \text{pattern}) = \frac{P(\text{rare} \cap \text{pattern})}{P(\text{pattern})}$$

203. **(E)** Probabilities are never less than 0.

204. **(D)** The probability of throwing a heads is 0.5. By the law of large numbers, the more times you flip a coin, the closer the relative frequency tends to be to this probability. With fewer tosses, there is more chance of wide swings in the relative frequency.

205. **(E)** Probability is associated with the area under the curve, and the area under the entire curve of a probability density function must be 1. In choice (A), the area under the graph is (1)(1) = 1. In choice (B), the area is (0.05)(20) = 1. In choice (C), the area is (0.3)(2) + (0.2)(2) = 1. In choice (D), the area is $\left(\frac{1}{2}\right)(1)(2) = 1$. However in choice (E), the area is $\left(\frac{1}{2}\right)(0.25)(1) = 0.125$.

206. **(A)** Let E be receiving an e-mail, and let T be receiving a text message. If $P(E \cap T) = P(E)P(T)$, then E and T are independent.

207. **(E)** Coins have no memory. So the probability that the next toss will be heads is 0.5 and the probability that it will be tails is 0.5. The law of large numbers says that as the number of tosses becomes larger, the cumulative proportion of tails tends to become closer to 0.5.

208. **(C)** $P(X \cap Y) = P(X \mid Y)P(Y) = (0.27)(0.41)$. Then

$$P(Y \mid X) = \frac{P(X \cap Y)}{P(X)} = \frac{(0.27)(0.41)}{0.36}.$$

209. **(B)** invNorm(0.9) = 1.282 and invNorm(0.2) = –0.842. Solving the system of equations

$$\begin{cases} \mu + 1.282\sigma = 37.333 \\ \mu - 0.842\sigma = 36.781 \end{cases}$$

gives $\mu = 37.000$.

210. **(B)** If E and F are mutually exclusive, $P(E \cap F) = 0$. Thus $P(E) + P(F) - P(E \cap F) = 0.25 + P(F) - 0 = 0.64$. So $P(F) = 0.64 - 0.25$.

211. **(D)** If E and F are independent, $P(E \cap F) = P(E)P(F)$. Thus $0.35 + P(F) - (0.35)P(F) = 0.73$, and so

$$P(F) = \frac{0.73 - 0.35}{1 - 0.35} = \frac{0.73 - 0.35}{0.65}.$$

212. **(B)** If A and B are independent $P(A \cap B) = P(A)P(B)$ and thus $P(A \cup B) = 0.32 + 0.45 - (0.32)(0.45) = 0.626$.

213. **(B)**

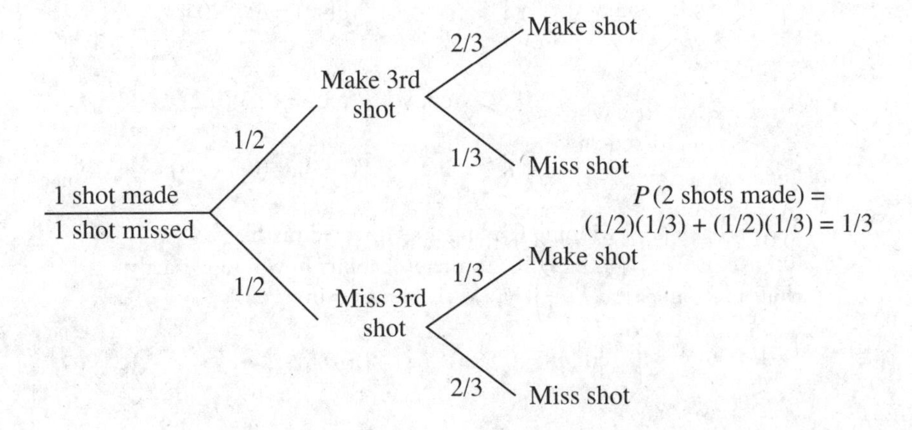

214. **(B)** $P(\text{healthy}) = 1 - 0.001 = 0.999$

0.001 Ebola $\underline{0.995}$ positive $P(\text{Ebola} \cap \text{positive}) = (0.001)(0.995)$

0.999 healthy $\underline{0.03}$ positive $P(\text{healthy} \cap \text{positive}) = (0.999)(0.03)$

$$P(\text{positive}) = (0.001)(0.995) + (0.999)(0.03)$$

$$P(\text{Ebola} \mid \text{positive} = \frac{P(\text{Ebola} \cap \text{positive})}{P(\text{positive})} = \frac{(0.001)(0.995)}{(0.001)(0.995) + (0.999)(0.03)}$$

215. **(B)** $P(E \cup F) = P(E) + P(F) - P(E \cap F)$, so $0.623 = 0.420 + 0.350 - P(E \cap F)$ and $P(E \cap F) = 0.147$. Since $P(E \cap F) \neq 0$, E and F are not mutually exclusive. However, $P(E \cap F) = 0.147 = (0.420)(0.350) = P(E)P(F)$, which implies that E and F are independent.

216. **(D)** The probabilities must sum to 1, and no individual probability can be negative.

217. **(C)**

$$P(\text{longest answer} \mid \text{doesn't guess}) = \frac{P(\text{longest answer})}{P(\text{doesn't guess})} = \frac{1 - 0.45 - 0.35}{1 - 0.45} = \frac{0.20}{0.55}$$

218. **(D)** $P(E \cap F) = P(E)P(F)$ only if E and F are independent. In this case, women live longer than men, and so the events are not independent.

219. **(B)**

0.7 — Box with 3 red and 2 blue: 0.6 red $P(\text{red}) = (0.7)(0.6)$; 0.4 blue $P(\text{blue}) = (0.7)(0.4)$

0.3 — Box with 1 red and 4 blue: 0.2 red $P(\text{red}) = (0.3)(0.2)$; 0.8 blue $P(\text{blue}) = (0.3)(0.8)$

The probability of winning (ending up with a red marble) is $(0.7)(0.6) + (0.3)(0.2) = 0.48$. So the probability of winning exactly twice in 4 games is

$$\binom{4}{2}(0.48)^2(0.52)^2 = 6(0.48)^2(0.52)^2.$$

220. **(D)** Independence gives

$$P(E)P(F) = P(E \text{ and } F),\ 0.4P(F) = 0.15,\ P(F) = 0.375, \text{ and}$$

$$P(E \text{ or } F) = P(E) + P(F) - P(E \text{ and } F) = 0.4 + 0.375 - 0.15 = 0.625.$$

221. **(B)** $P(\text{flooding} \mid \text{tornado}) = \dfrac{P(\text{flooding} \cap \text{tornado})}{P(\text{tornado})}$

222. **(D)** $P(\text{flosses} \cap \text{cavity}) = P(\text{flosses once} \cap \text{cavity}) + P(\text{flosses twice} \cap \text{cavity})$

$= P(\text{flosses once})P(\text{cavity} \mid \text{flosses once}) + P(\text{flosses twice})P(\text{cavity} \mid \text{flosses twice})$
$= (0.55)(0.04) + (0.30)(0.01)$

223. **(A)** $P(\text{white chip}) = \dfrac{1}{3}\cdot\dfrac{1}{5}+\dfrac{1}{3}\cdot\dfrac{2}{4}+\dfrac{1}{3}\cdot\dfrac{3}{4}=\dfrac{29}{60}$ so

$$P(\text{red chip}) = 1 - \frac{29}{60} = \frac{31}{60}.$$

To end with exactly \$10,000, one must either pick a white chip and then throw a "6" or pick a red chip and then don't throw a "6." So

$$P(\$10{,}000) = \frac{29}{60}\cdot\frac{1}{6}+\frac{31}{60}\cdot\frac{5}{6}=\frac{23}{45}.$$

224. **(A)** When events are independent, the probability of their intersection is the product of their probabilities. In this case, $\left(\dfrac{9}{10}\right)(0.10)(0.55) = 0.0495.$

225. **(D)** Independence implies $P(S \cap T) = P(S)P(T)$, while mutually exclusive implies $P(S \cap T) = 0$. We are given that $P(S)$ and $P(T)$ are nonzero, so it is impossible for S and T to be both independent and mutually exclusive.

226. **(B)** Events E and F are independent if $P(E \cap F) = P(E)P(F)$ and are mutually exclusive if $P(E \cap F) = 0$. In this context,

$$P(\text{state})P(\text{federal}) = (0.32)(0.15) = 0.048 \text{ and } P(\text{both}) = 0.048 \neq 0.$$

So the events are independent but are not mutually exclusive.

227. **(D)** A problem like this is most easily handled with a Venn diagram. Using

$$P(\text{dorm but not meal}) = 0.46 - 0.33 = 0.13 \text{ and}$$
$$P(\text{meal but not dorm}) = 0.65 - 0.33 = 0.32,$$

the resulting Venn diagram is the following:

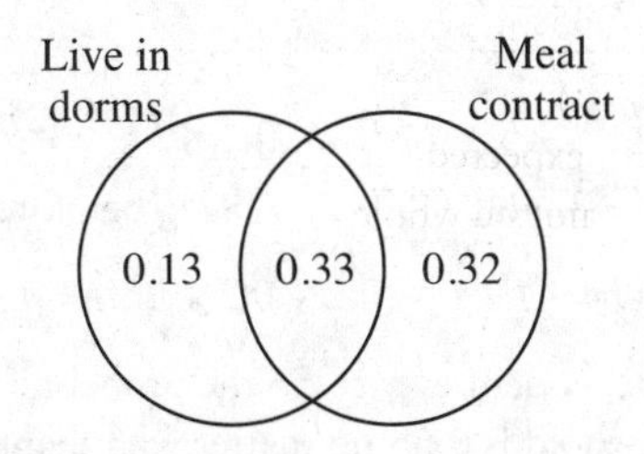

The Venn diagram gives that $P(\text{dorm or meal, not both}) = 0.13 + 0.32 = 0.45.$

228. **(D)** $P(\text{different}) = 1 - P(\text{same}) = 1 - (0.4^2 + 0.2^2 + 0.2^2 + 0.1^2 + 0.1^2)$
$= 1 - 0.26 = 0.74.$

229. **(D)** A tree diagram helps with the probability calculations.

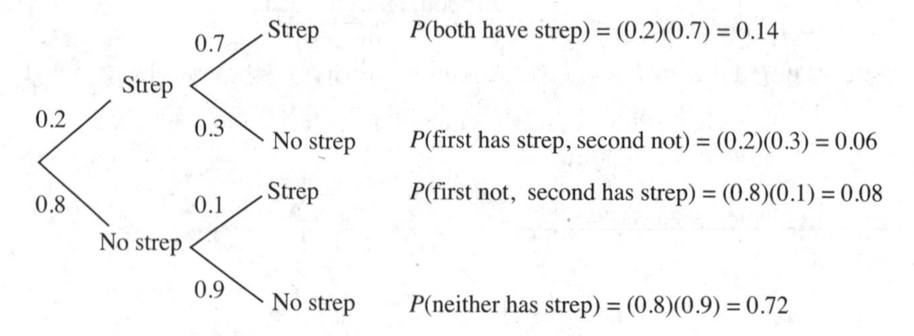

From the tree diagram, $P(0 \text{ strep}) = 0.72$, $P(1 \text{ strep}) = 0.06 + 0.08 = 0.14$, $P(2 \text{ strep}) = 0.14$.

230. **(C)** A tree diagram illustrates the relationship between the probabilities.

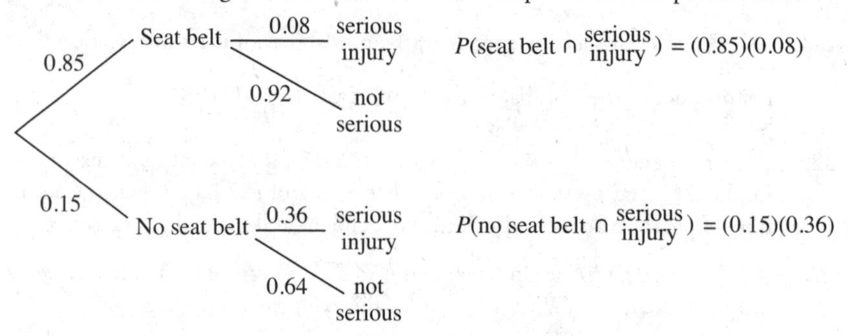

$$P(\text{serious injury}) = (0.85)(0.08) + (0.15)(0.36) = 0.122$$

231. **(B)** For events E and F, $P(E \cup F) = P(E) + P(F) - P(E \cap F)$. In this context, $0.62 = 0.16 + 0.53 - P(\text{both})$ and so $P(\text{both}) = 0.16 + 0.53 - 0.62$.

232. **(C)** Based on information in the probability distribution,

$$P(X = 3) = 1 - (0.4 + 0.1) = 0.5.$$

By independence,

$$P(X = 3, Y = 2) = P(X = 3)P(Y = 2).$$

So $0.1 = (0.5)P(Y = 2)$ and $P(Y = 2) = 0.2$. Therefore

$$P(Y = 4) = 1 - (0.2 + 0.2 + 0.3) = 0.3.$$

233. **(A)** These are independent events. So the probability of a successful contact for a selected household is 0.65 no matter what happened on the attempts to contact other households.

Binomial and Geometric Probabilities (pages 84–88)

234. **(B)**

$$P(X < 2) = P(X = 0) + P(X = 1)$$
$$= (0.3)^4 + {}_4C_1(0.7)(0.3)^3 = (0.3)^4 + 4(0.7)(0.3)^3$$

235. **(C)** With $p = 0.1$, there are low probabilities of a greater number of successes so the histogram is skewed to the right (to the higher values). No matter what p and n are, the binomial distribution is unimodal. No matter what p is, the larger the value of n, the more the histogram will appear symmetric and the closer it will come to a normal distribution.

236. **(D)**

$$P(X \geq 2) = P(X = 2) + P(X = 3)$$
$$= {}_3C_2(0.09)(0.91)^2 + (0.91)^3$$
$$= 3(0.09)(0.91)^2 + (0.91)^3$$

237. **(D)** $P(\text{at least one}) = 1 - P(\text{none}) = 1 - (0.94)^5$

238. **(C)** This is a binomial with $n = 15$ and $p = 0.28$. So the mean is

$$np = (15)(0.28) = 4.2.$$

239. **(B)** In this binomial example, the probability that a backseat passenger does wear a seat belt is $1 - 0.2 = 0.8$, the probability that all of the backseat passengers wear seat belts is $(0.8)^5$, and the probability that exactly one doesn't wear a seat belt is $5(0.2)(0.8)^4$. Thus the probability that at least two don't wear seat belts is $1 - [(0.8)^5 + 5(0.2)(0.8)^4]$.

240. **(A)** ${}_{12}C_2(1/6)^2(5/6)^{10} = 0.296$ while ${}_{120}C_{20}(1/6)^{20}(5/6)^{100} = 0.0973$.

You can also use

binompdf(12, 1/6, 2) = 0.296 while binompdf(120, 1/6, 20) = 0.0973.

The law of large numbers talks about relative frequency in the long run becoming close to an expected probability but not about getting that exact probability.

241. **(D)** $P(X \geq 1) = 1 - P(X = 0) = 1 - (0.95)^3$

242. **(E)** This is a binomial with $p = 1 - 0.41 = 0.59$ and $n = 5 + 8 = 13$. So the mean is $np = 13(0.59) = 7.67$.

243. **(A)** In choice (B), there is not a fixed probability p of success. In choice (C), there is not a fixed number of trials n. In choice (D), the probability of rain on a given day is not independent of whether or not there was rain on a previous day.

244. **(B)** With $B(12, 0.35)$ and $P(\text{not African-American}) = 1 - 0.35 = 0.65$, $P(X < 2) = P(X = 0) + P(X = 1) = (0.65)^{12} + 12(0.35)(0.65)^{11}$.

245. **(C)** This is a binomial with $n = 6$ and $p = 1/3$. So

$$P(X \geq 4) = 1 - P(X \leq 3) = 1 - \text{binomcdf}(6, 1/3, 3) = 1 - 0.8999 = 0.1001.$$

246. **(D)** A binomial with $n = 10$ and $p = 0.3$ has mean $np = 10(0.3) = 3$ and standard deviation $= \sqrt{np(1-p)} = \sqrt{10(0.3)(0.7)} = 1.45$.

247. **(D)** The probability that a person tests negative in the rapid test is $1 - 0.02 = 0.98$. A person doesn't have Zika either if he/she tests negative in the rapid test or he/she tests positive in the rapid test but the follow-up more accurate test shows he/she does not have Zika. So $P(\text{don't have Zika}) = 0.98 + (0.02)(0.06) = 0.9812$ or $1 - P(\text{Zika}) = 1 - (0.02)(0.94) = 0.9812$. In a random sample of three people, the probability of none of them having Zika is $(0.9812)^3 = 0.945$.

248. **(D)** This is a geometric setting, and the probability of two losses followed by a win is $(0.93)(0.93)(0.07) = (0.07)(0.93)^2$.

249. **(D)** This is a cumulative geometric setting with probability

$$0.15 + (0.85)(0.15) + (0.85)^2(0.15) + (0.85)^3(0.15) = 0.4780.$$

This can be calculated more simply with $1 - (0.85)^4 = 0.4780$.

250. **(B)** The probability an e-mail is not spam is $1 - 0.786 = 0.214$. The probability of none of the first three e-mails being spam is $(0.214)^3$. The probability the next e-mail is spam is 0.786. For the entire sequence to happen, the probability is $(0.214)^3(0.786)$.

Random Variables (pages 89–95)

251. **(C)** The mean is an unchanging, fixed number. The law of large numbers gives that the average of observed X-values tends to be closer and closer to the mean after more and more repetitions. The standard deviation is the positive square root of the variance. The only way that $\text{var}(X) = \Sigma(x_i - \mu_x)^2 p_i$ can be 0 is if every $x_i = \mu_x$.

252. **(C)** The mean is $\mu = \Sigma xP(x) = 0(0.8) + 25(0.16) + 50(0.04) = 6$ and the variance is

$$\Sigma(x - \mu)^2 P(x) = (0 - 6)^2(0.8) + (25 - 6)^2(0.16) + (50 - 6)^2(0.04).$$

253. **(C)** $25{,}000(0.01) + 5{,}000(0.04) = 450$, so the company can expect to make $750 - 450 = \$300$ per policy or $25(300) = \$7{,}500$ for 25 policies.

254. **(E)** $20\left(\frac{3}{4}\right) + 100\left(\frac{1}{4}\right) = 40$ and $10\left(\frac{100}{400}\right) + 50\left(\frac{300}{400}\right) = 40$. So all boxes have the same expected winning.

255. **(B)** The probability of an application being turned down is $1 - 0.08 = 0.92$. So the expected value of a binomial with $n = 50$ and $p = 0.92$ is $np = 50(0.92)$.

256. **(B)** The expected value for number of sales is

$$[0(0.37) + 1(0.28) + 2(0.25) + 3(0.10)].$$

The real estate agent receives $6,500 commission per sale.

257. **(D)** Option I gives the highest expected return: 50,000 is greater than both 70,000(0.50) = 35,000 and 90,000(0.10) = 9,000. Option II gives the best chance (0.50) of paying off the $60,000 loan, and Option III gives the only chance of paying off the $80,000 loan. The moral is that the greatest expected value is not automatically the "best" answer.

258. **(C)** The expected winnings in one game are $5(0.25) + 2(0.5) - 3 = -0.75$. For six games, the expected outcome is $6(-0.75) = -\$4.50$.

259. **(D)** The expected number of days served is

$$\Sigma xP(x) = 1(0.3) + 2(0.4) + 3(0.15) + 4(0.1) + 5(0.05) = 2.2.$$

At $40 per day, this gives a mean expected pay of 2.2($40) = $88.

260. **(B)** Expected value = $1.75 - 8(0.25) = -\$0.25$.

Variance = $[0.50]^2 + [2(0.25)]^2 = 0.50$. Standard deviation = $\sqrt{0.50} = 0.71$.

261. **(D)** The probabilities of winning different numbers of points is a random variable:

x:	1	2	3	4
$P(x)$:	0.4	0.2	0.2	0.2

Mean = $\Sigma xP(x) = 1(0.4) + 2(0.2) + 3(0.2) + 4(0.2) = 2.2$

Variance = $\Sigma(x - \mu)^2 P(x)$

$= 1.2^2(0.4) + 0.2^2(0.2) + 0.8^2(0.2) + 1.8^2(0.2) = 1.36$

262. **(E)** The variance of

$$X - Y \text{ is } \text{var}(X - Y) = \text{var}(X) + \text{var}(Y) = 3.4^2 + 2.9^2 = 19.97.$$

The mean of $X - Y$ is $E(X - Y) = E(X) - E(Y) = 28.1 - 23.7 = 4.4$. The standard deviation of $X - Y$ is $\sqrt{19.97} \approx 4.47$. The median and range of $X - Y$ cannot be determined from the given information.

263. **(D)** Variances add, so the new variance is $12^2 + 5^2 = 169$. The new standard deviation is $\sqrt{169} = 13$.

264. **(D)** Adding a constant to every value increases the expected value by that constant, so $E(X + c) = E(X) + c$. The sum of the probabilities of the x_i equals 1, but the sum of the x_i themselves can equal anything. $E(X) = \Sigma x_i P(x_i)$ and $\text{var}(X) = \Sigma (x_i - \overline{x})^2 P(x_i)$. Multiplying each value by a constant increases the standard deviation by that multiple and increases the variance by the multiple of the constant squared, so $\text{var}(aX) = a^2 \text{var}(X)$.

265. **(C)** There are 200 sales. The expected value for a single sale can be shown by the following equation.

$$\Sigma xP(x) = 11.98(0.10) + 1.47(0.60) + 5.40(0.30)$$

So the expected value of one day's sales is

$$200[11.98(0.10) + 1.47(0.60) + 5.40(0.30)].$$

266. **(E)** $E(X + Y) = 29 + 35 = 64$. However, without independence we cannot determine $\text{var}(X + Y)$ from the information given.

267. **(B)** For *independent* random variables, variances always add:

$$\text{var}(X \pm Y) = \text{var}(X) + \text{var}(Y).$$

So the variance of the random variable $X - Y$ is $4 + 3$.

268. **(A)** $X - Y$ is normally distributed with a mean of $45 - 35 = 10$ and a standard deviation of $\sqrt{4^2 + 3^2} = 5$. Then

$$P(X - Y > 0) = P\left(z > \frac{0 - 10}{5}\right) = P(z > -2).$$

269. **(D)** Multiplying by a constant increases both the mean and the standard deviation by that same multiple. Adding a constant increases the mean by that same constant but does not change the standard deviation. So the mean is $5 + 3(15) = 50$, and the standard deviation is $3(10) = 30$.

270. **(B)** $\mu_{X-Y} = \mu_X - \mu_Y$ and $\sigma^2_{X-Y} = \sigma^2_X + \sigma^2_Y$. So $\mu_Y = 25 - 9 = 16$ and $\sigma_Y = \sqrt{10^2 - 8^2} = 6$.

271. **(D)** The mean of the sum (or difference) of two random variables is equal to the sum (or difference) of the individual means. *If two random variables are independent*, the variance of the sum (or difference) of the two random variables is equal to the sum of the two individual variances.

272. **(C)** For independent random variables, the variance of the sum or difference of the random variables is the sum of the individual variances: $\frac{5}{4} + \frac{5}{4}$. The standard deviation is the square root of the sum of the variances: $\sqrt{\frac{5}{4} + \frac{5}{4}}$.

273. **(C)** Multiplying every value by a constant increases the mean, median, IQR, and standard deviation by the same multiple and increases the variance by the multiple of the constant squared. Adding the same constant to every value increases the mean and median by that constant but doesn't change the IQR, standard deviation, or variance. In this problem, the new variance is $4^2 \times 5 = 80$.

Normal Probabilities (pages 96–102)

274. **(C)** In both cases, 15 pounds is $\frac{5}{3}$ standard deviations from the mean with a right tail probability of 0.04779.

275. **(D)** The area under any probability distribution is equal to 1. The mean of a normal curve determines the value around which the curve is centered, while the standard deviation determines the height and spread of the curve.

276. **(A)** The *z*-score of 12 is $\frac{12-11}{2}$, and we are asked for the probability of a score greater than this.

277. **(A)** The critical *z*-score is invNorm(0.3) = –0.5244. Thus, $200 - 210 = -0.5244\sigma$. Solving for σ gives $\sigma = \frac{200-210}{-0.5244}$.

278. **(B)** Choice (B) is true by symmetry. Choice (A) is false because 0.4772 is not 2 times 0.3413. Choice (C) is false because 2 times 0.67 is not 3. Less than 10^{-20} (a very small number!) values are greater than a standardized *z*-score of 10. The 68-95-99.7 rule (the empirical rule) applies to *all* normal (actually all bell-shaped) distributions.

279. **(E)** The two critical *z*-scores are $\frac{50{,}000 - 58{,}760}{6{,}500}$ and $\frac{60{,}000 - 58{,}760}{6{,}500}$.

280. **(E)** All normal distributions have about 68% of their observations within one standard deviation of the mean.

281. **(D)** From the shape of the normal curve, the answer is in the middle. The middle $\frac{2}{3}$ (leaving $\frac{1}{6}$ in each tail) is between *z*-scores of $\pm\text{invNorm}\left(\frac{5}{6}\right) = \pm 0.9674$, and $8 \pm 2(0.9674)$ gives (6.1, 9.9).

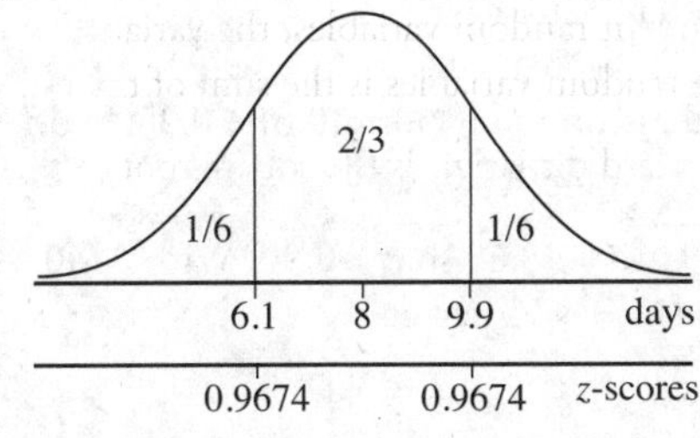

282. **(E)** $P(X > 10)$ = normalcdf(10, 1000, 12, 3.5) = 0.7161

$P(10 < X < 15)$ = normalcdf(10, 15, 12, 3.5) = 0.5205

$$P(X < 15 \mid X > 10 = \frac{P(10 < X < 15)}{P(X > 10)} = \frac{0.5205}{0.7161} = 0.727$$

283. **(E)** When we standardize a variable, the shape of the distribution doesn't change; standardized scores have a normal distribution only if the original distribution is normal. Because of symmetry, the median and mean are identical for all normal distributions. Many bell-shaped curves are *not* normal curves. For example, the t-distributions are not normal distributions. For all normal curves, 99.7% of the area is within three standard deviations of the mean.

284. **(B)** The critical z-score is invNorm(0.25) = –0.6745. We have $\mu - 0.6745(2.5) = 13.2$. Rearranging gives $\mu = 13.2 + 0.6745(2.5)$.

285. **(B)** With $\mu = 15{,}000$, $\sigma = 3{,}600$, and a normal model,

$P(X > 20{,}000) = P\left(z > \dfrac{20{,}000 - 15{,}300}{3{,}600}\right)$. The probability for two individuals is this answer squared: $\left[P\left(z > \dfrac{20{,}000 - 15{,}300}{3{,}600}\right)\right]^2$.

286. **(A)** The critical z-score is invNorm(0.10) = –1.282. Then $x = \mu + z\sigma = 275 + (-1.282)(40)$.

287. **(D)** 22.5 is $\dfrac{22.5 - 20}{4} = \dfrac{2.5}{4}$ standard deviations from the women's mean. However, 22.5 is $\dfrac{26 - 22.5}{5} = \dfrac{3.5}{5}$ standard deviations from the men's mean. Because $\dfrac{2.5}{4} = 0.625 < \dfrac{3.5}{5} = 0.7$, the footprint is closer to the women's mean as measured by standard deviations from the mean.

288. **(D)** $P(\text{over } 250{,}000 \mid \text{under } 300{,}000) = \dfrac{P(\text{over } 250{,}000 \cap \text{under } 300{,}000)}{P(\text{under } 300{,}000)}$.

For a normal distribution with $\mu = 275{,}000$ and $\sigma = 35{,}000$,

$P(250{,}000 < X < 300{,}000) = 0.5249$ and $P(X < 300{,}000) = 0.7625$.

Then $\dfrac{0.5249}{0.7625} = 0.688$.

289. **(D)** The critical z-score is invNorm(0.6) = 0.25, and $x = \mu + z\sigma = 27.1 + (0.25)(4.6)$.

290. **(C)** $40.44 - 1.944\sigma = 38.69$ so $\sigma = 0.90$. With $N(40.44, 0.90)$, $P(X > 42) = 0.0415 = 4.15$ percent.

291. **(D)** The probability of laying at least 2.5 miles on a good weather day is normalcdf(2.5, 1000, 3, 0.35) = 0.9234, while the probability of laying at least 2.5 miles on a bad weather day is

normalcdf(2.5, 1000, 1.9, 0.4) = 0.06681.

The probability of laying at least 2.5 miles on a random day is

(0.7)(0.9234) + (0.3)(0.06681) = 0.6664.

292. **(C)** The z-score of x is $\frac{x-28}{3.4}$, which should be set equal to

invNorm(0.80) = 0.84.

293. **(B)** The first quartile is the 25th percentile, and

normalcdf(–100, –0.6) = 0.27

or the 27th percentile. So $m < l < n$.

294. **(D)** 51 is $\left|\frac{51-28}{10}\right| = 2.3$ standard deviations from the mean for winter temperatures, while 51 is $\left|\frac{51-66}{6}\right| = 2.5$ standard deviations from the mean for summer temperatures. It's more unusual to be 2.5 than 2.3 standard deviations from a mean.

Sampling Distributions (pages 102–113)

295. **(D)** The sampling distribution of $\bar{x}$ is approximately normal with a mean of 10.00 and a standard deviation of $\frac{0.50}{\sqrt{5}}$. The probability that the average amount exceeds $\frac{52}{5} = 10.40$ is $P\left(z > \frac{10.40 - 10.00}{\left(\frac{0.50}{\sqrt{5}}\right)}\right)$.

296. **(B)** Use either 1 – binomcdf(1000, 0.067, 80) or

normalcdf$\left(0.08, 1, 0.067, \sqrt{\frac{(0.067)(0.933)}{1000}}\right)$ to get 0.05.

297. **(A)** The larger the sample size n is, the smaller the spread is in the sampling distribution. For example, the sampling distribution of a population mean or proportion has a standard deviation inversely proportional to $\sqrt{n}$. Provided that the population size is significantly greater than the sample size, the spread of a sampling distribution depends on the *sample* size, not the population size. A sampling distribution is *unbiased* if its *mean* is equal to the associated population parameter. A sampling distribution shows the distribution of a statistic over all possible samples of the same size. The larger

a sample is, the more the sample shows the distribution traits of the whole population.

298. **(D)** Both $np = 200(0.15) = 30 \geq 10$ and $n(1 - p) = 200(0.85) = 170 \geq 10$. The sampling distribution of $\hat{p}$ is approximately normal with a mean of 0.15 and a standard deviation of $\sigma_{\hat{p}} = \sqrt{\frac{(0.15)(0.85)}{200}}$. The critical z-score is $\frac{0.18 - 0.15}{\sqrt{\frac{(0.15)(0.85)}{200}}}$, and we are calculating the probability that z is greater than this critical value.

299. **(E)** For the central limit theorem to apply, the observations must be independent. It is always true that the sampling distribution of $\bar{x}$ has mean μ and standard deviation $\frac{\sigma}{\sqrt{n}}$. In addition, the sampling distribution will be normal if the population is normal, or it will be approximately normal, by the central limit theorem, if n is large.

300. **(C)** We note that $n_1p_1 = 115(0.73)$, $n_1(1 - p_1) = 115(0.27)$, $n_2p_2 = 80(0.61)$, and $n_2(1 - p_2) = 80(0.39)$ are all ≥ 10. The sampling distribution of $\hat{p}_1 - \hat{p}_2$ is approximately normal with a mean of $0.73 - 0.61 = 0.12$ and a standard deviation of $\sqrt{\frac{(0.73)(0.27)}{115} + \frac{(0.61)(0.39)}{80}}$.

The critical z-score is $\frac{0.10 - 0.12}{\sqrt{\frac{(0.73)(0.27)}{115} + \frac{(0.61)(0.39)}{80}}}$.

301. **(C)** $\frac{\bar{x} - \mu}{\left(\frac{s}{\sqrt{n}}\right)}$ has a t-distribution with $df = n - 1$. Since $n = 8$, $n - 1 = 7$.

302. **(D)** As the sample size n becomes larger, the sample distribution becomes closer to the population distribution. (The *sampling distribution* may become closer to a normal distribution.)

303. **(D)** The sample proportion is a random variable with a probability distribution called the sampling distribution of $\hat{p}$. The *sampling distribution* of $\hat{p}$ has a mean equal to the population proportion p, has a standard deviation equal to $\sqrt{\frac{p(1 - p)}{n}}$, and is considered close to normal provided that both np and $n(1 - p)$ are large enough (greater than 5 or 10 are standard guidelines). The sampling distribution of $\bar{x}$ is usually close to normal when $n \geq 30$.

304. **(A)** Variances can be added; standard deviations cannot. The sample sizes n_1 and n_2 do not have to be the same. Without independence, our methods of inference do not apply.

305. **(D)** The sampling distribution of $\bar{x}$ is approximately normal with a mean of $\mu_{\bar{x}} = 23.1$ and a standard deviation of $\sigma_{\bar{x}} = \frac{0.6}{\sqrt{10}}$. For the total to be between 228 and 232, the average of 10 must be between 22.8 and 23.2. The critical z-scores are $\frac{22.8 - 23.1}{\left(\frac{0.6}{\sqrt{10}}\right)}$ and $\frac{23.2 - 23.1}{\left(\frac{0.6}{\sqrt{10}}\right)}$.

306. **(E)** All are unbiased estimators for the corresponding population parameters. In other words, the means of their sampling distributions are equal to the population parameters.

307. **(D)** The sampling distribution of $\hat{p}$ is approximately normal with mean $\mu_{\hat{p}} = 0.6$ and standard deviation $\sigma_{\hat{p}} = \sqrt{\frac{(0.6)(0.4)}{150}} = 0.04$. Only the graph in choice (D) shows the correct mean and standard deviation.

308. **(C)** The mean of the set of sample means is equal to the mean of the population; it does not vary with the size of the samples. The variance of the set of sample means varies inversely as the size of the samples and directly as the variance of the original population. One must take a sample *four* times as large in order to cut the standard deviation of $\bar{x}$ in half.

309. **(E)** The sampling distribution of $\hat{p}$ is roughly normal with a mean of $\mu_{\hat{p}} = p = 0.3$ and a standard deviation of $\sigma_{\hat{p}} = \sqrt{\frac{p(1-p)}{n}} = \sqrt{\frac{(0.3)(0.7)}{200}}$. Note that the sample size is 200, not 100.

310. **(C)** The sampling distribution of $\hat{p}_1 - \hat{p}_2$ is approximately normal with a mean of $\mu_{\hat{p}_1 - \hat{p}_2} = p_1 - p_2$ and a standard deviation of $\sigma_{\hat{p}_1 - \hat{p}_2} = \sqrt{\frac{p_1(1-p_1)}{n_1} + \frac{p_2(1-p_2)}{n_2}}$. In this example, $p_1 = 0.09$, $p_2 = 0.12$, and $n_1 = n_2 = 100$. (Note that the sample sizes are 100, not 400.)

311. **(D)** The sampling distribution for $\bar{x}$ should be roughly bell-shaped with a mean around 19 and a standard deviation of approximately $\frac{2.31}{\sqrt{12}} \approx 0.67$. The graph in choice (D) is reasonable.

312. **(B)** The standard deviation of the sampling distribution of $\hat{p}$ is $\sigma_{\hat{p}} = \sqrt{\frac{p(1-p)}{n}}$. This will be larger when n (in the denominator) is smaller

or when p is closer to 0.5. (Note that $y = x(1 - x)$ is a parabola opening downward with its max at $x = 0.5$.)

313. **(C)** This follows from the central limit theorem.

314. **(C)** The sampling distribution of $\hat{p}_1 - \hat{p}_2$ has a mean of

$$\mu_{\hat{p}_1 - \hat{p}_2} = p_1 - p_2 = 0.48 - 0.44 = 0.04$$

and a standard deviation of

$$\sigma_{\hat{p}_1 - \hat{p}_2} = \sqrt{\frac{p_1(1-p_1)}{n_1} + \frac{p_2(1-p_2)}{n_2}} = \sqrt{\frac{(0.48)(0.52)}{100} + \frac{(0.44)(0.56)}{100}} = 0.07.$$

This is shown in the graph in choice (C).

315. **(E)** With this large a sample size, $n = 100$, the CLT gives that the sampling distribution of $\bar{x}$ is approximately normal with a mean equal to the population mean, $\mu_{\bar{x}} = \mu = 37$, and the standard deviation equal to the population standard deviation divided by the square root of the sample size, $\sigma_{\bar{x}} = \frac{\sigma}{\sqrt{n}} = \frac{0.4}{\sqrt{100}} = 0.04$. Note that the sample size is 100, not 500.

316. **(D)** The sampling distributions of proportions, means, and slopes are unbiased. In other words, they are centered on the population parameters. However, the sampling distribution of sample maxima cannot be centered on the population maximum as all the sample maxima are less than or equal to the population maximum.

317. **(D)** The population {1, 2, 3, 4, 5, 6} has mean $\mu = 3.5$ and standard deviation $\sigma = 1.708$. The sampling distribution of $\bar{x}$ has mean $\mu_{\bar{x}} = \mu = 3.5$ and standard deviation $\sigma_{\bar{x}} = \frac{\sigma}{\sqrt{n}} = \frac{1.708}{\sqrt{25}}$. Note that the sample size is $n = 25$, not 400.

318. **(B)** The sampling distribution for the difference of means has the standard deviation $\sqrt{\frac{\sigma_1^{\ 2}}{n_1} + \frac{\sigma_2^{\ 2}}{n_2}}$, which in this example is $\sqrt{\frac{(0.6)^2}{100} + \frac{(0.5)^2}{100}}$.

319. **(C)** The sampling distribution of $\bar{x}$ is approximately normal with a mean of $\mu_{\bar{x}} = \mu = 39.7$ and a standard deviation of $\sigma_{\bar{x}} = \frac{\sigma}{\sqrt{n}} = \frac{3.8}{\sqrt{200}}$.

320. **(D)** The standard deviation of the sampling distribution of $\hat{p}$ is

$$\sigma_{\hat{p}} = \sqrt{\frac{p(1-p)}{n}}.$$

321. **(C)** With $\mu_{MLB-NFL} = 26.8 - 25.6 = 1.2$ and

$$\sigma_{MLB-NFL} = \sqrt{\frac{3.8^2}{35} + \frac{2.9^2}{30}} = 0.8324,$$

then

$$P(\bar{x}_{MLB} - \bar{x}_{NFL} > 0) = P\left(z > \frac{0 - 1.2}{0.8324}\right) = 0.9253.$$

322. **(E)** The sampling distributions of proportions, means, and slopes are unbiased. In other words, they are centered on the population parameters.

323. **(D)** The *t*-distributions are symmetric, bell-shaped, and unimodal. They also have more, not less, spread than the normal distribution. The larger the *df* value, the closer the *t*-distribution is to the normal distribution.

324. **(E)** Like the *t*-distribution, there is a separate χ^2 curve for each *df* value. Like the *t*-distribution, for large *df* the chi-square distribution becomes closer to a normal curve. Every χ^2 curve represents a probability distribution, so the area is 1. Degrees of freedom for χ^2 distributions depend upon the number of categories, not on sample size.

325. **(A)** The larger the number of degrees of freedom, the closer the curve is to the normal curve. Although around the 30 level is often considered a reasonable approximation to the normal curve, it is not the normal curve. The degrees of freedom depend on the *sample* size. For any *df*, the probability that $t > 1.96$ is greater than the probability that $z > 1.96$ in a normal distribution. (Tails are fatter in the *t*-distribution than in the normal distribution.)

CHAPTER 4

MC STATISTICAL INFERENCE

Confidence Intervals for Proportions (pages 115–123)

326. **(D)** $\hat{p} = \frac{525}{1{,}500} = 0.35$, the critical *z*-scores for 95% are ±1.96, and the standard error is $\sqrt{\frac{\hat{p}(1-\hat{p})}{n}} = \sqrt{\frac{(0.35)(0.65)}{1{,}500}}$.

327. **(E)** Using a measurement from a sample, we are never able to say exactly what a population proportion is; rather, we always say we have a certain confidence that the population proportion lies in a particular interval. In this case, that interval is 29% ± 3% or between 26% and 32%.

328. **(E)** There is no guarantee that 17% is anywhere near any particular confidence interval, so none of the statements are true.

329. **(E)** We have $\hat{p} = \frac{330}{750} = 0.44$ and $\sigma_{\hat{p}} = \sqrt{(0.44)(0.56)/750}$. The z-scores of 0.40 and 0.48 are $\frac{0.40 - 0.44}{\sigma_{\hat{p}}}$ and $\frac{0.48 - 0.44}{\sigma_{\hat{p}}}$.

330. **(C)** $\sigma_{\hat{p}} = \sqrt{\frac{(0.28)(0.72)}{530}} = 0.0195$. Then $(z)(0.0195) = 0.03$ gives $z = 1.538$. $P(-1.538 \leq z \leq 1.538) = \text{normalcdf}(-1.538, 1.538) = 0.876 = 87.6\%$.

331. **(E)** This is an example of a voluntary response survey, which is a typically strongly biased survey that completely violates the simple random sample assumption.

332. **(E)** We have a random sample. (It is not an SRS, but there is no reason to think that it is not representative.) Given this procedure where each student has a 1% chance of being picked, it is reasonable to assume that the sample size is approximately 1% of 25,000 = 250 students. Therefore, both $np \approx (250)(0.86) = 215$ and $n(1 - p) \approx (250)(0.14) = 35$ are greater than 10. Since approximately 1% of the students are included in the sample, the sample size is less than 10% of the population.

333. **(A)** The confidence interval is $\hat{p} \pm z^* \sigma_{\hat{p}}$. Lower levels of confidence lead to smaller z^* and thus narrower confidence intervals. Also, $\sqrt{p(1 - p)}$, and thus $\sigma_{\hat{p}} = \sqrt{\frac{p(1 - p)}{n}}$ is larger when p is closer to 0.5. Among the choices, the lowest confidence level is 91%, and the furthest p from 0.5 is 0.18.

334. **(E)** The midpoint of the confidence interval is the sample proportion $\hat{p} = 0.04$.

335. **(B)** The critical z-scores will go from ±1.812 to ±2.17, resulting in an increase in the interval size: $\frac{2.17}{1.812} = 1.20$ or an increase of 20 percent.

336. **(C)** The men's point estimate is $\frac{0.095 + 0.115}{2} = 0.105$ so the women's point estimate is lower. Comparing standard errors results in $\sqrt{\frac{(0.1)(0.9)}{n}} < \sqrt{\frac{(0.105)(0.895)}{n}}$, so the women's interval is also narrower.

337. **(D)** Choice (A) is a correct interpretation of confidence *interval,* and choice (B) is a common misinterpretation of confidence interval. Choices (C) and (E) are common misinterpretations of confidence *level.*

338. **(E)** $\hat{p} = \dfrac{240}{800} = 0.4$, the critical z-score is invNorm(0.95) = 1.645, and the standard error is $\sqrt{\dfrac{\hat{p}(1-\hat{p})}{n}} = \sqrt{\dfrac{(0.3)(0.7)}{800}}$. The confidence interval is $\hat{p} \pm z^*SE(\hat{p}) = 0.3 \pm 1.645\sqrt{\dfrac{(0.3)(0.7)}{800}}$.

339. **(C)** Ninety-nine percent of all confidence intervals will contain the true population proportion. So the probability that the pollster will construct a confidence interval containing the population proportion is 0.99. Once he constructs a confidence interval, the probability that that particular interval contains the population proportion is either 0 or 1. The sample proportion is always in the confidence interval.

340. **(C)** The entire confidence interval is below 0.50. So it does provide evidence that less than half of all adults feel the death penalty is applied fairly by the courts. We never conclude an exact number like 46 percent, only intervals. What is important is the sample size, not the population size.

341. **(B)** For a given confidence level, halving the margin of error requires a sample *four times* as large. For a given sample size, more certainty means less precision and more precision leads to less certainty.

342. **(B)** The equation for margin of error gives $z\sqrt{\dfrac{(0.75)(0.25)}{450}} = 0.023$. Solving for z gives $z = 1.1268$. Then calculate using

normalcdf(–1.1268, 1.1268) = 0.74 = 74%.

343. **(A)** The critical z-score is invNorm(0.985) = 2.17, the standard error is $\sigma_{\hat{p}} = \sqrt{\dfrac{\hat{p}(1-\hat{p})}{n}}$, and the sample size is $n = 190$, not 12,000. The confidence interval is $\hat{p} \pm z^*SE(\hat{p}) = 0.22 \pm 2.17\sqrt{\dfrac{(0.22)(0.78)}{190}}$.

344. **(A)** $\hat{p}_1 = \dfrac{486}{1{,}080} = 0.45$ and $\hat{p}_2 = \dfrac{546}{1{,}050} = 0.52$. The critical z-scores for 90% are ±1.645, and the standard error is $\sqrt{\dfrac{\hat{p}_1(1-\hat{p}_1)}{n_1} + \dfrac{\hat{p}_2(1-\hat{p}_2)}{n_2}}$. The confidence interval is

$$\hat{p}_1 - \hat{p}_2 \pm z^*SE(\hat{p}_1 - \hat{p}_2) = (0.45 - 0.52 \pm 1.645\sqrt{\frac{(0.45)(0.55)}{1{,}080} + \frac{(0.52)(0.48)}{1{,}050}}.$$

345. **(C)** The critical z-score is invNorm(0.995) = 2.576, and the standard error is $\sqrt{\dfrac{\hat{p}_1(1-\hat{p}_1)}{n_1} + \dfrac{\hat{p}_2(1-\hat{p}_2)}{n_2}} = \sqrt{\dfrac{(0.7)(0.3)}{500} + \dfrac{(0.65)(0.35)}{500}}$.

346. **(E)** Confidence intervals come from sample statistics but are intended to capture population parameters, in this case the quotient of population proportions.

347. **(D)** When using 2-PropZInt on a calculator, the values of x_1 and x_2 must be integers. For example, putting in 0.4×756 (which equals 302.4) for x_1 or putting in 0.49×825 (which equals 404.25) for x_2 will result in an error message. 2-PropZInt, using $x_1 = 302$ and $x_2 = 404$, gives the interval (–0.141, –0.039). Zero is not in the interval (the whole interval is less than 0). So there is sufficient evidence of a difference in proportions of senior men and senior women who suffer from some form of arthritis.

Confidence Intervals for Means (pages 123–132)

348. **(B)** The wider interval corresponds to the higher (99%) confidence level ($58.3 - 52.7 = 5.6$ is less than $59.2 - 51.8 = 7.4$).

349. **(E)** In those rare cases where the population standard deviation is positively known, z-scores are appropriate. In the presence of strong skewness or outliers, the t-procedures are contraindicated. The t-procedures assume normality of the original population. When the population standard deviation σ is unknown and the standard error $\frac{s}{\sqrt{n}}$ is substituted for $\frac{\sigma}{\sqrt{n}}$, t-scores rather than z-scores are the proper choice.

350. **(E)** In determining confidence intervals, one uses sample statistics to estimate population parameters. If the data are actually the whole population, making an estimate has no meaning.

351. **(D)** $\sigma_{\bar{x}} = \frac{3.7}{\sqrt{60}} = 0.4777$ and $\frac{24.4 - 23.4}{0.4777} = 2.093$. With $df = 60 - 1 = 59$,

$P(-2.093 \le t \le 2.093) = \text{tcdf}(-2.093, 2.093, 59) = 0.959 = 96\%$.

352. **(E)** We have $\bar{x} = 55{,}000$ and $s = 5{,}477$. With $df = 4 - 1 = 3$, the critical t-score for 95% confidence is invT(0.975, 3) = 3.182. The standard error is $\frac{s}{\sqrt{n}} = \frac{5{,}477}{\sqrt{4}}$. The confidence interval is

$$\bar{x} \pm t^* SE(\bar{x}) = 55{,}000 \pm 3.182\left(\frac{5{,}477}{\sqrt{4}}\right).$$

353. **(C)** Narrower intervals result from smaller standard deviations and from larger sample sizes. Note that I is the narrower interval.

354. **(E)** The 99% refers to the method; 99% of all intervals obtained by this method will capture μ. Nothing is sure about any particular set of 100 intervals. For any particular interval, the probability that it captures μ is either 1 or 0 depending on whether μ is or isn't in the interval.

355. **(A)** Since 615 is the mean SAT Math score of the entire population, the margin of error is 0.

356. **(D)** With $df = 18 - 1 = 17$ and with 0.02 in each tail, the critical t-value is invT(0.98, 17) = 2.224. The confidence interval is

$$\bar{x} \pm t^*SE(\bar{x}) = 14.5 \pm 2.224\left(\frac{3.9}{\sqrt{18}}\right).$$

357. **(D)** We are 95% confident that the population mean is within the interval calculated using the data from the sample.

358. **(C)** invT(0.975, 74) = 1.993 and the standard error of $\bar{x}$ is $\frac{640}{\sqrt{75}}$.

The question asks for a confidence interval for the *total*, not the mean, paid by all 26,000 families. The confidence interval for the total is

$$26{,}000[\bar{x} \pm t^*SE(\bar{x})] = 26{,}000\left[1{,}950 \pm 1.993\left(\frac{640}{\sqrt{75}}\right)\right].$$

359. **(D)** With $df = n - 1 = 15$, we have invT(0.975, 15) = 2.131. The 95% confidence interval is

$$\bar{x} \pm t^*\left(\frac{s}{\sqrt{n}}\right) = 101 \pm 2.131\left(\frac{2.4}{\sqrt{16}}\right) = 101 \pm 2.131\left(\frac{2.4}{4}\right).$$

360. **(D)** The critical t-score is invT(0.95, 20) = 1.7247. Then

$$\bar{x} \pm t^*(SE) = 7.4 \pm (1.7247)(0.1091) = 7.4 \pm 0.1882.$$

361. **(C)** The confidence interval is about a mean, not about individuals. One particular sample mean doesn't say anything about other sample means. The confidence interval is always about the population, not the sample, and the confidence interval is not about a probability.

362. **(E)** Choice (D) is an interpretation of the confidence *level*, while choice (E) is an interpretation of the confidence *interval*. The other choices are typical misinterpretations.

363. **(C)** The critical t-scores are ± 1.645 and $\bar{x}_1 - \bar{x}_2 = 1.95 - 1.05 = 0.9$ and the standard error is $\sqrt{\frac{s_1^2}{n_1} + \frac{s_2^2}{n_2}}$. The confidence interval is

$$(\bar{x}_1 - \bar{x}_2) \pm t^*SE(\bar{x}_1 - \bar{x}_2) = 0.9 \pm 1.645\sqrt{\frac{(0.65)^2}{500} + \frac{(0.4)^2}{450}}.$$

364. **(E)** Using a measurement from a sample, we are never able to say *exactly* what a population mean is. Rather, we say we have a certain *confidence* that the population mean (or in this case, the difference in population means) lies in a particular *interval*.

365. **(C)** The standard error is $\sqrt{\frac{s_1^2}{n_1}+\frac{s_2^2}{n_2}}$, and we are given $s_1 = 0.4$, $n_1 = 240$, $s_2 = 0.5$, and $n_2 = 347$.

366. **(D)** When the population standard deviations are unknown with appropriate conditions satisfied, $\frac{(\bar{x}_1 - \bar{x}_2)-(\mu_1 - \mu_2)}{SE(\bar{x}_1 - \bar{x}_2)}$ follows a *t*-distribution.

367. **(E)** With a tail probability of 0.04, the area above the interval $(0, \bar{x}_1 - \bar{x}_2)$ must be 0.46. So a confidence interval centered at $\bar{x}_1 - \bar{x}_2$ with width $2(0.46) = 0.92$ would go right up to 0. Note that $0.95 > 0.92$ and $0.99 > 0.92$ but that $0.90 < 0.92$.

368. **(D)** The standard error is $\sqrt{\frac{s_1^2}{n_1}+\frac{s_2^2}{n_2}}$, and we are given $s_1 = 1.5$, $n_1 = 10$, $s_2 = 1.1$, and $n_2 = 8$.

369. **(D)** A necessary assumption of the *t*-test is that the distribution of sample means is approximately normal for each population. If the sample sizes were large enough, this assumption would follow from the central limit theorem. With small sample sizes, though, the other way of satisfying this assumption is for the population distributions themselves to be approximately normal.

Confidence Intervals for Slopes (pages 132–135)

370. **(C)** The sample slope b is 11.8204, and the standard error for b is 1.848. With a tail probability of 0.05 and $df = 25 - 2$, the critical *t*-scores are ± 1.714. The 90% confidence interval for the slope of the regression line is $11.8204 \pm 1.714(1.848)$.

371. **(E)** The critical *t*-scores for 96% confidence with

$$df = 18 \text{ are } \pm \text{invT}(0.98, 18) = \pm 2.2137.$$

From the computer printout, the standard error of the slope is 1,678. The 96% confidence interval for the slope of the regression line is $15{,}675 \pm 2.214(1.678)$.

372. **(E)** The critical *t*-scores for 90% confidence with $df = 13 - 2 = 11$ are $\pm$ invT(0.95, 11) = ± 1.796. The standard error of the slope is 1.135. The 90% confidence interval for the slope of the regression line is $5.52979 \pm 1.796(1.135)$.

373. **(C)** With $df = 18$, invT(0.975, 18) = 2.101. The confidence interval is $10.292 \pm 2.101(0.8192) = 10.292 \pm 1.721$ or from 8.57 to 12.01. An explanation of the coefficient of determination is not an interpretation of the slope.

374. **(E)** Regression is a line of *averages*! Choices (B) and (C) are common misinterpretations of confidence intervals.

375. **(E)** The sample slope is at the middle of the interval, so

$$b = \frac{-0.142 + 1.036}{2} = 0.447.$$

The correlation coefficient has the same sign as the sample slope. The sum and the mean of the residuals is always 0.

Determining Sample Sizes (pages 135–139)

376. **(C)** Decreasing the sample size by a factor of d multiplies the interval estimate by $\sqrt{d}$.

377. **(D)** $1.881\left(\frac{0.56}{\sqrt{n}}\right) \le 0.15$, $\sqrt{n} \ge 7.0224$, and $n \ge 49.314$. So choose $n = 50$.

378. **(E)** $2.326\left(\frac{0.5}{\sqrt{n}}\right) \le 0.045$, $\sqrt{n} \ge 25.844$, and $n \ge 667.9$. Only choice (E), 700, is ≥ 667.9.

379. **(C)** Larger samples (so $\frac{\sigma}{\sqrt{n}}$ is smaller) and lower confidence (so the critical z or t is smaller) both result in narrower intervals.

380. **(D)** Note that $\sqrt{p(1-p)\left(\frac{1}{n_1} + \frac{1}{n_2}\right)} \le (0.5)\sqrt{\frac{1}{n_1} + \frac{1}{n_2}}$. So if $n_1 = n_2 = n$, we have $(0.5)\sqrt{\frac{1}{n} + \frac{1}{n}} = (0.5)\sqrt{\frac{2}{n}} = \frac{0.5\sqrt{2}}{\sqrt{n}}$.

381. **(E)** To divide the interval estimate by d without affecting the confidence level, increase the sample size by a multiple of d^2. In this case, $9(200) = 1{,}800$.

382. **(D)** $\sigma_{\bar{x}_1 - \bar{x}_2} = \sqrt{\frac{(0.93)^2}{n} + \frac{(0.93)^2}{n}}$ and invNorm(0.995) = 2.576. To find the sample size, use $2.576\sqrt{\frac{0.93^2}{n} + \frac{0.93^2}{n}} \le 0.1$.

383. **(B)** Multiplying a sample size by a factor of m decreases the margin of error by a divisor of $\sqrt{m}$. Because sample T is 4 times the size of sample S, the margin of error using S is 2 times the margin of error of T.

384. **(D)** $1.96\frac{0.1}{\sqrt{n}} \le 0.008$ gives $\sqrt{n} \ge 24.5$ and $n \ge 600.25$, so choose $n = 601$.

385. **(D)** The critical z-score for 98% confidence is invNorm(0.99) = 2.326. Then we want $2.326\sqrt{\frac{p(1-p)}{n}} \le 0.035$. Since $p(1-p) \le (0.5)(0.5)$, it is sufficient to solve for n in $2.326\sqrt{\frac{(0.5)(0.5)}{n}} \le 0.035$ or $2.326\frac{0.5}{\sqrt{n}} \le 0.035$.

386. **(A)** The critical z-score is invNorm(0.95) = 1.645, and the standard deviation of the sampling distribution of $\bar{x}$ is $\frac{\sigma}{\sqrt{n}} = \frac{8}{\sqrt{n}}$. So a margin of error of size 5 gives $1.645\frac{8}{\sqrt{n}} \le 5$.

387. **(D)** The critical z-score is invNorm(0.98) = 2.054. The standard deviation of the sampling distribution of $\hat{p}$ is $\sqrt{\frac{p(1-p)}{n}}$. If p is completely unknown, we use 0.5 as the most conservative value in the calculation of n, but here we have $p \approx 0.20$. The equation $2.054\sqrt{\frac{(0.20)(0.80)}{n}} \le 0.03$ gives $\sqrt{n} \ge 27.387$, $n \ge 750.04$. Among the given choices, the smallest that is still greater than 750.04 is 800.

Logic of Significance Testing (pages 139–144)

388. **(D)** We attempt to show that the null hypothesis is unacceptable by showing that it is improbable; however, we cannot show that it is definitely true or false. If there is interest in deviations in only one direction, the alternative hypothesis is expressed using either < or >. When a true parameter value is further from the hypothesized value, it becomes easier to reject the null hypothesis. Increasing the sample size makes it easier to conclude that an observed difference between observed and hypothesized values is significant.

389. **(D)** Hypotheses are always about population parameters, not about sample statistics. The P-value does not give the probability that the null hypothesis is true; it gives the probability of such an extreme value or greater, assuming the null hypothesis is true. The smaller the P-value is, the stronger the evidence against the null hypothesis. If the P-value is small enough, we have evidence in support of the alternative hypothesis, but this does not prove for a fact that the alternative hypothesis is true. The null and alternative hypotheses are decided on before the data come in.

390. **(A)** The P-value is the smallest value of the significance level α for which the null hypothesis would be rejected. Equivalently, if the null hypothesis

is assumed to be true, the *P*-value of a sample statistic is the probability of obtaining a result as extreme as or more extreme than the one obtained. The *P*-value does not give the probability that the null hypothesis is true. Depending on the sample we happen to choose, we may mistakenly reject a true null hypothesis, in which case we commit a Type I error. A large *P*-value simply means that we do not have sufficient evidence to reject the null hypothesis; it does not mean we have evidence that the null hypothesis is true. The smaller the *P*-value is, the greater the evidence of something statistically significant.

391. **(A)** There is nothing to hypothesize about if the population parameter is already known. The *P*-value is a probability, so it is never negative. The *P*-value is based on a specific test statistic, and the test statistic is not known until after the experiment is conducted. If a *P*-value is smaller than a specified value α, the data are statistically significant at that level. The *P*-value is a probability calculated when assuming that the null hypothesis is true.

392. **(E)** If the sample statistic is far enough away from the claimed population parameter, we say there is sufficient evidence to reject the null hypothesis. In this case the null hypothesis is H_0: $\mu = 32.8$. The *P*-value is the probability of obtaining a sample statistic as extreme or more extreme than the one obtained if the null hypothesis is assumed to be true. The smaller the *P*-value, the more significant the difference between the null hypothesis and the sample results. With $P = 0.0416 < 0.05$, there is sufficient evidence to reject H_0.

393. **(A)** The *P*-value is *not* the probability that the null hypothesis is true, Rather it is the probability of obtaining a value as extreme or more extreme than the observed statistic *given that* the null hypothesis is true. Thus, it is a conditional probability. Power is the probability of not committing a Type II error, that is, the probability of rejecting the null hypothesis when a given alternative is true. *Small P*-values are evidence against the null hypothesis.

394. **(E)** With such small populations, censuses instead of samples are used, and there is no resulting probability statement about the difference.

395. **(E)** The proper conclusion is that we are 90% confident that the difference between the proportion of families not living near chemical plants who have children with leukemia and the proportion of families living near chemical plants who have children with leukemia is between –0.037 and –0.003. It is significant to note that 0 is not in this interval.

396. **(E)** For a confidence interval with $n = 11$, we use $df = n - 1 = 10$. For a two-population hypothesis test of the mean, we use the conservative

$$df = \min(n_1 - 1, n_2 - 1) = 11$$

or sometimes we can use $df = (n_1 - 1) + (n_2 - 1) = 22$. Instead, a calculator will give a fractional *df*. If a contingency table has 4 rows and 3 columns, $df = (4 - 1)(3 - 1) = 6$.

397. **(E)** The hypotheses are always about population parameters, never about sample statistics.

398. **(A)** A straightforward *z*-test is called for with H_0: $p = 0.85$ versus H_a: $p < 0.85$.

399. **(E)** With a two-sided test, half the *P*-value is in each tail. In this problem, 0.05 would be in each tail. If the test statistic is in the direction of the alternative hypothesis, the new *P*-value would be 0.05. If the test statistic is in the opposite direction of the alternative, though, the new *P*-value would be 0.95.

400. **(E)** It is bad practice for researchers to fish around in their data and keep trying different tests until they find something with the right *P*-value that clears an arbitrary threshold such as 0.05.

401. **(D)** Independence in sampling is necessary. When we sample without replacement (because we don't want to pick the same subject twice), independence is technically violated. However, if the sample isn't too large (less than 10 percent of the population), this dependence among observations is small enough to be acceptable for the test to proceed.

402. **(D)** The new *P*-value depends on the direction of the alternative hypothesis. With a two-sided test, half the *P*-value is in each tail. In this problem, 0.16 is in each tail. If the test statistic is in the direction of the alternative hypothesis, the new *P*-value is 0.16. However, if the test statistic is in the opposite direction of the alternative, the new *P*-value is 0.84.

403. **(E)** All three of these statements follow once it is given that there is actually no difference in support for the politician between male and female voters. This illustrates the important point that if enough tests are performed at a given significance level, some of them will result in small *P*-values simply by sample variability.

Hypothesis Tests for Proportions (pages 144–152)

404. **(A)** We have H_0: $p = 0.07$ and H_a: $p > 0.07$. We also have

$$\sigma_{\hat{p}} = \sqrt{\frac{(0.07)(0.93)}{200}} = 0.0180 \text{ and } \hat{p} = \frac{23}{200} = 0.115 \text{ with a } z\text{-score}$$

of $\frac{0.115 - 0.07}{0.0180} = 2.5$. Thus the *P*-value is normalcdf(2.5, 1000) = 0.0062. (1-PropZTest quickly gives $P = 0.0063$.) This low a *P*-value is strong evidence against H_0. The truant officer has strong evidence that the percentage of students skipping school during the World Series is even greater than the previously claimed 7%.

405. **(C)** $\hat{p} = \frac{57}{150} = 0.38$. With H_0: $p = 0.35$, the standard deviation of the test statistic is $\sigma_{\hat{p}} = \sqrt{\frac{p(1-p)}{n}} = \sqrt{\frac{(0.35)(1-0.35)}{150}}$. So the appropriate test statistic is $z = \frac{0.38 - 0.35}{\sqrt{\frac{(0.35)(1-0.35)}{150}}}$.

406. **(D)** $\hat{p} = \frac{280}{500} = 0.56$, and the standard deviation of the test statistic is $\sigma_{\hat{p}} = \sqrt{\frac{p(1-p)}{n}} = \sqrt{\frac{(0.54)(1-0.54)}{500}}$. Since this is a two-sided test, the *P*-value will be twice the tail probability of the test statistic. So the *P*-value is $2P\left(z > \frac{0.56 - 0.54}{\sqrt{\frac{(0.54)(1-0.54)}{500}}}\right)$.

407. **(B)** A two-sided test with $\alpha = 0.01$ has critical *z*-values of ±2.576, while a one-sided test with $\alpha = 0.01$ to the left has a critical *z*-value of –2.326. The only answer further out than –2.326 but not as far out as –2.576 is $z = -2.4$.

408. **(B)** In a hypothesis test for a population proportion, the standard error (*SE*) is $\sqrt{\frac{p_0(1-p_0)}{n}}$, where p_0 is the proportion claimed in the null hypothesis. The test statistic is $z = \frac{\hat{p} - p_0}{SE}$. The test statistic for the appropriate test is $\frac{0.66 - 0.75}{\sqrt{\frac{(0.75)(0.25)}{500}}}$.

409. **(E)** The *P*-value is 0.18, and the significance level is $\alpha = 0.10$. Since $0.18 > 0.10$, there is not sufficient evidence to reject H_0. In other words, there is not sufficient evidence to say that the proportion of American adults who are functionally illiterate is now less than 22 percent.

410. **(B)** The tail probability is normalcdf(1.15, 100) = 0.125. This is a two-sided test, so the *P*-value is twice the tail probability, or 0.25.

411. **(E)** We are comparing two population proportions. Thus the correct test involves H_0: $p_1 - p_2 = 0$ and H_a: $p_1 - p_2 > 0$, where p_1 and p_2 are the proportion of Europeans and Japanese, respectively, who believe bluefin tuna need protection.

412. **(B)** This is a two-proportion z-test with H_0: $p_1 - p_2 = 0$, H_a: $p_1 - p_2 > 0$, $\hat{p}_1 = \frac{315}{500} = 0.63$, $\hat{p}_2 = \frac{220}{400} = 0.55$, $\hat{p} = \frac{315 + 220}{500 + 400} = 0.594$, and $\sigma_d = \sqrt{(0.594)(0.406)(1/500 + 1/400)} = 0.0329$. The z-score of the difference is $\frac{0.63 - 0.55}{0.0329} = 2.432$ with $P = P(z > 2.432) = 0.0075$. (2-PropZTest quickly gives $P = 0.0076$.) With this small of a P-value, $0.0076 < 0.01$, there is very strong evidence that the proportion of STEM majors who are more interested in being challenged than in receiving high grades is greater than the proportion of business majors who are more interested in being challenged than in receiving high grades.

413. **(B)** We have H_0: $p_1 - p_2 = 0$, H_a: $p_1 - p_2 \neq 0$, $\hat{p}_1 = \frac{48}{53}$, $\hat{p}_2 = \frac{251}{285}$, $\hat{p} = \frac{48 + 251}{53 + 285}$, and $\sigma_d = \sqrt{\left(\frac{48 + 251}{53 + 285}\right)\left(1 - \frac{48 + 251}{53 + 285}\right)\left(\frac{1}{53} + \frac{1}{285}\right)}$.

The z-score of the difference is $\frac{\frac{48}{53} - \frac{251}{285}}{\sigma_d}$. This is a two-sided test, so the P-value is twice the tail probability.

414. **(B)** We have H_0: $p_1 - p_2 = 0$, H_a: $p_1 - p_2 < 0$, $\hat{p}_1 = \frac{560}{800} = 0.7$, and $\hat{p}_2 = \frac{450}{600} = 0.75$. Since the null hypothesis is that $p_1 = p_2$, we call this common value p and use it in calculating σ_d. Then $\hat{p} = \frac{560 + 450}{800 + 600} = 0.7214$, and $\sigma_d \approx \sqrt{(0.7214)(1 - 0.7214)(1/800 + 1/600)}$.

So the test statistic is $z = \frac{0.7 - 0.75}{\sqrt{(0.7214)(1 - 0.7214)(1/800 + 1/600)}}$.

415. **(E)** "The probability of observing a difference at least as large as the sample difference, if the two population proportions are the same" is the definition of the P-value. H_0: $p_M = p_W$ and H_a: $p_M > p_W$. 2-PropZTest gives $z = 1.599$ with a P-value of 0.0549. With this large of a P-value, $0.0549 > 0.05$, there is not sufficient evidence to reject H_0. In other words, there is not sufficient evidence to say that the proportion of men who are satisfied with their physical attractiveness is greater than the proportion of women who are satisfied with their physical attractiveness.

416. **(D)** 2-PropZTest gives $z = -2.08$ and $P = 0.0373$. With this small of a P-value, $0.0373 < 0.05$, there is sufficient evidence to reject H_0. In other words, there is sufficient evidence that the proportion of all Americans who

say they were born in another country is different from the proportion of all Canadians who say they were born in another country.

417. **(B)** The proper procedure is to compare the difference between the proportion of students at the first booth who picked fine Swiss chocolates (or Hershey chocolate kisses) and the proportion of students at the second booth who picked fine Swiss chocolates (or Hershey chocolate kisses).

418. **(B)** There is only a single sample! The two given proportions are not from two independent samples.

419. **(E)** The *P*-value is a conditional probability. In this case, there is a 0.028 probability of an observed difference in sample proportions as extreme as or more extreme than the one obtained if the null hypothesis is assumed to be true.

Hypothesis Tests for Means (pages 152–167)

420. **(D)** Since the sample sizes are small, the samples must come from normally distributed populations. Although the samples should be independent, np and $n(1-p)$ refer to conditions for tests involving sample proportions, not means.

421. **(D)** We have H_0: $\mu = 62.4$ and H_a: $\mu \neq 62.4$. The standard deviation of the sample means is $\frac{10.3}{\sqrt{32}}$. With $df = 32 - 1 = 31$, and with a two-sided test so that the *P*-value is twice the tail probability, we have the *P*-value equals $2P\left(t > \frac{65.0 - 62.4}{\left(\frac{10.3}{\sqrt{32}}\right)}\right)$.

422. **(C)** With $df = 20 - 1 = 19$, the critical *t*-scores for the 0.05 and 0.01 tail probabilities are 1.729 and 2.539, respectively. We have $2.615 > 1.729$ and $2.615 > 2.539$. So 2.615 is significant at both the 5% and the 1% levels. Alternatively, we could calculate $P(t > 2.615)$ = tcdf(2.615, 1000, 19) = 0.0085, and $0.0085 < 0.05$ and $0.0085 < 0.01$.

423. **(C)** The margin of error varies directly with the critical *z*-value and directly with the standard deviation of the sample but inversely with the square root of the sample size.

424. **(D)** We have H_0: $\mu = 1{,}250$ and H_a: $\mu < 1{,}250$. A *t*-test gives $t = -1.777$ and $P = 0.0516$. (Alternatively, $t = \frac{1{,}092 - 1{,}250}{\left(\frac{308}{\sqrt{12}}\right)} = -1.777$ with a *P*-value of $P(t < -1.777)$ = tcdf(−1000, −1.777, 11) = 0.0516.)

425. **(C)** If the true mean number of books read is 43, the probability curve should be centered at 43. The rejection region is $\bar{x} > 40$, so the failure to reject region will be $\bar{x} < 40$. With a standard error of $\frac{12}{\sqrt{100}}$, $P(\bar{x} < 40) = P\left(t < \frac{40 - 43}{\left(\frac{12}{\sqrt{100}}\right)}\right)$.

426. **(C)** We have H_0: $\mu = 9$, H_a: $\mu > 9$, and $\sigma_{\bar{x}} \approx \frac{3}{\sqrt{64}} = 0.375$. The t-score of 9.55 is $\frac{9.55 - 9}{0.375} = 1.467$, which gives a P-value of $P(t > 1.467) =$ tcdf(1.467, 1000, 63) = 0.0737. (Using the calculator, T-Test quickly gives $P = 0.0737$.) A P-value between 0.05 and 0.10 indicates some evidence against H_0. The program developer has some evidence to dispute the dietician's claim that the new weight loss program will result in an average loss of 9 pounds in the first month.

427. **(B)** A calculator gives $\bar{x} = 5.21$ and $s = 0.0652$. Since we are not told that the inspector suspects that the average weight is over or under 5.25 ounces and since a baseball that is either underweight or overweight clearly should be brought to the manufacturer's attention, this is a two-sided test. Thus the P-value is twice the tail probability obtained using the t-distribution with $df = n - 1 = 4$. So the P-value of this test is $2P\left(t < \frac{5.21 - 5.25}{\left(\frac{0.0652}{\sqrt{5}}\right)}\right)$.

428. **(D)** The standard deviation of the test statistic is $\sigma_{\bar{x}} = \frac{\sigma}{\sqrt{n}} \approx \frac{s}{\sqrt{n}} = \frac{38{,}000}{\sqrt{10}}$.

Since this is a two-sided test, the P-value will be twice the tail probability of the test statistic. However, the test statistic itself is not doubled. The test statistic in performing a hypothesis test of H_0 is $t = \frac{245{,}000 - 258{,}000}{38{,}000/\sqrt{10}}$.

429. **(B)** The particular result of one particular sample has nothing to do with other samples. The first statement comes directly from the definition of a P-value. With $P < 0.05$, there is sufficient evidence to reject the null hypothesis at the 5% significance level. The t-distribution has fatter tails than the normal distribution. In a two-sided test, we are interested in deviations in either direction, and so half the P-value is in each tail.

430. **(D)** We have H_0: $\mu = 3{,}700$ and H_a: $\mu < 3{,}700$. Here

$$\sigma_{\bar{x}} \approx \frac{s}{\sqrt{n}} = \frac{1{,}120}{\sqrt{30}} = 204.5.$$

The t-score of 3,490 is $\frac{3{,}490 - 3{,}700}{204.5} = -1.027$. With $df = 30 - 1 = 29$, the

P-value is tcdf(−1000, −1.027, 29) = 0.156. (Using the calculator, T-Test quickly gives $P = 0.156$.) With such a large P-value, there is little evidence against H_0. The teacher should not reject the salesperson's claim.

431. **(D)** The standard error of the test statistic is $SE(\bar{x}) = \frac{s}{\sqrt{n}} = \frac{7.5}{\sqrt{30}}$. The test statistic is $t = \frac{\bar{x} - x_0}{SE(\bar{x})} = \frac{104 - 100}{\left(\frac{7.5}{\sqrt{30}}\right)}$.

432. **(D)** The distance from $\bar{x}$ to 20 is the same as from 20 to $\bar{x}$. Since 20 is in the 95 percent interval centered at $\bar{x}$, $\bar{x}$ must be in the 95 percent interval centered at 20.

433. **(E)** With $df = n - 1 = 6 - 1 = 5$, the tail probability was

tcdf(1.641, 100, 5) = 0.08086.

This is a two-sided test. So the *P*-value is twice the tail probability or 0.1617.

434. **(B)** We're given that the Walmart mean is known, so we are not testing its mean in the hypotheses. The null hypothesis is always an equality statement. The interest is in whether the Target mean is greater than the Walmart mean of $75.

435. **(E)** The hypotheses are H_0: $\mu = 3.09$ versus H_a: $\mu > 3.09$. If the *P*-value is less than the significance level, there is sufficient evidence to reject the null hypothesis. In other words, there is sufficient evidence in favor of the alternative hypothesis.

436. **(A)** We have H_0: $\mu_1 - \mu_2 = 0$, H_a: $\mu_1 - \mu_2 > 0$. With fractional *df*, 2-SampTTest gives $P = 0.00115$.

437. **(C)** Two-sample tests require that the two samples being compared be independent of each other. However, in this case, the data occur in related pairs, namely, the cursive and printing speeds of each individual student. The proper procedure is to run a one-sample test on the single variable consisting of the difference between the cursive and printing speeds of each student.

438. **(A)** We have H_0: $\mu_1 - \mu_2 = 0$, H_a: $\mu_1 - \mu_2 > 0$, and $\sigma_{\bar{x}_1 - \bar{x}_2} \approx \sqrt{\frac{1.8^2}{45} + \frac{1.8^2}{45}}$. So $\alpha = P\left(z > \frac{1 - 0}{\sqrt{\frac{1.8^2}{45} + \frac{1.8^2}{45}}}\right)$. Note that with known population standard deviations, we can use a normal model.

439. **(A)** The standard error of the test statistic is $SE(\bar{x}_1 - \bar{x}_2) = \sqrt{\frac{(7.4)^2}{230} + \frac{(5.6)^2}{185}}$. Since this is a two-sided test, the *P*-value will be twice the tail probability of the test statistic; however, the test statistic itself is not doubled. The test statistic is $t = \frac{\bar{x}_1 - \bar{x}_2}{SE(\bar{x}_1 - \bar{x}_2)} = \frac{76.5 - 78.8}{\sqrt{\frac{(7.4)^2}{230} + \frac{(5.6)^2}{185}}}$.

440. **(E)** The two-sample hypothesis test should be used only when the two sets are independent. In this case, there is a clear relationship between the data, which occur in pairs. This relationship is completely lost in the procedure for the two-sample test. The proper procedure is to run a one-sample test on the single variable consisting of the differences from the paired data.

441. **(D)** H_0: $\mu_1 - \mu_2 = 0$ and H_a: $\mu_1 - \mu_2 \neq 0$. On the calculator, 2-SampTTest gives $P = 0.067$. Since $0.067 > 0.05$, there is not sufficient evidence to reject H_0 at the 5% significance level. In other words, there is not sufficient evidence that the mean winnings using the two strategies are different.

442. **(E)** Two-sample tests require that the two samples being compared be independent of each other. However, in this case, the data occur in related pairs, namely, the prices of the same 50 items on the two sites. The proper procedure is to run a one-sample test on the single variable consisting of the difference between online prices for each item.

443. **(E)** Using a measurement from a sample, we are never able to say exactly what a population mean is. Rather, we say we have a certain confidence that the population mean lies in a particular interval. In this case, we are 95% confident that the difference in the mean clotting time increase between Coumadin and Plavix is between 2.1 and 3.7 seconds.

444. **(B)** The data come in pairs, two ratings from each player, hence the matched pairs design. The *t*-test is run on a single sample, the set of 12 score differences.

445. **(B)** The null hypothesis is that the difference of population means is 0. Thus if the magnitude of the difference of the two sample means is larger, it would be further from 0, which makes the *P*-value smaller. If the sample standard deviations are smaller, the standard error (SE) of the differences is smaller and any observed difference will have a greater absolute value *t*-score. So again the *P*-value would be smaller.

446. **(D)** Two-sample *t*-tests of the difference between two population means are appropriate when the two samples being compared are independent of each other. In all the answer choices except for choice (D), the data are selected in naturally related pairs and the proper procedure is a one-sample analysis on the set of differences.

447. **(D)** The standard error of $\bar{x}_1 - \bar{x}_2$ is $\sqrt{\frac{s_1^2}{n_1} + \frac{s_2^2}{n_2}}$ and $t = \frac{\bar{x}_1 - \bar{x}_2 - 0}{SE(\bar{x}_1 - \bar{x}_2)}$.

So the *t*-statistic is $t = \frac{75.4 - 71.3}{\sqrt{\frac{9.8^2}{175} + \frac{8.2^2}{160}}}$.

448. **(A)** This was an experiment because treatments were applied (putting tees for sale into stores). A cause-and-effect conclusion is possible because the two treatments were randomly assigned. The fact that the 60 stores were not randomly selected could limit the stores to which the conclusion can be generalized.

449. **(D)** The data come in pairs, and the two-sample test does not apply the knowledge of what happened to each individual driver (the condition of independence of the two samples is violated). This is a two-sided test as we are simply testing for a difference in stopping times.

450. **(B)** Although both lead to the same conclusion, this did not have to happen because the computer printout shows a 2-sided confidence interval but a 1-sided hypothesis test.

451. **(D)** The data came in pairs, by months, so a match paired test is appropriate rather than a two-sample test. The interest is in whether there are more admissions on Friday the 13ths, that is, in whether the differences (6ths minus 13ths) are negative.

452. **(B)** The study compared the means of two independent random samples.

Type I and II Errors, Powers (pages 168–174)

453. **(B)** We have H_0: $\mu = 29{,}500$ and H_a: $\mu < 29{,}500$. The standard deviation of the sample means is $\sigma_{\bar{x}} = \dfrac{2{,}500}{\sqrt{40}}$. A Type I error is a mistaken rejection of a true claim. $P(\bar{x} < 29{,}000) = P\left(z < \dfrac{29{,}000 - 29{,}500}{\left(\dfrac{2{,}500}{\sqrt{40}}\right)}\right)$. Note that the population standard deviation is known, so we use the normal model instead of a t-model.

454. **(E)** A Type I error means that the null hypothesis is correct (the weather remains dry) but that you reject it (thus you needlessly carry around an umbrella). A Type II error means that the null hypothesis is wrong (it rains) but you fail to reject it (thus you get drenched).

455. **(C)** Power = $1 - \beta$ where β is the probability of a Type II error and where β is smallest when α and n are greater. Among the choices, the greatest α is 0.05, and the greatest n is 45.

456. **(C)** A Type II error is a mistaken failure to reject a false null hypothesis or, in this case, a failure to realize that the machine is turning out spinners with low spin times.

457. **(C)** Increasing the sample size allows us to decrease the probabilities of both Type I and Type II errors and to increase the power. The significance level α

is the same as the probability of a Type I error, and 1 – power is the same as the probability of a Type II error.

458. **(B)** We have H_0: $\mu = 6.0$ and H_a: $\mu < 6.0$. The standard deviation of sample means is $\sigma_{\bar{x}} = \frac{\sigma}{\sqrt{n}} = \frac{0.75}{\sqrt{50}}$. Since the population standard deviation is known, we use a normal distribution rather than a *t*-distribution. The probability the inspector will commit a Type I error is $P\left(z < \frac{5.5 - 6.0}{\left(\frac{0.75}{\sqrt{50}}\right)}\right)$.

459. **(C)** In general, there is one probability of a Type I error that is at the significance level and there is a different probability of a Type II error associated with each possible alternative answer. Although Type I and Type II errors are related, their sum has no specific value. Power is 1 minus the probability of a Type II error given a specific alternative. Power is the probability of rejecting a false null.

460. **(B)** If the alternative is true, the probability of failing to reject H_0 and thus committing a Type II error is 1 minus the power, that is, 1 – 0.8 = 0.2.

461. **(A)** *P*(Type I error) = *P*(reject H_0 | H_0 is true). Type I and Type II errors are caused by variability in data collection processes, not by human error. The probabilities of Type I and Type II errors are related. For example, lowering the probability of a Type I error increases the probability of a Type II error. A Type I error can be made only if the null hypothesis is true, while a Type II error can be made only if the null hypothesis is false.

462. **(A)** The significance level is the probability of a Type I error, not of a Type II error. When holding the significance level fixed, increasing the sample size reduces the probability of a Type II error and raises the power. Increasing the significance level makes it easier to reject a null hypothesis, which in turn decreases the probability of failing to reject a wrong null hypothesis.

463. **(C)** In the production of surgical gowns, the more serious concern is a Type II error, which is that the gowns are not performing correctly (they leak!), yet the check does not pick up this. At the blanket manufacturing plant, the more serious concern is a Type I error, which is that the equipment is performing correctly yet the check causes a production halt.

464. **(C)** There is a different probability of Type II error for each possible correct value of the population parameter, and 1 minus this probability is the power of the test against the associated correct value. Increasing the sample size makes the standard deviations smaller and thus can decrease the probabilities of Type I and Type II errors and so increase the power. Increasing the significance level α lowers the probability of a Type II error and thus also increase the power. Both decreased variability in the data and a greater difference

between the true and hypothesized parameters make it more likely to detect that the null hypothesis is false.

465. **(C)** This is a hypothesis test with H_0: bungee cord strength is within specifications, and H_a: bungee cord strength is below specifications. A Type I error is committed when a true null hypothesis is mistakenly rejected, or in this case, bungee cord strength is within specifications but the production process is halted.

466. **(D)** Given a particular alternative, power equals 1 minus the probability of a Type II error, and a Type II error is the probability of mistakenly failing to reject the false null hypothesis. In this case, the probability of a Type II error is $1 - 0.97 = 0.03 = 3$ percent.

467. **(A)** A Type I error means that the null hypothesis is correct (the smart phone will sell for only $900) but you reject H_0 (thus the company produces a nonprofitable smart phone.) A Type II error means that the null hypothesis is wrong (the smart phone will sell for over $900) but you fail to reject H_0 (thus you fail to produce a profitable smart phone).

468. **(D)** Power is the probability of correctly rejecting a false null hypothesis. A false null hypothesis in this context means that the employees are abusing the sick day policy. Correctly rejecting this means correctly accusing employees who are abusing the policy.

469. **(C)** The probability of committing a Type I error is precisely the significance level. Thus, if the significance level is kept at 0.10, the probability of committing a Type I error stays the same.

470. **(B)** Power is the probability of correctly rejecting a false null hypothesis. The further the true proportion is from the false claimed proportion in the direction of the alternative, the easier it is to show the null is false and so the greater the power. Also, the larger the sample size is, the smaller the standard error of the sampling distribution. Therefore, the more meaningful the difference is between the true proportion and the false claimed proportion, again resulting in the greater power.

471. **(B)** A Type II error is a mistaken failure to reject a false null hypothesis. In this context, a false null hypothesis means the subject does have cancer. A failure to recognize this means the cancer was not recognized. Note that choice (D) is a Type I error, while the probability of choice (A) is the power of the test.

472. **(E)** There is a different probability of Type II error for each possible correct value of the population parameter.

P-Values from Simulations (pages 175–176)

473. **(E)** The dotplot shows the differences in proportions we would expect to see by random chance if the null hypothesis of no difference was true. In $18 + 11 + 6 + 3 + 1 + 1 = 40$ out of the 200 simulations, the difference was equal to or greater than 0.1106. The estimated *P*-value is $\frac{40}{200} = 0.20$. With this large of a *P*-value, $0.20 > 0.05$, there is not sufficient evidence to reject the null hypothesis. In other words, there is not sufficient evidence of a difference in the proportion of happy days experienced by husbands and wives in marriages where the wife works and the husband is a stay-at-home dad.

474. **(E)** The control check has hypotheses: H_0: the machinery is operating properly, and H_a: the machinery is not operating properly. In the simulation with the machine operating properly, there were 3 values out of 100 that were 24 or greater. This gives an estimated *P*-value of 0.03. With this small of a *P*-value, $0.03 < 0.05$, there is sufficient evidence to reject H_0. In other words, there is sufficient evidence to necessitate a recalibration of the machinery.

Chi-Square Tests (pages 177–187)

475. **(B)** We have H_0: support and gender are independent, and H_a: support and gender are not independent. Putting the data into a matrix and using χ^2-test gives $\chi^2 = 5.550$ and $P = 0.062$. With this *P*-value, $0.05 < 0.062 < 0.10$, there is evidence at the 10% but not at the 5% significance level to reject H_0. In other words, there is weak evidence of a relationship between gender and support for the initiative.

476. **(A)** The expected values given independence can be calculated from the row and column totals: $\frac{(70)(83)}{150} = 38.7$, $\frac{(70)(67)}{150} = 31.3$, $\frac{(80)(83)}{150} = 44.3$, and $\frac{(80)(67)}{150} = 35.7$. Then

$$\chi^2 = \sum \frac{(\text{obs} - \exp)^2}{\exp}$$

$$= \frac{(46 - 38.7)^2}{38.7} + \frac{(24 - 31.3)^2}{31.3} + \frac{(37 - 44.3)^2}{44.3} + \frac{(43 - 35.7)^2}{35.7}.$$

477. **(C)** Although the observed frequencies are whole numbers, the expected frequencies, as calculated by the rule in choice (E), may or may not be whole numbers.

478. **(B)** This is a goodness of fit test for a uniform distribution. The expected value of each cell for a uniform distribution is $\frac{12+5+9+4+15}{5} = \frac{45}{5} = 9$. Then

$$\chi^2 = \sum \frac{(\text{obs} - \text{exp})^2}{\text{exp}}$$

$$= \frac{(12-9)^2}{9} + \frac{(5-9)^2}{9} + \frac{(9-9)^2}{9} + \frac{(4-9)^2}{9} + \frac{(15-9)^2}{9}.$$

479. **(C)** Since 1 + 6 + 9 = 16, according to the geneticist, the expected number of fruit flies of each species is $\frac{1}{16}(320) = 20$, $\frac{6}{16}(320) = 120$, and $\frac{9}{16}(320) = 180$.

We calculate $\chi^2 = \frac{(25-20)^2}{20} + \frac{(100-120)^2}{120} + \frac{(195-180)^2}{180} = 5.8333$.

With $df = 3 - 1 = 2$, the P-value is χ^2cdf(5.8333, 1000, 2) = 0.0541. (Alternatively, putting {25, 100, 195} in L1 and {20, 120, 180} in L2 and then using χ^2GOF-Test gives the same result.) With this P-value, 0.0541 > 0.05, there is not sufficient evidence to reject the geneticist's claim at the 5% significance level.

480. **(E)** Three of the *expected* cell counts (for the "Grades D, F" cells) are all 3. With this number of cell counts less than 5, it is not proper to use a chi-square test of homogeneity.

481. **(B)** This is a chi-square test of independence. The expected values given independence can be calculated from the row and column totals: $\frac{(110)(65)}{170} = 42.1$, $\frac{(110)(55)}{170} = 35.6$, $\frac{(110)(50)}{170} = 32.4$, $\frac{(60)(65)}{170} = 22.9$, $\frac{(60)(55)}{170} = 19.4$, and $\frac{(60)(50)}{170} = 17.6$. (With a calculator, you can use Matrix and χ^2-Test to find the expected values.) Then

$$\chi^2 = \sum \frac{(\text{obs} - \text{exp})^2}{\text{exp}}$$

$$= \frac{(55-42.1)^2}{42.1} + \frac{(40-35.6)^2}{35.6} + \frac{(15-32.4)^2}{32.4} + \frac{(10-22.9)^2}{22.9}$$

$$+ \frac{(15-19.4)^2}{19.4} + \frac{(35-17.6)^2}{17.6}$$

and df = (rows – 1)(columns – 1) = (2 – 1)(3 – 1) = 2.

482. **(D)** Chi-square tests are for categorical data. In particular, the chi-square test of independence looks at one group of subjects cross-classified by two categorical variables.

483. **(C)** The expected values are 0.21(80) = 16.8, 0.26(80) = 20.8, 0.12(80) = 9.6, and 0.41(80) = 32.8. We calculate

$$\chi^2 = \sum \frac{(\text{obs} - \text{exp})^2}{\text{exp}}$$

$$= \frac{(23 - 16.8)^2}{16.8} + \frac{(14 - 20.8)^2}{20.8} + \frac{(10 - 9.6)^2}{9.6} + \frac{(33 - 32.8)^2}{32.8}$$

The degrees of freedom are 4 – 1 = 3.

484. **(E)** First note the row and column totals. The proportion of Republicans is $\frac{170}{340}$, so $\frac{170}{340}(93) = \frac{(170)(93)}{340}$ is the expected number of Republicans who answer Willard. Alternatively, the proportion of people answering Willard is $\frac{93}{340}$, so $\frac{93}{340}(170) = \frac{(93)(170)}{340}$ gives the expected number of Republicans who answer Willard.

485. **(A)** This is a goodness of fit test for a uniform distribution. If the admissions were equally spread out, there would have been $\frac{35 + 77 + 53}{3} = 55$ on each day. We then calculate

$$\chi^2 = \frac{(35 - 55)^2}{55} + \frac{(77 - 55)^2}{55} + \frac{(53 - 55)^2}{55} = 16.15.$$

With $df = 3 - 1 = 2$, the P-value is χ^2cdf(16.15, 1000, 2) = 0.0003. (Using lists, putting {35, 77, 53} in L1 and {55, 55, 55} in L2, χ^2GOF-Test gives the same result.) With such a small P-value, there is very strong evidence that the distribution is not uniform. In other words, the number of admissions is different on the different days depending on phases of the moon.

486. **(E)** The chi-square test of independence is the proper procedure to test for an association between two categorical variables when there is one random sample cross-classified on the two variables.

487. **(E)** Chi-square measures differences. The greater the differences are, the greater the chi-square statistics. The sample sizes do not have to be the same for a test of homogeneity. The samples of men and women should be independent, which would not be the case if married couples were used. As long as all expected cells are at least five (and we are given that all conditions for inference are met), the sample sizes are not too small to run this chi-square

test. The null hypothesis is that the proportion of people preferring each of the four TV models is the same for men and for women.

488. **(C)** This is a chi-square goodness of fit test. We have 3 + 2 + 3 + 4 = 12. Thus the expected numbers according to the engineer's claim are $\frac{3}{12}(3{,}600) = 900$, $\frac{2}{12}(3{,}600) = 600$, $\frac{3}{12}(3{,}600) = 900$, and $\frac{4}{12}(3{,}600) = 1{,}200$. Using chi-square to test for goodness of fit, we calculate

$$\chi^2 = \frac{(920-900)^2}{900} + \frac{(570-600)^2}{600} + \frac{(700-900)^2}{900} + \frac{(1{,}410-1{,}200)^2}{1{,}200}$$

with $df = 4 - 1 = 3$.

489. **(E)**

$$\text{Expected cell count} = \frac{(\text{Row total})(\text{Column total})}{\text{Grand total}} = \frac{(37+52+764)(32+764)}{920}$$

490. **(B)** With about 68% of the values within 1 standard deviation of the mean in a normal distribution and with a sample size of 200, the expected numbers are as follows: 16% of 200 = 32, 34% of 200 = 68, 34% of 200 = 68, and 16% of 200 = 32. Then

$$\chi^2 = \sum \frac{(\text{obs} - \text{exp})^2}{\text{exp}} = \frac{(21-32)^2}{32} + \frac{(84-68)^2}{68} + \frac{(67-68)^2}{68} + \frac{(28-32)^2}{32}.$$

491. **(D)** The degrees of freedom are (rows – 1)(columns – 1) = (2 – 1)(3 – 1). The totals don't count for numbers of rows and columns. Expected cell values should all be greater than 5; χ^2-Test gives that the smallest expected cell value is 5.07. The null hypothesis in a chi-square test of independence is always *no association.* Statements such as in choice (E) cannot be concluded from a chi-square test of independence. For example, perhaps the explanation of this test result is that police tend to search certain types of cars and these cars happen to be driven by drivers of certain races.

492. **(B)** The chi-square test of homogeneity compares samples from two or more populations (adults and children in this example) with regard to a single categorical variable (ice cream preference in this example). A chi-square test of independence tests for evidence of an association between two categorical

variables from a single sample. (There are two independent samples in this example.) The difference between these two tests is in the *design* of the study.

493. **(A)** The binomial distribution with $n = 2$ and $p = 0.8$ is $P(0) = (0.2)^2 = 0.04$, $P(1) = 2(0.2)(0.8) = 0.32$, and $P(2) = (0.8)^2 = 0.64$, resulting in expected values of $0.04(200) = 8$, $0.32(200) = 64$, and $0.64(200) = 128$. Thus,

$$\chi^2 = \sum \frac{(\text{obs} - \text{exp})^2}{\text{exp}} = \frac{(12-8)^2}{8} + \frac{(75-64)^2}{64} + \frac{(113-128)^2}{128}.$$

494. **(E)** With a P-value this small (less than 0.05), there is sufficient evidence in support of the alternative hypothesis H_a: the distribution of reading preferences is different, that is, it differs for at least one of the proportions.

Hypothesis Tests for Slopes (pages 188–190)

495. **(B)** The computer output for regression is always for a *two*-sided test. In this case, $2P = 0.0478$ and $P = 0.0239$.

496. **(E)** A pattern in the residual plot is a strong indicator that there is a better fit out there (although a linear fit might still be good). Even with a low r, a least squares line might give the best fit. The mean of the residuals is always 0. A low P-value indicates a good fit. A high r^2 means that when using the linear model, a high percentage of the variation in one variable is explained by variation in the other variable.

497. **(D)** From the computer printout, we immediately see that the P-value is 0.0053, and $0.001 < 0.0053 < 0.01$.

498. **(E)** For inference with regard to the slope of a least squares regression line, $df = n - 2$. In this case, $df = 20 - 2 = 18$, and invT(0.975, 18) = 2.101.

499. **(C)** With a small P-value in a significance test on the slope, H_0: $\beta = 0$ and H_a: $\beta > 0$, there is sufficient evidence of a positive linear relationship. In other words, a one-unit (one-year) increase in the x-variable corresponds to a constant positive increase in the y-variable (ages among women) on average.

500. **(D)** $t = \dfrac{b - 0}{SE(b)} = \dfrac{-8.73}{0.74}$

FREE-RESPONSE

EXPLORATORY ANALYSIS

One-Variable Data Analysis (pages 193–200)

501. (a)

Stem	Leaves
0	0 4 8
1	2 3 5 5 8
2	3 4
3	
4	0 3 7
5	0 3 7 9
6	1 1 5

(6|1 means 61 texts)

(Note that a key must be included, and the line corresponding to 30 texts should not be deleted even though it is empty.)

(b) A complete answer must address shape, center, spread, and any unusual features. The answer must also mention context in at least one of the responses.

Shape: Bimodal with two distinct clusters and a gap between 24 and 40

Center: The median number of text messages is $\frac{24 + 40}{2} = 32$

Spread: The range is $65 - 0 = 65$ text messages

(c) A boxplot would give less information. A boxplot would not give the individual values as does the stemplot, and a boxplot would completely miss showing the clusters and gap.

502. (a) A complete answer must address shape, center, spread, and any unusual features. The answer must also mention context in at least one of the responses.

The distribution of candy bag weights for the given 1-minute interval is skewed left with a gap between 15.3 and 15.5 ounces and a possible outlier between 15.2 and 15.3. The median weight is between 15.9 and 16.0 ounces. The bag weights vary from a minimum between 15.2 and 15.3 ounces to a maximum between 16.1 and 16.2 ounces.

(b) If the 15.85-ounce bag weighed 15.75 ounces, the mean would decrease by 0.10 divided by 21. The median would not change because the current

CHAPTER 5

median is between 15.9 and 16.0 ounces and both 15.85 and 15.75 ounces are smaller than that.

503. (a) A complete answer must address shape, center, spread, and outliers. The answer must also mention context in at least one of the responses.

Shape: Skewed right with a possible outlier around 23

Center: The median number of malware intrusions is around 5

Spread: The range is 23 – 1 = 22 malware intrusions

(b) In a skewed right distribution, the mean is usually greater than the median, so the quotient $\frac{\text{mean intrusions}}{\text{median intrusions}}$ is probably greater than 1. (Note: You would receive no credit without an explanation.)

(c) Yes, the conditions for inference are met. The sample is given to be random. The sample size, $n = 109 \geq 30$ is large enough for the central limit theorem to apply.

504. (a) A complete answer compares shape, center, and spread. The answer also mentions context in at least one of the responses.

Shape: The University *A* starting salary distribution is roughly symmetric and bell-shaped, while the University *B* starting salary distribution is skewed left (toward the lower values).

Center: The center of the distribution of University *A* starting salaries (at around 52 or 53) is less than the center of the distribution of University *B* starting salaries (at around 67).

Spread: The spreads of the two distributions are roughly the same. The range of the University *A* starting salaries (70 – 35 = 35) equals the range of the University *B* starting salaries (75 – 40 = 35).

(b) Because of the roughly symmetric distribution, the mean and median starting salaries for University *A* are about the same. Because of the skewed left distribution, the mean starting salary of University *B* is less than the median starting salary of University *B*. The graphs show that the mean starting salary of University *B* is thus closer to the mean and median starting salaries of University *A*. We conclude that the difference between the two university median starting salaries is greater than the difference between the two university mean starting salaries.

505.

HS students		College students
6	1	
7 3	2	2
9 6 2	3	4 5 5 8 8
8 5 3 0	4	2 7 9 9
7 7 4 1	5	0 4 8
9 4 0	6	0 5 9
5 2	7	5 9
2	8	7
	9	8

2|8|7 indicates ratings of 82 and 87 of a high school students and a college student, respectively.

(Note the column labeling and the presence of a key.)

A complete answer compares shape, center, spread, and any unusual features. The answer also mentions context in at least one of the responses.

Shape: The HS students' distribution of music website ratings is roughly symmetric and bell-shaped. The college students' distribution of music website ratings is skewed right (toward higher values).

Center: The center of the high school students' distribution is about the same as the center of the college students' distribution, both about 50.

Spread: The spread of the high school students' distribution (with a range of 82 – 16 = 66) is less than the spread of the college students' distribution (with a range of 98 – 22 = 76).

506. A complete answer compares shape, center, and spread. The answer also mentions context.

Shape: Country *A*, for which the cumulative frequency plot rises steeply at first, has more lower A1C level readings and thus a distribution skewed to the right (toward the higher A1C level readings). Country *C*, for which the cumulative frequency plot rises slowly at first and then steeply toward the end, has more higher A1C level readings and thus a distribution skewed to the left (toward the lower A1C level readings). Country *B*, for which the cumulative frequency plot rises slowly at each end and steeply in the middle, has a more bell-shaped distribution.

Center: Considering the center to be a value separating the area under a histogram roughly in half, the centers correspond to a cumulative frequency of 0.5. Reading across from 0.5 to the intersection of each graph and then down to the *x*-axis shows the center (median) of the Country *A* distribution to be lowest and the center (median) of the Country *C* distribution to be the highest.

Spread: The spreads of the A1C level readings in all three countries are the same: from 5.0 to 8.0.

507. (a) $Q_1 = 88$, Med = 89, and $Q_3 = 90$. So the

$$\textit{trimean} = \frac{88 + 2(89) + 90}{4} = 89.$$

(b) The mean of {88, 88, 89, 89, 90, 90} is 89.

(c) The median trims the upper 49+% and the lower 49+% of the scores.

(d) The trimean and a trimmed mean are "resistant" to extreme values, while the mean is "sensitive" to extreme values. So the trimean and trimmed mean are useful measurements when the extreme values are in some way suspicious or when we want to diminish their effect.

508. (a) Using a little algebra to solve for °C gives the equation

$$°C = \left(\frac{1}{1.8}\right)(°F) - \left(\frac{32}{1.8}\right).$$

Adding (or subtracting) the same constant to every value in a set increases (or decreases) the mean by the same amount but leaves the standard deviation unchanged. Multiplying (or dividing) every value by the same constant multiplies (or divides) both the mean and the standard deviation by that amount. So the new mean will be $\left(\frac{1}{1.8}\right)(1{,}760) - \left(\frac{32}{1.8}\right) = 960°C$.

The new standard deviation will be $\left(\frac{1}{1.8}\right)(27) = 15°C$.

(b) The size of the sample is extraneous information. Changing units does not involve the sample size. The answer is the same whether $n = 10$ or $n = 20$.

509. (a) For each set the mean is 2. For Set *A*, we calculate

$$\begin{aligned}
\text{MAD} &= \frac{1}{n}\Sigma|x - \bar{x}| \\
&= \frac{1}{9}\Sigma|x - 2| \\
&= \frac{1}{9}(2 + 1 + 1 + 0 + 0 + 0 + 1 + 1 + 2) \\
&= \frac{8}{9}.
\end{aligned}$$

For Set B, we calculate

$$\begin{aligned}\text{MAD} &= \frac{1}{n}\Sigma|x - \bar{x}| \\ &= \frac{1}{11}\Sigma|x - 2| \\ &= \frac{1}{11}(2 + 2 + 2 + 1 + 1 + 0 + 1 + 1 + 2 + 2 + 2) \\ &= \frac{16}{11}.\end{aligned}$$

(b) We note that the MAD of Set B is greater than the MAD of Set A. Yes, this is what was expected because Set B looks more spread out than Set A.

510. (a) $\bar{x} = \frac{3 + 4 + 2(5) + 3(6) + 4(7) + 5(8) + 5(9)}{21} = \frac{148}{21} = 7.048$, while the median of the 21 numbers is 7.

(b) This is *not* what was expected. When data are skewed to the left (toward the lower numbers), the mean is usually less than the median.

511. (a)

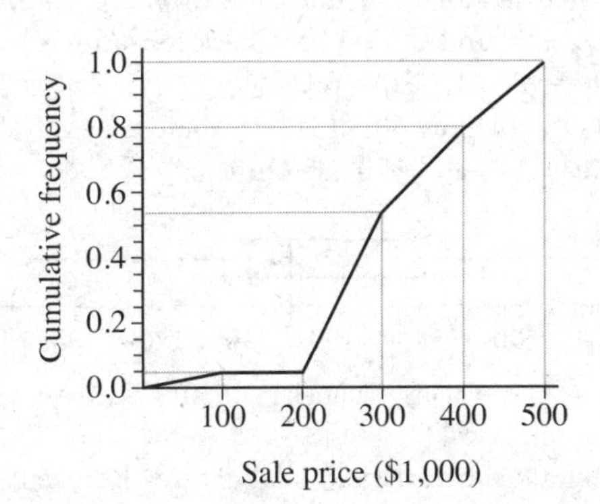

(b) The median is more useful than the mean when outliers are present because the median is "resistant" to extreme values. In our example, we most likely want to diminish the effect of the one sale below $100,000, which is not at all representative.

512. (a) For medication A pressure differences, 1.5(IQR) = 1.5(10 – 0) = 15. So any pressure differences below $Q_1 - 15 = -15$ or above $Q_3 + 15 = 25$ are outliers; there are none. For medication B pressure differences, 1.5(IQR) = 1.5(5 – 3) = 3. So any pressure differences below $Q_1 - 3 = 0$ or above $Q_3 + 3 = 8$ are outliers; this includes –2 and 10.

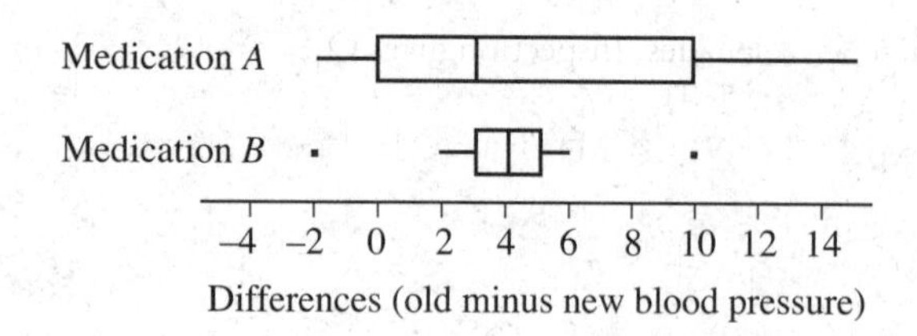

(b) For blood pressures using medication *A*, $Q_1 = 0$. For medication *B*, all but one value is > 0. So medication *B* has one fewer than 100 percent of the subjects improving, while medication *A* has only 75% of the subjects improving. Choose medication *B*.

(c) The distribution of blood pressures under medication *B* appears roughly symmetric so the mean should be close to the median value of 4. For medication *A*, the distribution appears sharply skewed right. So the mean is probably considerably greater than the median of 3. Choose medication *A* for the greatest likelihood of the greater mean improvement.

513. (a) The median is the 46th score from either side. Adding column totals quickly gives the median to be 105. With 45 scores on either side of the median, the quartiles are the 23rd scores from each end. Adding column totals gives $Q_1 = 90$ and $Q_3 = 110$. Check for outliers by calculating $Q_1 - 1.5(IQR) = 90 - 1.5(20) = 60$ and $Q_3 + 1.5(IQR) = 110 + 1.5(20) = 140$. There are no values below 60. However, there are two values above 140. *These* two values, 145 and 150, are outliers.

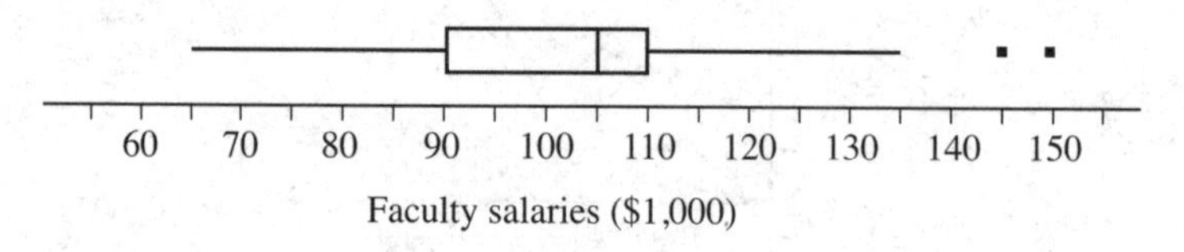

(b) The histogram shows the bimodal nature of the data, something lost in the boxplot.

(c) The boxplot more clearly distinguishes the outliers than does the histogram.

514. (a) There are 11 values among the boy times. Inspection gives $Q_1 = 58$, median = 59, $Q_3 = 60$, and IQR = 60 – 58 = 2. Any value below $Q_1 - 1.5(IQR) = 58 - 3 = 55$ or above $Q_3 + 1.5(IQR) = 60 + 3 = 63$ is an outlier. Thus 64 is the only outlier. Similarly, there are 7 values among the girl times. Inspection gives $Q_1 = 59$, median = 62, $Q_3 = 63$, and IQR = 63 – 59 = 4. Any value below $Q_1 - 1.5(IQR) = 59 - 6 = 53$ or above $Q_3 + 1.5(IQR) = 63 + 6 = 69$ is an outlier. Thus there are no outliers.

(b) There are 18 values. Inspection gives $Q_1 = 58$, $Q_3 = 62$, and $IQR = 62 - 58 = 4$. There are no values below $Q_1 - 1.5(IQR) = 58 - 6 = 52$ or above $Q_3 + 1.5(IQR) = 62 + 6 = 68$. So the combined set has no outliers.

(c) The outlier, 64, in the boy distribution was on the high side. However, the girl distribution had several values around 64. So 64 was no longer an outlier in the combined set.

515. (a) The median is a value with 50 percent of the values to each side. Adding the given proportions, for either males or females, for Species *B* shows the median to be in the 40–49 interval.

(b) For Species *A*, a greater proportion of males than females are found in their respective 90+ and 80–90 intervals. So males appear to have longer life expectancies. For Species *B*, a greater proportion of females than males are found in their respective 90+, 80–90, and 70–80 intervals. So females appear to have longer life expectancies.

(c) In both species, there was a greater proportion of male than female births during the last 20 days of the study.

516. (a) The entire low-end price distribution is less than the entire high-end price distribution. (The maximum low-end price is less than the minimum high-end price.) The low-end price distribution has two outliers. The median price of the low-end chairs is more than $150 less than the median price of the high-end chairs. The IQR of the low-end price distribution is less than the IQR of the high-end price distribution. However because of the outliers, the range of the low-end price distribution is greater than the range of the high-end price distribution.

(b) The shape of a single histogram made of all the lounge chair prices would depend on the width of the bins. Assuming the width is not too large (or too small), there would be two clusters, one centered just less than $200 with the other centered around $350. There would be a solitary value at $100 and another solitary value at $250 (in what otherwise would be a gap between $225 and $275).

517. (a) 20% of the scores are below 50.

(b) No, the plot shows relative frequencies, not actual numbers. So an estimate cannot be made.

(c) Going over from 0.5 on the vertical axis and then down to the horizontal axis gives a score of 70 for the median.

(d) 0.7 or 70 percent are below 80; so 0.3 or 30 percent are above 80.

(e) No one scored between 80 and 90.

518. (a) The median fiber content of all three cereals is about the same, 10 grams per serving. The ranges are different, with the boxes from Cereal *A* having the greatest range and the boxes from Cereal *B* having the smallest range.

(b) We are looking at only samples. However, in these samples, there is no overlap among the three sugar boxplots. So analyzing sugar content would distinguish among the cereals.

(c) For Cereal *B* the three minimums in the boxplots total 13 + 9 + 9 = 31 grams. For Cereal *C*, the three maximums in the boxplots total 7 + 13 + 9 = 29 grams. For Cereal *A*, the three minimums total less than 30 and the three maximums total more than 30 grams. With your analysis showing a total of 30 grams, you probably have Cereal *A*!

Two-Variable Data Analysis (pages 201–204)

519. (a) Yes, a linear model is appropriate for these data. The rent scatterplot is approximately linear, and there is no apparent pattern in the residuals plot.

(b) The slope of the regression line is 1.077. This means that for each additional square foot in size, the *average* rise in rent is $1.077.

(c) The coefficient of determination is $r^2 = 94.7\%$, meaning that 94.7% of the variation in weekly rent is explained by the linear model of weekly rent and size.

520. (a) The correlation $r = \sqrt{0.779} = 0.883$ suggests a moderately strong, positive, linear relationship. A linear regression *t*-test yields $t = 5.32$ with $P = 0.0007$, indicating very strong evidence of a linear association. However, the absolute values of the residuals decrease with increasing *x*, indicating a model other than linear might be more appropriate.

(b) The slope is 2.74, indicating that on average, the calorie content increases by 2.74 for each additional gram of fat.

(c) The *y*-intercept, 78.3, refers to the average calorie content for potato chips with 0 grams of fat. It's not clear if this is possible!

(d) For 10 grams of fat, the predicted calories are 78.3 + 2.74(10) = 105.7.

(e) From the residual plot, the residual is 4 for $x = 10$. Since residual = actual – predicted, the actual calories are 105.7 + 4 = 109.7.

521. For male students, the regression equation is ESP = 50.89 + 0.04768(SAT). For female students, the regression equation is ESP = 22.19 + 0.09615(SAT). In the male student equation, 50.89 + 0.04768(689) = 83.7. In the female student equation, 22.19 + 0.09615(689) = 88.4. The actual ESP test result of 84 is much closer to the prediction for male students, and thus we guess the student's gender is male.

522. (a)

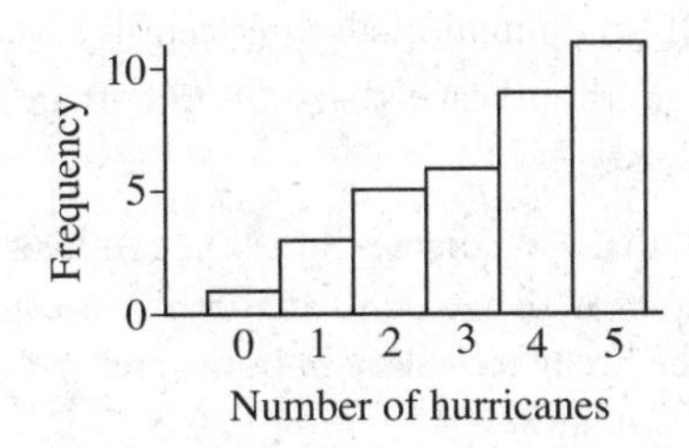

(b) The scatterplot shows a roughly linear trend, with a positive slope, associated with the number of hurricanes each year. This feature is not shown in the histogram.

(c) The histogram shows that the distribution of number of hurricanes is skewed left (to the lower values). This feature can be seen in the scatterplot but not as clearly as in the histogram.

523. (a) *Linear* means that as distance from the front increases by 1 foot, exam average changes by a constant amount, on average. *Negative* means that students sitting at greater distances from the front tend to have lower exam averages. *Strong* means that the data points fall close to the line of best fit (the actual exam averages are close to the predicted exam averages).

(b) The slope of −1.15 means that if a student is sitting 1 foot farther from the front than another student, his/her exam average is predicted, on average, to be 1.15 lower than the student closer to the front.

(c) The student's predicted exam average was 97.5 − 1.15(12) = 83.7. Actual exam average = Predicted exam average + Residual, so the student's exam average was 83.7 − 4.7 = 79.

524. (a) A linear model is not appropriate for this data: the percent of variation in numbers sold explained by price (r^2) is only 6.1%, the residual plot has a strong pattern, and a linear regression t-test would show a large P-value of 0.5556.

(b) Given the above, or reading from the computer output, the linear regression line would be almost flat. Thus the residual plot shows the approximate actual shape of the scatterplot of numbers sold versus price. From this we conclude that the best cost to achieve the most sales is $40,000.

525. (a) 99.8% of the variation in weight loss is explained by the linear model of weight loss and the square root of the dosage of the experimental drug.

(b) Comparing the residual plots, the transformed data $(\sqrt{x}, y)$ shows higher r^2 and has a residual plot with far less of a pattern than that of the other two regression lines, which are both indicators of a better linear fit.

526. (a) Salary versus years of college is a better linear fit than salary versus age, as can be seen in the two residual plots. The salary versus age plot shows a

distinct curved pattern going from negative to positive and back to negative. Such a distinct pattern indicates that a fit other than linear is probably a better model.

(b) Residual = Actual – Predicted. Since the residuals for a salary with 4 years of college and for an age of 40 are both positive, the actual salary must be greater than the predicted salary in both cases. So each of the regression lines gives underestimates.

FR COLLECTING AND PRODUCING DATA (PAGES 205–208)

527. (a) Interview a random sample of people as to whether or not they sit within five feet when watching TV and whether or not they have eye problems. Compare the proportion of people who sit within five feet of the TV who have eye problems to the proportion of people who sit over five feet from the TV who have eye problems.

(b) From a group of volunteers without eye problems, use chance to pick half to sit within five feet of the TV for some period of years while the remaining half are instructed to sit over five feet from the TV for the same period of time. At the end of the time period, compare the proportion of people in each group who developed eye problems.

(c) The experimental approach is better in that the observational study may have confounding variables. For example, perhaps those who choose to sit closer than five feet are those who already have eye problems. However, a choice of observational study is a reasonable answer if the argument is made that an experiment would be unethical.

(d) If the researcher notices that members of the group instructed to sit within five feet of the TV are coming down with eye problems faster than members of the other group, the experiment should be stopped even if the desired time period has not been completed.

528. (a) Because of random sampling, we should be able to generalize. So conclusions will apply to adult nondieters with high A1C levels who are not taking medications.

(b) The design should incorporate random assignment of the adults to four treatment groups: medication and diet, medication and no diet, placebo and diet, placebo and no diet. The design should include a measurement and comparison of A1C levels before and after. For example, for each participant we can use a random number generator to pick an integer between 1 and 4, with each number standing for one of the four groups. (This method will not guarantee equal numbers in each group, but that wasn't asked for.)

(c) There are many possible answers, but each should be explained in the context of the problem. For example, you could block on gender or block on age because men versus women or old versus young might respond differently to medication or diet. In any case, give the scheme of first splitting participants into separate blocks (such as gender or age) and then randomly assigning subjects within each block to the four treatment groups.

(d) Blinding is incorporated through the use of a placebo if there is a possibility some participants might be able to influence their A1C levels subconsciously if they know for sure they are taking the metformin.

529. (a) This is an observational study because no treatment is being imposed on anyone.

(b) There are many possible confounding variables. For example, it is possible that the employees who eat under 1,000 calories are those who sleep in late, have to rush to work without eating, have not fully awakened, and thus have lower productivity.

(c) No, it is not reasonable. Cause-and-effect conclusions cannot be drawn from observational studies.

(d) Use random sampling to pick subjects for the experiment. (For example, number the employees starting with 1. Then use a random number generator, throwing out repeats, until the desired number of subjects are chosen.) This will allow for generalization of any results to all of the company employees. Then use random assignment to place the subjects into two groups. (For example, put cards in a box, half labeled ">1,000" and the other half labeled "<1,000." Have the subjects reach in and pick cards without replacement.) One group is told to eat under 1,000 calories at breakfast, while the other group is told to eat over 1,000 calories at breakfast. Compare productivity records after a given time period.

530. There is a real problem with possible response bias in that parishioners may not answer the pastor honestly. The assumption of a random sample from the population of all parishioners is not met as the sample was taken from those coming to church one Sunday, and they may not be representative of all the parishioners. There is the assumption that the population is roughly normally distributed. However, the sample data seem to indicate a right skewed distribution: $\text{Max} - Q_3 = 15.8$ is much larger than $Q_1 - \text{Min} = 5.2$ and $Q_3 - \text{Med} = 14.2$ is much larger than $\text{Med} - Q_1 = 8.3$. Also, the minimum value is only $\frac{13.5 - 31.2}{13.2} = -1.34$ standard deviations below the mean and the maximum is only $\frac{57 - 31.2}{13.2} = 1.95$ standard deviations above the mean. Both of these are less than what would be expected with a roughly normal distribution.

531. (a) In each block, two of the tables will be randomly assigned to receive one type of soldering iron, while the remaining two tables in the block will receive the other type of iron. Randomization of irons to tables within each block should reduce bias due to confounding variables associated with the tables that might be related to sources of error. In particular, the randomization in blocks in Scheme *A* should even out the effect of the distance tables are from the manager's desk.

(b) Scheme *A* creates homogeneous blocks with respect to window exposure. Scheme *B* creates homogeneous blocks with respect to distance from the manager's desk. Randomization of soldering irons to tables within blocks in Scheme *A* should even out the effects of distance to the manager's desk. Randomization of soldering irons to tables within blocks in Scheme *B* should even out the effects of window exposure.

532. (a) The response variable is the exam scores. The treatments are the two classrooms. The experimental units are the two algebra classes (not the students themselves!).

(b) Yes, the two classrooms (the treatments) were randomly assigned to the two algebra classes (the experimental units).

(c) No, there was no replication in this study. Each treatment was applied to only one experimental unit.

(d) If a difference in exam scores is noted between the two classes, it is not known if this is due to the difference in classrooms or the difference in teachers. That is, the treatments (classrooms) are confounded with the teachers.

533. (a) Both surveys lead to voluntary response samples, based on individuals who offer to participate, resulting in bias as too much emphasis is typically given to people with strong opinions. Both surveys also lead to selection bias as readers of *VFW* magazine and readers of the *Washington Post* are not representative of the American public.

(b) The word "retreat" is loaded with emotion, while the phrase "phased troop deployment" is too evasive. A more neutral wording would simply ask about support for "pulling troops out of Afghanistan."

534. (a) Number the employees 1 through 250,000. Use a random number generator to pick integers between 1 and 250,000, throwing out repeats, until 250 unique integers have been picked. The sample will be the 250 employees whose numbers correspond to the picked integers. A disadvantage would be the difficulty of listing and numbering all 250,000 employees.

(b) Method 2 is a stratified sample where the U.S. employees and non-U.S. employees are the two strata. An advantage of stratified sampling here is that this method guarantees good numbers of both U.S. and non-U.S. employees will be in the sample. You want both groups well represented because they might have very different views about new coffee drinks that their clientele will enjoy and purchase.

(c) Method 3 is a cluster sample where the 25 stores are clusters. An advantage of cluster sampling here is the ease and low costs in implementing this method. Employees throughout the company do not have to be listed. The research team only has to pick the 25 stores randomly and then simply use all full-time employees in those stores for their sample.

535. (a) In each hospital, number the doctors 1 through 50. Then use a random number generator to pick two unique integers between 1 and 50. (Throw out a repeat.) The two doctors with numbers corresponding to the picked integers will be in the sample. Repeat for each hospital to obtain a total of 50 doctors in the sample.

(b) Number the hospitals 1 through 25. Then use a random number generator to pick a single integer between 1 and 25. The sample will be all 50 doctors who work at the hospital with the number corresponding to the picked integer.

(c) The doctors working at different hospitals might be very different and have different opinions. For example, different hospitals might specialize in treating different medical conditions. Stratified sampling guarantees that some doctors from every hospital will be represented in the sample. In this context with cluster sampling, the sample will be made up of doctors from only a single hospital.

(d) In this context, a cluster sample is much easier to obtain as only a single hospital will have to be visited and the doctors do not have to be numbered. This is different than a stratified sample, where every hospital must be visited and in each hospital every doctor must be numbered.

536. (a) Number the students from 1 to 60. Then use a random number generator to pick numbers between 1 and 60, throwing out repeats, until a set of 30 unique numbers has been selected. The students corresponding to these numbers read from a physical book, while the remaining group of 30 students read the same passages on an iPad for the same predetermined length of time. The mean number of words read by each of the two groups can be compared using a two-sample t-test for the difference between two means or by using a confidence interval for the difference between two means.

(b) Randomly assign the 60 students into two groups of 30 as above. The first group reads from a physical book, while the remaining group of 30 students reads the same passages on an iPad for the same predetermined length of time. Then each student should switch, physical book to iPad and iPad to physical book. Again all students should read the same set of passages for the same predetermined length of time. The difference in number of words read (physical book – iPad) should be calculated for each student. Then use a one-sample t-test on this set of differences (with null hypothesis that the mean difference is 0). Alternatively, determine a confidence interval of this set of differences and see if 0 is in the interval.

FR PROBABILITY (PAGES 209–216)

537. (a) $(0.6)^2 = 0.36$.

(b) Player A can win the match in exactly 3 games if he/she loses one of the first two games and wins the other two, if he/she wins one game and draws the other two, or if he/she wins one of the first two games, draws the other, and then wins the third game. $P(\text{L}) = 1 - 0.9 = 0.1$ and $P(\text{WLW or LWW or WDD or DWD or DDW or DWW or WDW}) =$ $2(0.6)^2(0.1) + 3(0.6)(0.3)^2 + 2(0.3)(0.6)^2 = 0.45$.

(c) $P(\text{first game was draw} \mid A \text{ wins in 3 games})$

$$= \frac{P(\text{first game draw} \cap A \text{ wins in 3})}{P(A \text{ wins in 3})} = \frac{P(\text{DWD or DDW})}{P(A \text{ wins in 3})}$$

$$= \frac{2(0.6)(0.3)^2 + (0.3)(0.6)^2}{0.45} = \frac{0.216}{0.45} = 0.48.$$

(d) Since the game outcomes are independent, $P(\text{W} \mid \text{D}) = P(\text{W}) = 0.6$.

538. Clearly something is wrong. If you follow this line of reasoning, you should always switch envelopes. However, there is a probability of 0.5 that you picked the envelope with the larger money to start with. The correct line of reasoning is as follows. One envelope has \$X while the other has \$2X. If you stick with your original choice, your expected payoff is (0.5)(\$X) + (0.5)(\$2X) = \$1.5X. If, however, you switch envelopes, your expected payoff is (0.5)(\$2X) + (0.5)(\$X) = \$1.5X. There is the same expected payoff whether or not you switch envelopes!

539. (a) This is a geometric probability with $p = 0.06$ and $k = 5$. The probability of the next homer on the 5th at bat is

$$(1 - 0.06)^4(0.6) = (0.94^4(0.06) = 0.0468.$$

(b) This is a binomial probability with $n = 5$, $p = 0.06$, and $k = 1$. The probability of exactly one homer in 5 at bats is

$$\binom{5}{1}(0.06)(1 - 0.06)^4 = 5(0.06)(0.94)^4 = 0.2342.$$

(c) With $B(10, 0.06)$, the mean is $np = (10)(0.06) = 0.6$. The mean number of homers for every 10 at bats is 0.6.

(d) The mean in this geometric setting is $\frac{1}{p} = \frac{1}{0.06} = 16.67$. The mean number of at bats before Rizzo hits a homer is 16.67.

540. (a) With $N(7.0, 0.8)$, $P(X < 5.5) = P\left(z < \frac{5.5 - 7.0}{0.8}\right) = 0.0304$.

(b)

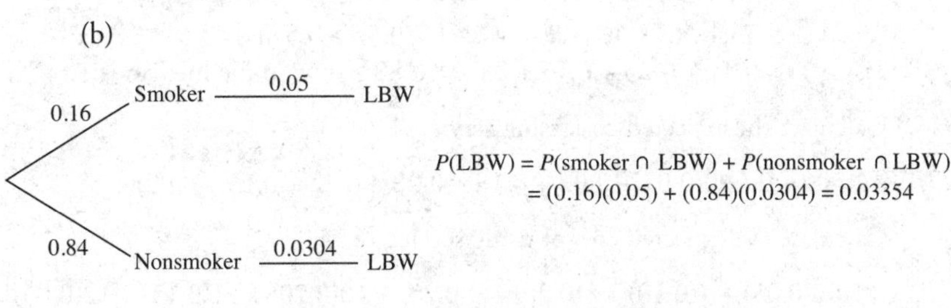

(c) $P(\text{nonsmoker} \mid \text{LBW}) = \frac{(0.84)(0.0304)}{0.03354} = 0.761$.

541. (a) $E(\text{charge}) = (2.50)(0.30) + (5.00)(0.40) + (7.50)(0.20) + (10.00)(0.05) + (0.00)(0.05) = \$4.75$.

(b) Note that $P(x \geq 5.00) = 0.40 + 0.20 + 0.05 = 0.65$ and $P(x \leq 5.00) = 0.30 + 0.40 + 0.05 = 0.75$. Since both $P(x \geq 5.00) \geq 0.5$ and $P(x \leq 5.00) \geq 0.5$, \$5.00 is the median charge.

542. (a) For die A, $P(5) = \frac{1}{2}$, $P(3) = \frac{1}{3}$, and $P(1) = \frac{1}{6}$. For die B, $P(2) = \frac{1}{3}$ and $P(4) = \frac{2}{3}$. If the player rolls a 5 with die A or if he/she rolls a 3 with die A and the other player rolls a 2 with die B, then the player rolling die A wins. If the player rolls a 2 with die B and the other player rolls a 1 with die A or if he/she rolls a 4 with die B and the other player rolls either a 3 or 1 with die A, then the player rolling die B wins. So

$$P(A \text{ wins}) = \frac{1}{2} + \left(\frac{1}{3}\right)\left(\frac{1}{3}\right) = \frac{11}{18}$$

$$\text{or } P(B \text{ wins}) = \left(\frac{1}{3}\right)\left(\frac{1}{6}\right) + \left(\frac{2}{3}\right)\left(\frac{1}{2}\right) = \frac{7}{18}.$$

So rolling die A gives a greater chance of winning.

(b) $E(X) = \Sigma x P(x)$, so

$$E(A) = 5\left(\frac{1}{2}\right) + 3\left(\frac{1}{9}\right) = 2\frac{5}{6} \text{ and } E(B) = 2\left(\frac{1}{18}\right) + 4\left(\frac{1}{3}\right) = 1\frac{4}{9}.$$

543. (a) $\mu = \Sigma xP(x)$ (0.12) + 1(0.41) + 2(0.25) + 3(0.15) + 4(0.07) = 1.64. On the average, college students get a good night's sleep on 1.64 days of the week.

(b) The average for the first sample was $\frac{20}{10} = 2$. It is expected that the average based on 50 students will be closer to 1.64 than was the average based on only 10 students. The variability for sample averages based on 50 students is smaller than for sample averages based on 10 students.

(c) We see that $P(X \le 1) = 0.12 + 0.41 = 0.53 \ge 0.5$ and $P(X \ge 1) = 0.41 + 0.25 + 0.15 + 0.07 = 0.88 \ge 0.5$. So the median is 1.

544. Calculate the expected cost using service *A*:

$$E(A) = 2{,}500 + 2{,}000(0.01 + 0.02 + 0.18 + 0.28 + 0.22 + 0.13 + 0.02) = \$4{,}220.$$

Calculate the expected cost of using service *B*:

$$E(B) = 450[6(0.05) + 7(0.10) + 8(0.20) + 9(0.25) + 10(0.20) + 11(0.15) + 12(0.05)]$$

$$= 450(9.1) = \$4{,}095.$$

So the school should choose service *B*. (Don't forget to answer the question!)

545. (a)

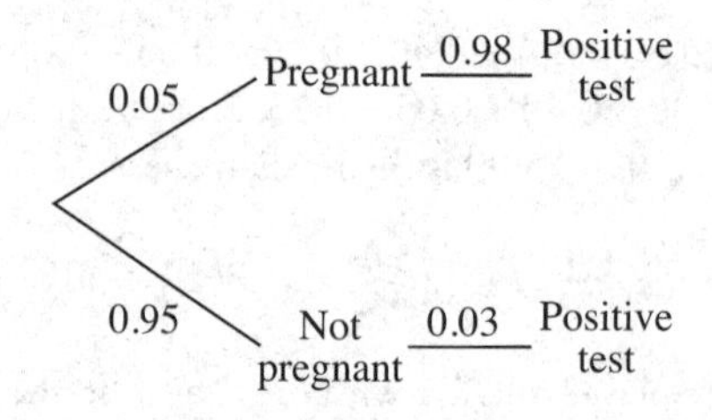

$P(\text{pregnant} \cap \text{positive test}) = (0.05)(0.98) = 0.049$

$P(\text{not pregnant} \cap \text{positive test}) = (0.95)(0.03) = 0.0285$

$P(\text{positive test}) = 0.049 + 0.0285 = 0.0775$

(b)

$$P(\text{pregnant} \mid \text{positive test}) = \frac{P(\text{pregnant} \cap \text{positive test})}{P(\text{positive test})} = \frac{0.049}{0.0775} = 0.6323$$

(c) With $B(2, 0.6323)$, we calculate

$$P(X = 1) = 2(0.6323)(1 - 0.6323) = 0.4650.$$

546. (a) The probability a computer will fail within 18 months is

$$P\left(z < \frac{18 - 24}{3}\right) = \text{normalcdf}(0, 18, 24, 3) = 0.02275.$$

So the expected cost per computer is (0.02275)($350) = $7.96.

(b) If the company is willing to warranty 0.02275 + 0.05 = 0.07275 of the computers, the new warranty should be for invNorm(0.07275, 24, 3) = 19.6 months.

547. For young adults, the minimum of 13 and the maximum of 37 are both only $\frac{12}{10}$ = 1.2 standard deviations from the mean, an unlikely occurrence for a roughly normal distribution. For older adults, the min, Q_1, Q_3, and max have z-scores of $-\frac{12}{4} = -3$, $-\frac{3}{4} = -0.75$, $\frac{3}{4} = 0.75$, and $\frac{12}{4} = 3$, respectively. Under a normal curve, almost all values are between –3 and 3 and the quartiles have z-scores of ±0.67. Thus, there is evidence that the older adult distribution is roughly normal.

548. For young adults, $z = \frac{8{,}000 - 6{,}500}{1{,}200} = 1.25$, which gives 0.894 or 89.4% with debt less than $8,000. For older adults, $z = \frac{8{,}000 - 5{,}550}{1{,}800} = 1.361$, which gives 0.913 or 91.3% with debt less than $8,000. So with $8,000, a greater percentage of older adults will more likely be able to pay off their credit debt.

549. (a)

$$P(\text{grant}) = 1 - P(z > 1.5) = \text{normalcdf}(1.5, 100) = 0.0668$$

(b)

$$P(\text{at least 1 out of 3 receives grant}) = 1 - P(\text{none receive grant})$$
$$= 1 - (0.9332)^3 = 0.1873$$

(c) The quartiles are at ± invNorm(0.75) = ± 0.6745 standard deviations from the mean. Thus IQR = 2(0.6745) = 1.3490 standard deviations. The cutoff for the lab equipment is a z-score of 0.6745 + 1.5(1.3490) = 2.698.

$$P(\text{lab equipment}) = P(z > 2.698) = \text{normalcdf}(2.698, 100) = 0.0035$$

(d) The probability a scientist receives the grant but not the lab equipment is $P(1.5 < z < 2.698) = 0.0633$.

550. (a) There is a probability of 0.95 the crow can solve the puzzle. If it does, the probability the crow does so in under 1.5 minutes is $P\left(z < \frac{1.5 - 1.8}{0.4}\right) = 0.2266$. The probability of both occurrences is (0.95)(0.2266) = 0.2153.

(b) $0.05 + 0.95P\left(z > \frac{2.0 - 1.8}{0.4}\right) = 0.3431$ or

$$1 - 0.95P\left(z < \frac{2.0 - 1.8}{0.4}\right) = 0.3431$$

551. (a) $0.30 + 0.10 = 0.40$

(b) $2,000,000(0.30 + 0.30 + 0.20) + $5,000,000(0.10 + 0.05 + 0.05) = $2,600,000

(c) No, position and bonus are not independent. For example,

$$P(\$2{,}000{,}000 \text{ bonus}) = 0.30 + 0.30 + 0.20 = 0.80.$$

However,

$$P(\$2{,}000{,}000 \text{ bonus} \mid \text{pitcher}) = \frac{0.30}{0.30 + 0.10} = 0.75 \neq 0.80.$$

552. (a) If p is the probability that a box is under 16 ounces, the expected value of the fine for each of the two boxes is $10{,}000p$. Then $10{,}000p + 10{,}000p = 125$ gives $p = 0.00625$.

(b) $z = \text{invNorm}(0.00625) = -2.498$, and $m = 16 + 2.498(0.05) = 16.125$ ounces.

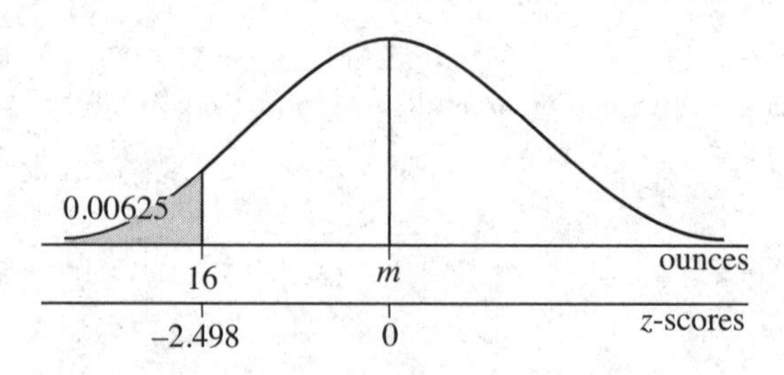

553. (a) The total weights are approximately normally distributed with a mean equal to the sum of the means of the 12 fruit and a variance equal to the sum of the variances of the 12 fruit.

Mean $= 6(8) + 6(6) = 84$ ounces

Variance $= 6(0.5)^2 + 6(0.4)^2 = 2.46$

Standard deviation $= \sqrt{2.46} = 1.568$ ounces

(b) Five pounds equals 80 ounces. With $N(84, 1.568)$,

$$P(X \geq 80) = P\left(z \geq \frac{80 - 84}{1.568}\right) = 0.995.$$

(c) Adding a constant to a random variable will change the mean by the same constant but will not change the variance or standard deviation. Thus the total weights are approximately normally distributed with a mean of $84 + 12 = 96$ ounces and a standard deviation of still 1.568 ounces.

554. (a) The sum of normal distributions is a normal distribution. So the times to complete the triathlon are roughly normally distributed because the individual leg times are roughly normally distributed. The means add. Because of independence, so do the variances: $\mu_{SUM} = 30 + 90 + 70 = 190$ min and $\sigma^2_{SUM} = 5^2 + 10^2 + 10^2 = 225$. So $\sigma_{SUM} = \sqrt{225} = 15$ min.

(b) $P\left(z < \frac{180 - 190}{15}\right) = 0.252$.

555. (a) Given independence, means and variances add.

$\mu_{SUM} = 6(175) + 4(130) = 1{,}570$

$\sigma^2_{SUM} = 6(15^2) + 4(10^2) = 1{,}750$

$\sigma_{SUM} = \sqrt{1{,}750} = 41.83$

(b) Since the individual distributions are normal, so is the distribution of the sums. Then $P\left(z > \frac{1{,}650 - 1{,}570}{41.83}\right) = 0.028$.

556. (a) $E(M + W) = E(M) + E(W) = 22.8 + 25.2 = 48.0$ seconds. Because of independence, $\text{Var}(M + W) = \text{Var}(M) + \text{Var}(W) = (0.9)^2 + (1.2)^2 = 2.25$. So $SD(M + W) = \sqrt{2.25} = 1.5$ seconds. If two independent random variables have normal distributions, so does their sum. Given $N(48.0, 1.5)$, $P(x > 50) = 0.0912$.

(b) $E(M - W) = E(M) - E(W) = 22.8 - 25.2 = -2.4$ seconds. Because of independence, $\text{Var}(M - W) = \text{Var}(M) + \text{Var}(W) = (0.9)^2 + (1.2)^2 = 2.25$. So $SD(M - W) = \sqrt{2.25} = 1.5$ seconds. If two independent random variables have normal distributions, so does their difference. Given $N(-2.4, 1.5)$, $P(x < 0) = 0.9452$.

557. (a) Since 102 is only $\frac{102 - 98.3}{3.2} = 1.15625$ standard deviations away from the mean, there is a probability of $P(z > 1.15625) = 0.124$ that Feigner would throw a ball 102 mph or faster, which is not a very unusual occurrence.

(b) The standard deviation of the sampling distribution of $\bar{x}$ is $s_{\bar{x}} \approx \frac{s}{\sqrt{n}} = \frac{3.2}{\sqrt{35}} = 0.541$. A sample mean of 102 is $\frac{102 - 98.3}{0.541} = 6.84$ standard deviations from the mean, which is a very large z-score. So the answer is "yes," it would have been highly unusual for Feigner to average 102 mph over 35 random fastballs. (Don't forget to answer the question!)

558. (a) The standard deviation of the sampling distribution of $\bar{x}$ is proportional to $1/\sqrt{n}$, which is smaller for larger n. Thus the probability of a sample mean more than 1 minute less than the population mean decreases for larger

n. So the sample of size 25 is more likely to have a mean more than 1 minute less than the population mean.

(b) We calculate $z = \frac{21 - 19}{0.8} = 2.5$ and $P(z > 2.5) = 0.0062$.

(c) The calculation would not change. By the central limit theorem, the sampling distribution of $\bar{x}$ is approximately normal for large n.

FR STATISTICAL INFERENCE (PAGES 217–234)

559. (a) *Procedure:* One-sample z-interval for a population proportion

Check conditions: Random sample (given), and both $n\hat{p} = 30 \geq 10$ and $n(1 - \hat{p}) = 120 \geq 10$

Mechanics: 1-PropZInt gives (0.116, 0.284)

(Alternatively, $0.2 \pm 2.576\sqrt{\frac{(0.2)(0.8)}{150}} = 0.2 \pm 0.084$.)

Conclusion in context: We are 99 percent confident that the proportion of all adults over the age of 45 who have played in a band at least one time in their lives is between 0.116 and 0.284.

(b) A 99 percent estimate for the number of adults sending in donations is between (0.116)(125,000,000) = 14,500,000 and (0.284)(125,000,000) = 35,500,000. At $10 each, a 99% confidence interval for the total donations from this group of adults to Musicians without Borders is $145,000,000 to $355,000,000.

560. (a) Number the students 1 through 12,500. Use a random number generator to generate integers between 1 and 12,500, ignoring repeats, until 500 unique integers between 1 and 12,500 have been generated. The sample will consist of the students corresponding to those integers.

(b) 0.6 is a conditional probability, P(parent did homework | someone other than student did homework), but the staff is interested in a confidence interval of the probability, P(parent did homework).

(c) $np = 12{,}500\left(\frac{90}{500}\right) = 2{,}250$

561. (a) We first note that the sample is given to be random, $n\hat{p} = 366$ and $n(1 - \hat{p}) = 549$ are both ≥ 10, and the sample size is clearly small compared to the total population of all adults. We calculate $\hat{p} = \frac{366}{915} = 0.4$ and $\sigma_{\hat{p}} = \sqrt{\frac{(0.4)(0.6)}{915}} = 0.0162$. For 95% confidence, the critical z-scores are ± 1.96. The margin of error is $\pm 1.96(0.0162) = \pm 0.032$.

(b) We have $(z)(0.0162) = 0.02$, which gives $z = 1.235$. The resulting confidence level is $P(-1.235 < z < 1.235) = 0.783$ or 78.3 percent.

562. (a) This is not an SRS since all possible samples of size 50 are not equally likely to be picked. For example, there is no chance of picking a sample that has 11 people from the same firm. This answer would not change if we knew that all the banking firms had equal numbers of employees. Although this would mean that all banking firm employees have an equal chance of being selected, it would still not mean that all possible samples of size 50 were equally likely to be picked.

(b) *Procedure:* One-sample z-interval for a population proportion

Check conditions: Although we don't have an SRS, the procedure did involve random selection. $n\hat{p} = 38$ and $n(1 - \hat{p}) = 33$ are both ≥ 10.

Mechanics: 1-PropZInt gives (0.642, 0.878).

(Alternatively, $\hat{p} = \frac{38}{50} = 0.76$ and $\sigma_{\hat{p}} = \sqrt{\frac{(0.76)(0.24)}{50}} = 0.0604$, giving $0.76 \pm 1.96(0.0604) = 0.76 \pm 0.12$.)

Conclusion in context: We are 95% confident that the proportion of banking assistants among this city's banking firm employees is between 0.642 and 0.878.

563. (a) This method is an example of a convenience sample and could be highly unrepresentative of the desired population. The socioeconomic status of the downtown area could significantly influence the data. For example, perhaps only well-to-do individuals work and shop in the downtown area, so the proportion earning over $7.25 per hour will seem higher than it really is.

(b) This method is an example of a voluntary response sample, which is based on individuals who offer to participate. It typically gives too much emphasis to people with strong opinions. Perhaps young adults making less than $7.25 are less (or more?) likely to be online.

(c) We have $1.96\left(\frac{0.5}{\sqrt{n}}\right) \leq 0.05$, giving $\sqrt{n} \geq \frac{1.96(0.5)}{0.05} = 19.6$ and $n \geq 384.16$. So we choose $n = 385$.

(d) Use a stratified random sample with two strata: males and females. Within each stratum, a random sample of young adults should be interviewed.

564. (a) invNorm(0.03) = −188. $1.88\sqrt{\frac{(0.65)(0.35)}{n}} \leq 0.035$ gives $\sqrt{n} \geq 25.62$ and $n \geq 656.4$, so 657 people must be surveyed. Note that we use 0.65 in the calculation because we believe $p \approx 0.65$. Otherwise, we would use 0.5.

(b) *Procedure:* Two-sided, one-sample z-test for a population proportion

Check conditions: The sample is given to be randomly chosen, and both $np = 700(0.65) = 455$ and $n(1 - p) = 700(0.35) = 245$ are ≥ 10.

Hypotheses: H_0: $p = 0.65$ and H_a: $p \neq 0.65$.

Test statistic and P-value: 1-PropZTest gives $z = 1.8226$ and $P = 0.0684$. (Alternatively, with $\hat{p} = \frac{432}{700} = 0.6171$, we get $z = \frac{0.6171 - 0.65}{\sqrt{\frac{(0.65)(0.35)}{700}}} = -1.825$.

This is a two-sided test, so the

$$P\text{-value} = P = 2P(z < -1.825) = 2(0.034) = 0.068.$$

Conclusion in context linked to the P-value: With this large of a P-value, $0.068 > 0.05$, there is not sufficient evidence that the approval rating has changed.

(c) With H_0: $p = 0.65$ and H_a: $p < 0.65$, $P = 0.034$. With $0.034 < 0.05$, there is sufficient evidence that the senator's approval rating has slipped.

(d) There is no contradiction. The tail probability of the test statistic is doubled when finding the P-value in a two-sided test. Thus, what might be a small enough tail probability to reject H_0 for a one-sided test might no longer be small enough when doubled for a two-sided test.

565. (a) Label the data values 01 to 50. Pick two digits at a time from the random table line, throwing out repeats and ignoring 00 and numbers over 50, until 10 unique numbers are selected. Note the data values corresponding to these numbers. In our case, this results in selecting the following.

77219 48190 20235 26836 23590 44492 14607 09431 75299 42662

The set {21, 20, 23, 36, 04, 44, 14, 31, 29, 26} corresponds to the data values {33, 40, 29, 40, 39, 33, 38, 32, 39, 32}, which becomes our sample.

(b) *Procedure:* Confidence interval for a population mean

Check conditions: We are given that all conditions for inference are met.

Mechanics: Putting the data in a list, TInterval gives (33.13, 37.87). (Alternatively, $35.5 \pm 1.833\frac{4.089}{\sqrt{10}} = 35.5 \pm 2.37$.)

Conclusion in context: We are 90 percent confident that the mean score on this exam for all the AP Statistics students at the school is between 33.13 and 37.87.

(c) The true population mean, 32.44, is not in the interval. However, this was not entirely unexpected since we were only 90 percent confident that our interval contained μ.

566. (a) The standard deviation, 1.5, represents, on average, how far from Team *A*'s mean score is a typical number of goals in any game. During most games, the team will score within ±1.5 goals of the mean number it scores.

(b) Team *A*, which scores 3, 4, or 5 goals in many games, appears to score more goals in any given game than Team *B*, which typically scores around 2 goals per game.

(c) Since 0 is in the confidence interval, there is not sufficient evidence to say that the mean number of goals scored by the two teams are different.

567. *Procedure:* Confidence interval for the difference of two population means

Check conditions: The samples are representative and independent (given). We are not given normal populations or large sample sizes. So we must check if the samples are roughly unimodal and symmetric with no extreme outliers. Using dotplots gives the following.

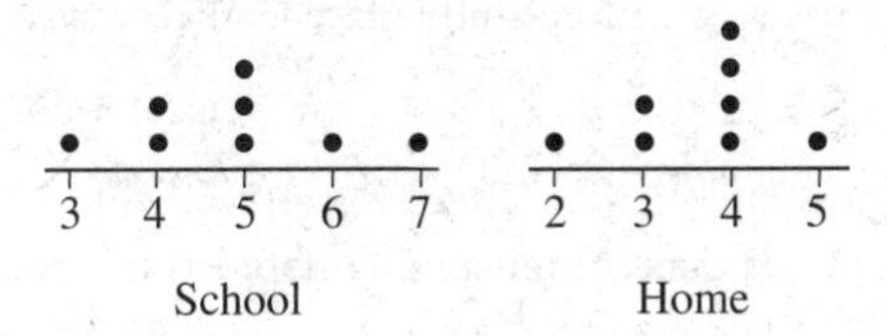

Based on the dotplots, it is not unreasonable to assume that the populations are roughly normal.

Mechanics: 2-SampTInt gives (0.281, 2.219).

Conclusion in context: We are 90% confident that the difference between the mean number of times a month this student eats processed meat at school and at home (school – home) is between 0.281 and 2.219.

568. (a) A graphing calculator gives $\hat{y} = 56.66 + 2.114x$, where x is the weight (in ounces) and y is the grade. This regression model gives a predicted grade for a term paper weighing $x = 16$ ounces of $\hat{y} = 56.66 + 2.114(16) = 90.48$.

(b) *Procedure:* Confidence interval for the slope

Check conditions: We are given that all conditions for inference are met.

Mechanics: LinRegTInt gives (0.6565, 3.5711).

Conclusion in context: We are 95 percent confident that the predicted mean increase in grade for each additional ounce in term paper weight is between 0.66 and 3.57.

569. (a) *Predicted number of cones sold* = 92.9805 + 0.459939(*Outside temperature*).

(b) The slope (0.46 cones per degree) gives an estimate for the predicted mean increase in number of cones sold for each additional degree of outside temperature.

(c) The margin of error for a confidence interval of the slope is $t_{n-2} \times \text{SE}(b)$. For a 95% confidence interval with $df = 10 - 2 = 8$, $t_{n-2} = \text{invT}(0.975, 8) = 2.306$. So the margin of error is $2.306 \times 0.1971 = 0.45$ cones per degree.

570. (a) No treatment is being imposed, so this is an observational study.

(b) Although this is a random sample of long-haul truckers, response bias could be a real problem as truckers might not want to admit to a lack of sleep.

(c) Although the residual plot shows no pattern and the histogram of residuals is approximately normal, a further assumption is that the scatterplot looks approximately linear. The scatterplot does not show this. Therefore, the necessary conditions for inference are not all met.

571. (a) $\hat{y} = 20.05 - 0.7809x$, where x is the urine THC level (100 ng/mL) and y is the number of objects successfully placed. Alternatively, the answer can be stated:

Predicted number of objects = 20.05 – 0.7809(*THC level*).

(b) *Procedure:* Confidence interval for a population slope

Check conditions: It is given that all conditions are satisfied.

Mechanics: With $df = 8$, the critical t-values are $\pm\text{invT}(0.975, 8) = \pm 2.306$. The confidence interval for slope is $-0.7809 \pm 2.306(0.0713) = -0.78 \pm 0.16$ or between –0.94 and –0.62.

Conclusion in context: We are 95% confident that the mean decrease in number of objects that would be successfully placed is between 0.62 and 0.94 for each additional 100 ng/mL increase in urine THC level.

(c) The confidence interval for the y-intercept is $20.05 \pm 2.306(0.969) = 20.05 \pm 2.23$. We are 95% confident that the mean number of objects that would be successfully placed by people with a urine THC level of 0 ng/mL (no THC registered in the urine) is between 17.82 and 22.28.

572. *Hypotheses:* H_0: $p = 0.10$ and H_a: $p < 0.10$, where p = the proportion of the university baseball players who go on to play professionally after graduation.

Procedure: A one-sample z-test for a proportion.

Check conditions: We note three items. First, we are given that the sample is random. Second, both $np_0 = 450(0.10) = 45 \geq 10$ and $n(1 - p_0) = 450(0.90) = 450 \geq 10$. Third, it is reasonable that 450 players is less than 10% of all of the university baseball players who graduated during the past 20 years.

Test statistic and P-value: 1-PropZTest gives $z = -2.0428$ and $P = 0.0205$. (Alternatively, calculate $\hat{p} = \frac{32}{450} = 0.0711$, $z = \frac{0.0711 - 0.10}{\sqrt{\frac{(0.10)(0.90)}{450}}} = -2.0435$.

So $P = P(z < -2.0435) = 0.0205$.)

Conclusion in context linked to the P-value: With this small of a P-value, $0.0205 < 0.05$, there is sufficient evidence to reject the null hypothesis. In other words, there is sufficient evidence that less than 10% of the university's baseball players go on to play professionally after graduation. Thus there is sufficient evidence to write an article disputing the university's claim.

573. (a) The sample size n must be large enough so that both np and $n(1 - p)$ are ≥ 10. In this case, $np = (150)(0.03) = 4.5$. (Note that some statisticians use 5 instead of 10.)

(b) For $np \geq 10$, n must be at least $\frac{10}{p} = \frac{10}{0.03} = 333.3$. So choose $n = 334$.

(c) *Hypotheses:* $H_0: p = 0.03$, $H_a: p > 0.03$

Procedure: z-test for a proportion

Check conditions: This is a random sample (given). Both $np = (334)(0.03) = 10.02$ and $n(1 - p) = (334)(0.97) = 323.98$ are ≥ 10. The sample is less than 10% of the population (seems reasonable here).

Test statistic and P-value: 1-PropZTest gives $z = 1.918$ and $P = 0.028$.

(Alternatively, calculate $z = \frac{\frac{16}{334} - 0.03}{\sqrt{\frac{(0.03)(0.97)}{334}}} = 1.918$ and

$P = P(z > 1.918) = 0.028$.)

Conclusion in context linked to the P-value: With this small of a P-value, $0.028 < 0.05$, there is sufficient evidence to reject H_0. In other words, there is sufficient evidence to say that the percentage of applicants lying about their salary is now over 3 percent.

574. (a) *Hypotheses:* $H_0: p_1 = p_2$ and $H_a: p_1 < p_2$, where p_1 is the divorce rate for men who ask a woman's father for permission before asking the woman and p_2 is the divorce rate for men who don't ask a woman's father for permission before asking the woman. Alternatively, $H_0: p_{\text{ask}} = p_{\text{don't ask}}$ and $H_a: p_{\text{ask}} < p_{\text{don't ask}}$.

Procedure: Two-sample z-test to compare population proportions

Check conditions: These are independent random samples (given). All the relevant counts, 143, 357, 169, and 331, are ≥ 10.

Test statistic and P-value: 2-PropZTest gives $z = -1.775$ and $P = 0.0380$.

Conclusion in context linked to the P-value: With this small of a P-value, $0.0380 < 0.05$, there is sufficient evidence to reject H_0. In other words, there is sufficient evidence that men who ask a woman's father for permission before asking the woman have marriages with lower divorce rates.

(b) The marriage counselor's conclusion is not appropriate based on the study. This was an observational study, not an experiment. With observational studies, cause and effect cannot be concluded. There could be a confounding variable. For example, men who ask a father's permission might be more thoughtful, and being more thoughtful might be the reason for the lower divorce rate.

575. (a) H_0: $p = 0.5$ and H_a: $p > 0.5$, where p is the proportion of correct answers.

(b) This is a binomial distribution with $n = 16$ and $p = 0.5$. Then

$$P(X = 12) = \binom{16}{12}(0.5)^{12}(0.5)^4 = 0.0278 \text{ (or use binompdf).}$$

(c) $B(16, 0.5)$ so $P(X \geq 12) = 1 - P(X \leq 11) = 1 -$ binomcdf(16, 0.5, 11) = 0.0384. If he was guessing, the probability of getting a result this extreme or more (12 or more correct out of 16) would be 0.0384.

(d) With this small of a P-value, $0.0384 < 0.05$, there is sufficient evidence to reject H_0. In other words, there is sufficient evidence that he can distinguish between marshmallows and mushrooms by taste alone better than simply by guessing.

576. (a) *Hypotheses:* H_0: $\mu = 35$; H_a: $\mu < 35$

Procedure: One-sample t-test for a mean

Check conditions: A random sample and a roughly normal population are given.

Test statistic and P-value: T-Test gives $t = -1.0725$ and $P = 0.1508$.

(Alternatively, calculate $t = \dfrac{34.82 - 35}{\left(\dfrac{0.65}{\sqrt{15}}\right)} = \dfrac{-0.18}{0.1678} = -1.073$; and with

$df = 15 - 1 = 14$, the P-value is 0.151.)

Conclusion in context linked to the P-value: With this large of a P-value, $0.151 > 0.05$, there is not sufficient evidence to reject H_0; that is, there is not sufficient evidence for the geologist to dispute the advertised claim of extraction of a mean 35 grams of dissolved salts per 1 liter of seawater.

(b) With $N(34.75, 0.83)$, $P(x \geq 35) = 0.3816$.

(c) We have a binomial distribution with $n = 10$ and $p = 0.3816$.

$P(x \geq 2) = 1 - [P(x = 0) + P(x = 1)] = 1 - [(0.6184)^{10} + 10(0.6184)^9(0.3816)]$
$= 1 - [0.00818 + 0.05047] = 0.9414$, or use 1 – binomcdf(10, 0.3816, 1).

577. (a) *Hypotheses:* H_0: $\mu_A - \mu_B = 0$, H_a: $\mu_A - \mu_B > 0$. Or, H_0: $\mu = 35$ and H_a: $\mu < 35$.

Procedure: Two-sample *t*-test for the difference in population means

Check conditions: Independent random samples (given) with large $n = 100$

Test statistic and P-value: 2-SampTTest gives $t = 1.666$ and $P = 0.049$.

Conclusion in context linked to the P-value: With this small of a *P*-value, $0.049 < 0.05$, there is sufficient evidence that the mean sale in Store *A* is greater than the mean sale in Store *B*.

(b) *Hypotheses:* H_0: $p_A - p_B = 0$, H_a: $p_A - p_B < 0$. Or, H_0: $p_A = p_B$, H_a: $p_A < p_B$.

Procedure: Two-sample *z*-test for the difference in population proportions

Check conditions: These are independent random samples (given). Although $n_A(1 - p_A) = 93$, $n_B p_B = 15$, and $n_B(1 - p_B) = 85$ are all ≥ 10, $n_A p_A = 7 \geq 5$ but ≤ 10, so proceed with caution.

Test statistic and P-value: 2-PropZTest gives $z = -1.808$ and $P = 0.035$.

Conclusion in context linked to the P-value: With this small of a *P*-value, $0.035 < 0.05$, there is sufficient evidence to reject H_0. In other words, there is sufficient evidence that the proportion of $50 sales in Store *A* is less than the proportion of $50 sales in Store *B*.

(c) Store *A* should emphasize mean sales prices, while Store *B* should emphasize a high proportion of $50 sales.

578. (a) Although the use of random samples (given) is good, some random method (such as a coin toss) must be employed for each secretary to decide which method of dictation, live or recorded, is timed first. The second time a letter is typed naturally might lead to faster typing times, giving an unfair advantage if recorded dictation is always timed second. The design should also recognize that this is a matched pair experiment.

(b) *Hypotheses:* H_0: $\mu_D = 0$, H_a: $\mu_D < 0$, where μ_D is the mean difference (live minus recorded) in dictation times among all the company secretaries

Procedure: Matched pairs *t*-test on the set of differences (live – recorded)

Check conditions: This is a random sample (given), and the population of differences is roughly symmetric with no outliers.

–4 –2 0 2

Differences
(Live-Recorded dictation times)

Test statistic and P-value: T-test on the set of differences gives $t = -1$ and $P = 0.187$.

Conclusion in context linked to the P-value: With this large of a P-value, $0.187 > 0.05$, there is not sufficient evidence to reject H_0. In other words, there is not sufficient evidence that recorded dictation leads to slower typing speeds.

579. (a) Given that we have a random sample and that the histogram is roughly normal, we can calculate a t-interval. TInterval gives (162.4, 167.1). (Alternatively, invT(0.975, 19) = 2.093 and calculate

$$164.75 \pm 2.093\left(\frac{5.02}{\sqrt{20}}\right) = 164.75 \pm 2.35$$

or (162.40, 167.10).) So, yes, 167 pounds is in the 95% confidence interval.

(b) *Hypotheses:* H_0: $\mu = 167$ and H_a: $\mu < 167$

Procedure: Use a t-test of a population mean.

Check conditions: There is a random sample (given), and the histogram of the data is roughly bell-shaped.

Test statistic and P-value: A t-test gives $t = -2.0044$ and $P = 0.0297$.

(Alternatively, calculate $t = \dfrac{164.75 - 167}{\left(\dfrac{5.02}{\sqrt{20}}\right)} = -2.0044$ $df = 20 - 1 = 19$

with $P(t < -2.0044) = 0.0297$.)

Conclusion in context linked to the P-value: With this small of a P-value, $0.0297 < 0.05$, there is sufficient evidence to reject H_0. In other words, there is sufficient evidence that the mean weight of all junior boxers is less than 167 pounds.

(c) There is no contradiction. The 95 percent (two-sided) confidence interval corresponds to a 2.5 percent one-sided hypothesis test. At the 2.5% significance level, there is not sufficient evidence that the mean weight is less than 167 pounds.

580. (a) One possible answer is the following.

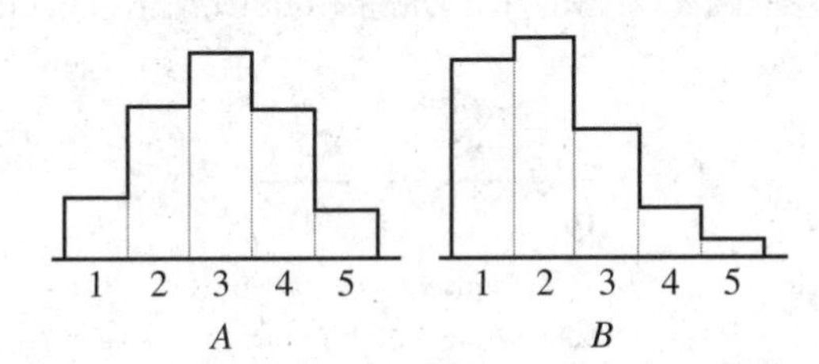

The comparison must address shape, center, and spread. The answer must mention context and use comparative language for center and spread.

The distribution of scores for Professor *A* is roughly symmetric and bell-shaped, while that for Professor *B* is skewed to the right. Both distributions have the same range with lows of 1 and highs of 5. However, both the mean and median scores for Professor *A* appear greater than those of Professor *B*.

(b) *Hypotheses:* H_0: $\mu_A - \mu_B = 0$; H_A: $\mu_A - \mu_B > 0$. (Or, H_0: $\mu_A = \mu_B$; H_A: $\mu_A > \mu_B$.)

Procedure: Two-sample *t*-test to compare population means

Check conditions: Random assignment is satisfied by the given scheme. By the central limit theorem and large sample sizes, use of the *t*-distribution is appropriate.

Test statistic and the P-value: 2-SampTTest gives $t = 6.552$ and $P = 0.0000$.

Conclusion in context linked to the P-value: With such a small *P*-value, $0.000 < 0.05$, there is very strong evidence to reject H_0. That is, there is very strong evidence of greater mean satisfaction with Professor *A* as compared to Professor *B*.

581. (a) $\mu_S = 3.68$ and $\mu_T = 3.86$.

(b) To perform a two-sample test, one condition is independence of the two samples, which is not satisfied here as the same five boards were used for each saw.

(c) *Hypotheses:* H_0: $\mu_D = 0$, H_a: $\mu_D \neq 0$, where μ_D is the mean difference in cutting times (Saw *S* minus Saw *T*) over all pieces of lumber.

Procedure: Matched pairs *t*-test on the set of differences.

Check conditions: The procedure involves random choice. A dotplot of the set of differences is very roughly symmetrical with no extreme skewness.

−0.4 −0.2 0 0.2

Differences
(Saw S time – Saw T time)

Test statistic and P-value: T-Test gives $t = -2.449$ and $P = 0.0705$.

Conclusion in context linked to the P-value: With this small of a P-value, $0.0705 < 0.10$, there is sufficient evidence (at the 10% significance level) to reject H_0. In other words, there is sufficient evidence that the observed difference in mean cutting times of the two saws is significant.

582. (a) *Hypotheses:* H_0: $\mu_1 - \mu_2 = 0$; H_a: $\mu_1 - \mu_2 > 0$ (or, H_0: $\mu_1 = \mu_2$; H_a: $\mu_1 > \mu_2$), where μ_1 is the mean score of all students at the school using the innovative program who take the state test and μ_2 is the mean score of all students at the school using the traditional approach who take the state test.

Procedure: Two-sample t-test

Check conditions: The scheme gives independent random samples, and the normal/sample size condition is satisfied by the large sample sizes (both 100 and 150 are ≥ 30).

Test statistic and P-value: 2-SampTTest gives $t = 1.891$ and $P = 0.030$.

Conclusion in context linked to the P-value: With this small of a P-value, $0.030 < 0.05$, there is sufficient evidence to reject H_0. In other words, there is sufficient evidence that students using the innovative approach have a higher mean score on the state exam than students using the traditional approach.

(b) Even with a small P-value, we cannot conclude causation because this is an observational study, not an experiment. For example, it may be that schools with the top teachers and students are the ones that choose to try the innovative approach.

(c) $1.96\sqrt{\frac{27^2}{n} + \frac{31^2}{n}} \leq 5$ gives $\sqrt{n} \geq 16.11$ and $n \geq 259.5$, so choose $n = 260$.

583. (a) *Hypotheses:* H_0: $\mu_D = 0$; H_a: $\mu_D > 0$, where μ_D is the mean difference in jump heights (with new shoes minus with old shoes) for professional high jumpers.

Procedure: Matched-pair t-test on the set of differences

Check conditions: Randomization of order is given. A dotplot of the differences is roughly symmetrical with no outliers or extreme skewness.

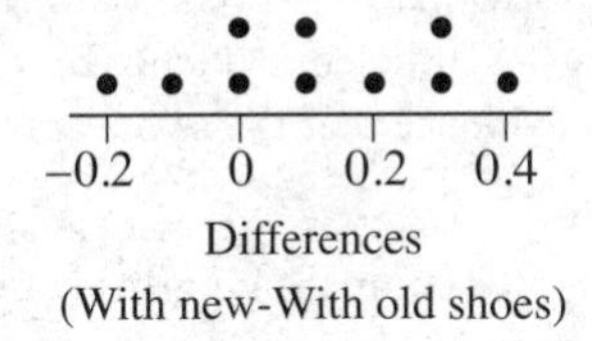

(b) *Hypotheses:* H_0: $\beta = 0$, H_a: $\beta \neq 0$ or H_0: $\beta = 0$, H_a: $\beta > 0$

Procedure: Use a t-test for regression on the matched pairs.

Check conditions: You must check that the scatterplot is roughly linear, that the residuals have no apparent pattern, and that the histogram of the residuals is roughly normal.

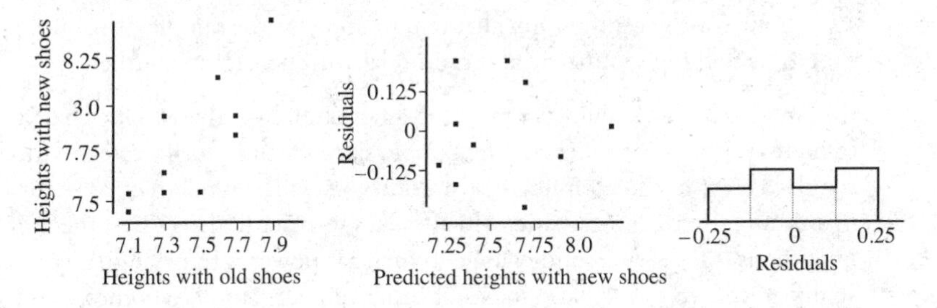

Test statistic and the P-value: LinRegTTest gives $t = 3.82$ and $P = 0.0051$ (with a one-sided test, $P = 0.0025$).

Conclusion in context linked to the P-value: With this small of a P-value, $0.0051 < 0.05$ (or $0.0025 < 0.05$ if a one-sided test), there is sufficient evidence to reject H_0. In other words, there is sufficient evidence that knowing a jumper's jump height with his old shoes does help predict, using a linear model, his jump height with the new shoes.

584. *Hypotheses:* H_0: $\mu_A - \mu_B = 0$ and H_a: $\mu_A - \mu_B \neq 0$ (or, H_0: $\mu_A = \mu_B$; H_a: $\mu_A > \mu_B$), where μ_A and μ_B are the mean slugging averages of the players on Teams A and B, respectively.

Procedure: Two-sample t-test

Check conditions: The independent random samples (given), and the large sample sizes (75) together with the central limit theorem allow us to use the t-distribution.

Test statistic and P-value: Using Lists [(1,2,3,4) in L1, (44,17,10,4) in L2, and (34,22,12,7) in L3] and 2-SampTTest gives $\bar{x}_A = 1.653$, $\bar{x}_B = 1.893$, and $t = -1.544$ with $P = 0.125$.

Conclusion in context linked to the P-value: With this large of a *P*-value, $0.125 > 0.05$, there is not sufficient evidence to reject H_0. In other words, there is not sufficient evidence that the two teams have different slugging averages.

585. (a) *Test:* A paired *t*-test for comparing means (a one-sample *t*-test on the set of differences, before video – after video, reaction times for each batter)

Hypotheses: H_0: $\mu_d = 0$ and H_a: $\mu_d > 0$, where μ_d is the mean difference between reaction times of batters who have not trained and those who have trained on the video game.

Conditions: The subjects of the test should be a random sample of batters. Unless it is somehow known that the set of all batter reaction times is roughly normally distributed, either the sample size has to be large enough, say $n \geq 30$, for the central limit theorem to apply or the sample distribution should be roughly symmetric and unimodal with no extreme outliers.

(b) Power is the probability of rejecting a false null hypothesis. One change to increase power is to increase the sample size, which in turn decreases the standard error of the sampling distribution. A smaller standard error results in a more extreme test statistic, which makes it easier to detect that the null hypothesis is false. A second change to increase power is to use a greater significance level, which also makes it easier to reject a null hypothesis.

586. (a) *Procedure:* One-sample confidence interval for a mean (of the differences, before additive – after additive)

Check conditions: Given that the normal/large sample condition is satisfied, we must assume we have a representative sample of luxury SUVs.

Mechanics: TInterval gives (–0.339, 1.439). (Alternatively, calculate

$$0.55 \pm 2.093\frac{1.90}{\sqrt{20}} = 0.55 \pm 2.093(0.42) = 0.55 \pm 0.88 \text{ or } (-0.33, 1.43).)$$

Conclusion in context: We are 95% confident that the mean difference in carbon dioxide emissions (for luxury SUVs) between before and after additive input is between –0.33 and 1.43 pounds per gallon.

(b) Because 0 is in the interval, there is not sufficient evidence that this additive reduces mean CO_2 emissions for luxury SUVs.

587. (a)

Region 1		Region 1
8	4	
9 2 1	5	7
8 7 3 3	6	6 9
2 0	7	0 3 4 7 8
	8	6
	9	1

1|5|7 means countries in Region 1 and 2 with life expectancies of 51 and 57 years, respectively

(b) Both distributions are unimodal and roughly symmetric. The median in the Region 2 distribution is a higher life expectancy than the median in the Region 1 distribution. The range in the Region 2 distribution (91 – 57 = 34) is greater than the range in the Region 1 distribution (72 – 48 = 24).

(c) With this small of a *P*-value, $0.0029 < 0.05$, there is sufficient evidence to reject H_0. In other words, there is sufficient evidence that the mean life expectancy of all countries in Region 1 is less than the mean life expectancy of all countries in Region 2.

588. (a) *Hypotheses:* $H_0: \beta_1 = 0$; $H_a: \beta_1 > 0$

Procedure: Hypothesis test for slope of regression line

Check conditions: There is a random sample (given). The scatterplot is roughly linear. There is no pattern in the residual plot. The distribution of residuals is roughly normal (as seen by linearity in the normal probability plot).

Test statistic and P-value: LinRegTTest gives $\hat{y} = -56.96 + 2.225x$, where x = waist and y = % fat, $t = 10.56$, $P = 0.0009$, and $s = 1.917$.

Conclusion in context linked to the P-value: With such a small *P*-value, $0.0009 < 0.05$, there is sufficient evidence to reject H_0. In other words, there is sufficient evidence of a positive linear relationship between waist size and percent body weight.

(b) $-56.96 + 2.225(37) = 25.37$ (percent body fat)

(c) An approximate range is 3 standard deviations (1.917) to either side of the point prediction (25.37). $\approx 25.37 \pm 3(1.917) = 25.37 \pm 5.75$ or between 19.62 and 31.12 (percent body fat).

589. (a) The null hypothesis is that the remission rate among child leukemia patients treated with the new therapy is equal to the remission rate of child leukemia patients treated with standard chemotherapy. The alternative

hypothesis is that the remission rate of those receiving the new therapy is greater than that of those receiving standard chemotherapy.

(b) A Type I error—that is, incorrectly rejecting a true null hypothesis—in this context means that the new treatment is not helping but the company believes it is. The serious consequence would be that children might take this new therapy, wrongly thinking that it will help. Instead, their disease will worsen because they are not taking the standard chemotherapy.

(c) A Type II error—that is, incorrectly failing to reject a false null hypothesis—in this context means that the new drug really works better than chemotherapy but the company doesn't realize this. The serious consequence would be that children will not take a new therapy that really is better than the old standard.

590. (a) A Type II error is a failure to reject a false null hypothesis. In this study, this would mean that the new cooler really does keep drinks cold for over 24 hours. However, the test does not give significant evidence of this. The company would not move to what is really a better cooler.

(b) A significance level of $\alpha = 0.10$ would result in a smaller probability of a Type II error because this makes it easier to reject a null hypothesis. The significance level and the probability of a Type II error are inversely related.

(c) Power is the probability of not committing a Type II error. So for this study, if the true mean is $\mu = 25$ hours, there is a 0.85 probability of correctly rejecting H_0 and concluding that there is sufficient evidence that the new cooler keeps drinks cold for over 24 hours.

591. A chi-square goodness-of-fit test is not proper for any of these examples.

(a) A chi-square goodness-of-fit test is for counts, not for quantitative measures like weights (ounces).

(b) To run a chi-square goodness-of-fit test, we need enough data. This is typically checked by the expectation of at least 5 in each cell. In this case, the expected values for the cells are 1.9, 3.8, 7.6, and 5.7.

(c) A chi-square goodness-of-fit test does not test whether or not a given observed cell is greater than expected. Rather, it tests if an overall pattern is followed.

592. (a) Using Lists and 1-Var Stats L_1, L_2 gives $\mu = 1.460$ and $s = 1.332$.

(b) By the central limit theorem, the sampling distribution of $\bar{x}$ is approximately normal with a mean of $\mu_{\bar{x}} = 1.460$ and a standard deviation of

$$\sigma_{\bar{x}} = \frac{1.332}{\sqrt{50}} = 0.1884.$$

(c) *Hypotheses:* H_0: all-nighters pulled by students during a semester are 0, 1, 2, 3, 4, 5 in the ratio 15:10:10:10:4:1; H_a: all-nighters pulled by students are 0, 1, 2, 3, 4, 5 in some ratio different from 15:10:10:10:4:1

Procedure: χ^2 goodness-of-fit test

Check conditions: There is a random sample (given). With

$$15 + 10 + 10 + 10 + 4 + 1 = 50,$$

the expected cells, $\frac{15}{50}(880) = 264$, $\frac{10}{50}(880) = 176$, $\frac{4}{50}(880) = 70.4$, and $\frac{1}{50}(880) = 17.6$, are all ≥ 5.

Test statistic and P-value: χ^2 GOF-Test gives $\chi^2 = 17.68$ and $P = 0.0034$.

Conclusion in context linked to the P-value: With this small of a P-value, $0.0034 < 0.05$, there is sufficient evidence to reject H_0. In other words, there is sufficient evidence that all-nighters pulled by students during a semester are 0, 1, 2, 3, 4, 5 in some ratio different from 15:10:10:10:4:1. (There is sufficient evidence to dispute the counselor's belief.)

593. (a) $P(\text{teaches history} \cap \text{very happy}) = \frac{20}{250} = \frac{2}{25}$

(b) $P(\text{mildy happy} \mid \text{teaches math}) = \frac{25}{10 + 25 + 30} = \frac{25}{65} = \frac{5}{13}$

(c) $P(\text{teaches English} \mid \text{unhappy}) = \frac{20}{10 + 20 + 20} = \frac{20}{50} = \frac{2}{5}$

(d) *Hypotheses:* H_0: happiness level and subject taught are independent; H_a: there is a relationship between happiness level and subject taught

Procedure: Chi-square test for independence

Check conditions: We are given a random sample of teachers, and the expected cell counts are all ≥ 5.

13	27.3	24.7
24	50.4	45.6
13	27.3	24.7

Test statistic and P-value: Running a chi-square test gives $\chi^2 = 7.975$. With $df = (r - 1)(c - 1) = 4$, we get $P = 0.0925$.

Conclusion in context linked to the P-value: With this P-value ($0.05 < P < 0.10$), we say that the data provide some (but not strong) evidence to reject H_0. We conclude that there is sufficient evidence at the 10% significance level, but not at the 5% significance level, of a relationship between happiness level and subject taught.

594. *Hypotheses:* H_0: the proportions of students who admit to cheating have not changed over the years 2014, 2016, 2018; H_a: the proportions of students who admit to cheating is not the same for the years 2014, 2016, 2018

Procedure: Chi-square test of homogeneity

Check conditions: There is a random sample (given). The expected cell values {70.5, 61.9, 63.6, 51.5, 45.1, 46.4}, which are calculated by putting the data into a matrix and using χ^2-Test, are all ≥ 5.

Test statistic and P-value: χ^2-Test gives $\chi^2 = 8.799$ and $P = 0.0123$

Conclusion in context linked to the P-value: With this small of a P-value, $0.0123 < 0.05$, there is sufficient evidence to reject H_0. In other words, there is sufficient evidence that the proportions of students who admit to cheating are not the same for the years 2014, 2016, and 2018.

595. (a) *Hypotheses:* H_0: good fit with 0.3, 0.5, 0.2 distribution for restaurant visits to Starbucks, Chick-fil-A, and Chipotle; H_a: pattern has changed from 0.3, 0.5, 0.2 distribution

Procedure: Chi-square test for goodness of fit

Check conditions: We are given a random sample, and the expected counts $(0.3)(250) = 75$, $(0.5)(250) = 125$, and $(0.2)(250) = 50$ are all ≥ 5.

Test statistic and P-value: χ^2GOF-Test gives $\chi^2 = 2.161$ and $P = 0.3394$.

(Alternatively, $\chi^2 = \frac{(71-75)^2}{75} + \frac{(136-125)^2}{125} + \frac{(43-50)^2}{50} = 2.161$, and with $df = 3 - 1 = 2$, we get $P = 0.3394$.)

Conclusion in context with linkage to the P-value: With this large of a P-value, $0.3394 > 0.05$, there is not enough evidence to say that the overall pattern of sales has changed from 30% Starbucks, 50% Chick-fil-A, and 20% Chipotle.

(b) Noting that $n\hat{p} = 71$ and $n(1 - \hat{p}) = 179$ are ≥ 10 and noting that we have a random sample, we calculate $\hat{p} = \frac{71}{250} = 0.284$ and the margin of error to be $\pm 1.96\sqrt{\frac{(0.284)(0.716)}{250}} = \pm 0.0559$.

(c) $n\hat{p} = 136$ and $n(1 - \hat{p}) = 114$ are ≥ 10 with a random sample. So we calculate $\hat{p} = \frac{136}{250} = 0.544$, $(z)\sqrt{\frac{(0.544)(0.456)}{250}} = 0.03$, and $z = 0.952$, which results in a confidence level of normalcdf(–0.952, 0.952) = 0.659 or 65.9%.

596. (a) The intended population is all high school students. However, the samples were gathered from a single teacher's classes. Any conclusion is applicable only to students of this particular teacher.

(b) The teacher is interested in a one-sided test, namely whether social studies students are more likely than other students to read newspapers. A chi-square test of homogeneity will indicate only whether students taking social studies and students taking other classes have different patterns of newspaper reading.

(c) As long as the sample of social studies students and the sample of other students were drawn randomly and independently, a *z*-test for differences of proportions can be performed.

(d) Segmented bar charts give a good initial impression of this data.

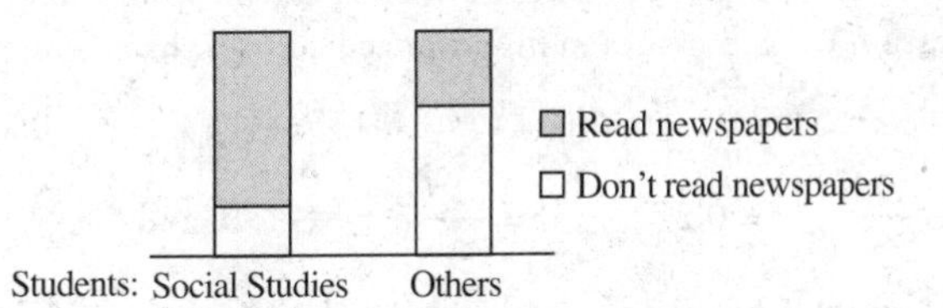

597. (a) *Hypotheses:* H_0: there is no association between whether or not people survive major heart attacks and their education levels; H_a: there is an association between whether or not people survive major heart attacks and their education levels.

Procedure: χ^2-test of independence

Check conditions: There is a random sample (given), and all expected counts are ≥ 5 (the smallest is given to be 9.9).

Test statistic and P-value: χ^2-Test gives $\chi^2 = 8.986$ and $P = 0.0112$.

Conclusion in context linked to P-value: With this small of a P-value, $0.0112 < 0.05$, there is sufficient evidence to reject H_0. In other words, there is sufficient evidence of an association between whether or not people survive major heart attacks and their education levels.

(b) Although the χ^2-Test test concludes there is sufficient evidence of an association, it does not give direction. The segmented bar charts further indicate that adults without a high school degree have roughly twice the probability of not surviving a major heart attack as do those with high school or college degrees.

598. (a) $df = (2 - 1)(3 - 1) = 2$, so the *P*-value is χ^2cdf(4.749, 1,000, 2) = 0.093. With this large of a *P*-value, 0.093 > 0.05, there is not sufficient evidence of an association between grades and course level. (Alternatively, with this small of a *P*-value, 0.093 < 0.10, there is sufficient evidence of an association between grades and course level.)

(b) *Hypotheses:* $H_0: p_{Int} - p_{Adv} = 0$; $H_a: p_{Int} - p_{Adv} < 0$ (or, $H_0: p_{Int} = p_{Adv}$; $H_a: p_{Int} < p_{Adv}$)

Procedure: Two-proportion *z*-test

Check conditions: We are given independent random sample, and $n_{Int}\hat{p}_{Int} = 35$, $n_{Int}(1 - \hat{p}_{Int}) = 65$, $n_{Adv}\hat{p}_{Adv} = 22$, and $n_{Adv}(1 - \hat{p}_{Adv}) = 18$ are all ≥ 10.

Test statistic and P-value: 2-PropZTest gives $z = -2.176$, and $P = 0.0148$.

Conclusion in context linked to the P-value: With such a small *P*-value, 0.0148 < 0.05, there is sufficient evidence that the professor gives a higher proportion of A, B grades in his advanced courses than in his intro courses.

599. (a) The probabilities are $P(0) = e^{-2} = 0.135$, $P(1) = 2e^{-2} = 0.271$,

$P(2) = \frac{(2)^2}{2}e^{-2} = 0.271$, $P(3) = \frac{(2)^3}{3!}e^{-2} = 0.180$, and

$P(4 \text{ or more}) = 1 - (0.135 + 0.271 + 0.271 + 0.180) = 0.143$.

(b) Multiplying each of these probabilities by 500 gives the number of patients expected to survive different numbers of years.

Survival Years	0	1	2	3	4 or more
Expected Number of Patients	67.5	135.5	135.5	90	71.5

(c) *Hypotheses:* H_0: good fit with the given distribution of survival years, H_a: not a good fit with the given distribution

Procedure: χ^2 goodness-of-fit test

Check conditions: All the expected values are ≥ 5, and we must assume a representative sample.

Test statistic and P-value: With {69, 106, 153, 98, 74} in L_1 and {67.5, 135.5, 135.5, 90, 71.5} in L_2, with $df = 5 - 1 = 4$, χ^2GOF-Test gives $\chi^2 = 9.515$ and $P = 0.0494$.

Conclusion in context linked to P-value: With this small of a *P*-value, 0.0494 < 0.05, there is sufficient evidence to reject H_0. In other words, there is sufficient evidence that the distribution of survival years follows a distribution different from the given Poisson distribution.

600. (a) The observed value of the test statistic is $\frac{s_8}{s_3} = \frac{5.4}{2.7} = 2.0$.

(b) The observed test statistic of 2.0 is very far to the right in the distribution of simulated ratios and so is unlikely to occur by random sampling alone if the null hypothesis of equal standard deviations is true. We conclude that there is sufficient evidence to reject H_0. In other words, there is sufficient evidence that there is more variability in the mean verbal scores of 8-year-olds than of 3-year-olds.

The tools you need to succeed on your AP Statistics exam

Barron's AP Statistics, 9th Ed.

Martin Sternstein, Ph.D.
Includes a diagnostic test and five full-length and practice exams with test questions answered and explained; the 35 best AP Stat exam hints found anywhere; 15 thorough chapter reviews; and more. Can also be purchased with an optional CD-ROM that presents two additional practice tests with automatic scoring, as well as a second CD-ROM introducing the TI-Nspire.
BONUS ONLINE PRACTICE TEST: Students who buy this book or package will also get FREE access to one additional full-length online AP Statistics test with all questions answered and explained.
Book only: ISBN 978-1-4380-0904-9, $18.99, *Can$23.99*
Book w/CD-ROM: ISBN 978-1-4380-7712-3, $29.99, *Can$37.50*

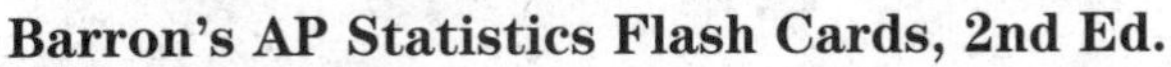

Barron's AP Statistics Flash Cards, 2nd Ed.

Martin Sternstein, Ph.D.
Questions and answers on this set of 450 flash cards encompass four general statistics-based themes: Exploratory Analysis, Planning a Study, Probability, and Statistical Inference. New to this edition are 50 extra multiple-choice questions that cover all topics. Includes a metal key-ring-style card holder.
BONUS ONLINE PRACTICE TEST: Students who purchase this set of flash cards will also get FREE access to one additional full-length online AP Statistics test with all questions answered and explained.
ISBN 978-1-4380-7401-6, $18.99, *Can$21.99*

Barron's AP Q&A Statistics
600 Questions and Answers

Martin Sternstein, Ph.D.
This all-new, handy AP test prep guide features 600 questions and thoroughly explained answers. Learn why your answer is correct—and the rationale behind why each other answer choice is incorrect, thereby reinforcing the facts you need to know in order to answer each question correctly on your AP exam. *AP Q&A Statistics* includes questions and answers on Exploratory Analysis, Collecting and Producing Data, Probability, and Statistical Inference.
ISBN 978-1-4380-1189-9, $16.99, *Can$21.50*

Ultimate AP Statistics

Martin Sternstein, Ph.D.
Find everything you need to score a 5 on your AP Statistics exam—and **save over 20% OFF** items when purchased separately! Includes *Barron's AP Statistics, Barron's AP Statistics Flash Cards*, and *Barron's AP Q&A Statistics*.
ISBN 978-1-4380-7921-9, paperback, $45.99, *Can$57.99*

Available at your local bookstore or visit **www.barronseduc.com**

Prices subject to change without notice.

Barron's Educational Series, Inc.
250 Wireless Blvd.
Hauppauge, N.Y. 11788
Order toll-free: 1-800-645-3476

In Canada:
Georgetown Book Warehouse
34 Armstrong Ave.
Georgetown, Ontario L7G 4R9

#319 R 5/18

Success on Advanced Placement Tests Starts with Help from BARRON'S

Each May, thousands of college-bound students take one or more Advanced Placement Exams to earn college credits—and for many years, they've looked to Barron's, the leader in Advanced Placement test preparation. Many titles offer practice tests online or on CD-ROM. You can get Barron's user-friendly manuals for 23 different AP subjects.

AP: Art History, 4th Ed., w/online exams
ISBN 978-1-4380-1103-5, $29.99, *Can$37.50*

AP: Biology, 6th Ed., w/optional CD-ROM
Book: ISBN 978-1-4380-0868-4, $18.99, *Can$23.99*
Book w/CD-ROM: ISBN 978-1-4380-7691-1, $29.99, *Can$37.50*

AP: Calculus, 14th Ed., w/optional CD-ROM
Book: ISBN 978-1-4380-0859-2, $18.99, *Can$23.99*
Book w/CD-ROM: ISBN 978-1-4380-7674-4, $29.99, *Can$37.50*

AP: Chemistry, 9th Ed., w/online tests
ISBN 978-1-4380-1066-3, $24.99, *Can$31.50*

AP: Chinese Language and Culture, 2nd Ed., w/MP3 CD
ISBN 978-1-4380-7388-0, $29.99, *Can$34.50*

AP: Comparative Government & Politics
ISBN: 978-1-4380-0615-4, $16.99, *Can$19.99*

AP: Computer Science A, 8th Ed., w/online tests
ISBN 978-1-4380-0919-3, $23.99, *Can$29.99*

AP: English Language and Composition, 7th Ed., w/optional CD-ROM
Book: ISBN 978-1-4380-0864-6, $16.99, *Can$21.50*
Book w/CD-ROM: ISBN 978-1-4380-7690-4, $29.99, *Can$37.50*

AP: English Literature and Composition, 7th Ed., w/online tests
ISBN 978-1-4380-1064-9, $22.99, *Can$28.99*

AP: Environmental Science, 7th Ed., w/online tests
ISBN 978-1-4380-0865-3, $24.99, *Can$31.50*

AP: European History, 9th Ed., w/online tests
ISBN 978-1-4380-1067-0, $23.99, *Can$29.99*

AP: French Language and Culture, 2nd Ed., w/MP3 CD
ISBN 978-1-4380-7603-4, $26.99, *Can$32.50*

AP: Human Geography, 7th Ed., w/online tests
ISBN 978-1-4380-1068-7, $22.99, *Can$28.99*

AP: Microeconomics/Macroeconomics, 6th Ed., w/online tests
ISBN 978-1-4380-1065-6, $21.99, *Can$27.50*

AP: Music Theory, 3rd Ed., w/downloadable audio
ISBN 978-1-4380-7677-5, $34.99, *Can$43.99*

AP: Physics 1, w/online tests
ISBN 978-1-4380-1071-7, $21.99, *Can$27.50*

AP: Physics 1 and 2, w/optional CD-ROM
Book w/CD-ROM: ISBN 978-1-4380-7379-8, $29.99, *Can$34.50*

AP: Physics 2, w/online tests
ISBN 978-1-4380-1123-3, $22.99, *Can$28.99*

AP: Physics C, 4th Ed.
ISBN 978-1-4380-0742-7, $18.99, *Can$22.99*

AP: Psychology, 8th Ed., w/online tests
ISBN 978-1-4380-1069-4, $22.99, *Can$28.99*

AP: Spanish Language and Culture, 9th Ed., w/MP3 CD and optional CD-ROM
Book w/MP3 CD: ISBN 978-1-4380-7682-9-0, $26.99, *Can$33.99*
Book w/MP3 CD and CD-ROM: ISBN 978-1-4380-7675-1, $34.99, *Can$43.99*

AP: Statistics, 9th Ed., w/optional CD-ROM
Book: ISBN 978-1-4380-0904-9, $18.99, *Can$23.99*
Book w/CD-ROM: ISBN 978-1-4380-7712-3, $29.99, *Can$37.50*

AP: U.S. Government and Politics, 11th Ed., w/online tests
ISBN 978-1-4380-1168-4, $22.99, *Can$28.99*

AP: U.S. History, 4th Ed., w/online tets
ISBN 978-1-4380-1108-0, $23.99, *Can$29.99*

AP: World History, 8th Ed., w/online tests
ISBN 978-1-4380-1109-7, $23.99, *Can$29.99*

Available at your local bookstore or visit
www.barronseduc.com

Barron's Educational Series, Inc.
250 Wireless Blvd.
Hauppauge, NY 11788
Order toll-free: 1-800-645-3476

In Canada:
Georgetown Book Warehouse
34 Armstrong Ave.
Georgetown, Ontario L7G 4R9
Canadian orders: 1-800-247-716

Prices subject to change without notic

(#93 R5/18)

The information you need in a flash!

BARRON'S AP FLASH CARDS

Study, prepare, and get ready to score a 5 on your AP subject test with these essential flash card sets. From Biology and Chemistry to Human Geography and Spanish, students will find the information they need to do their best on test day. Test takers will find questions and answers, as well as important terms to remember, vocabulary, explanations, definitions, and more. A corner punch hole on each card accommodates the enclosed metal key-ring-style card holder. Flash cards can be arranged in any sequence desired and are perfect for study on-the-go.

BONUS ONLINE PRACTICE TEST: Students who purchase most of these flash cards will also get ***FREE*** access to one additional full-length online AP subject test with all questions answered and explained.*

Each boxed set 4 7/8" x 5" x 3"

BARRON'S AP BIOLOGY FLASH CARDS, 3rd Edition
ISBN 978-1-4380-7611-9, $18.99, *Can$22.99*

BARRON'S AP CALCULUS FLASH CARDS, 2nd Edition
ISBN 978-1-4380-7400-9, $18.99, *Can$21.99*

BARRON'S AP CHEMISTRY FLASH CARDS, 2nd Edition
ISBN 978-1-4380-7419-1, $18.99, *Can$21.99*

BARRON'S AP COMPUTER SCIENCE A FLASH CARDS*
ISBN 978-1-4380-7856-4, $18.99, *Can$23.99*

BARRON'S AP ENVIRONMENTAL SCIENCE FLASH CARDS, 2nd Edition
ISBN 978-1-4380-7403-0, $18.99, *Can$21.99*

BARRON'S AP EUROPEAN HISTORY FLASH CARDS
ISBN 978-1-4380-7651-5, $18.99, *Can$22.99*

BARRON'S AP HUMAN GEOGRAPHY FLASH CARDS, 3rd Edition
ISBN 978-1-4380-7680-5, $18.99, *Can$22.99*

BARRON'S AP PSYCHOLOGY FLASH CARDS, 3rd Edition*
ISBN 978-1-4380-7713-0, $18.99, *Can$22.99*

BARRON'S AP SPANISH FLASH CARDS, 2nd Edition
ISBN 978-1-4380-7610-2, $18.99, *Can$22.99*

BARRON'S AP STATISTICS FLASH CARDS, 2nd Edition
ISBN 978-1-4380-7401-6, $21.99, *Can$27.50*

BARRON'S AP U.S. GOVERNMENT AND POLITICS FLASH CARDS, 2nd Edition
ISBN 978-1-4380-7402-3, $18.99, *Can$21.99*

BARRON'S AP UNITED STATES HISTORY FLASH CARDS, 3rd Edition
ISBN 978-1-4380-7609-6, $18.99, *Can$22.99*

BARRON'S AP WORLD HISTORY FLASH CARDS, 3rd Edition
ISBN 978-1-4380-7630-0, $18.99, *Can$22.99*

*Bonus Online Practice Test not available.

Available at your local bookstore or visit **www.barronseduc.com**

BARRON'S EDUCATIONAL SERIES, INC.
250 Wireless Blvd.
Hauppauge, NY 11788
Order toll-free: 1-800-645-3476
Prices subject to change without notice.

IN CANADA: Georgetown Book Warehouse
34 Armstrong Ave.
Georgetown, Ontario L7G 4R9
Canadian orders: 1-800-247-7160

(#93a) R4/18

College-bound students can rely on Barron's for the best in SAT Subject test preparation...

Every Barron's SAT Subject Test manual contains full-length practice exams with answers and explanations. All exams mirror the actual tests in length, format, and degree of difficulty. Manuals also present extensive review, study tips, and general information on the test. Many titles offer practice tests online. Foreign language manuals include MP3 files for listening comprehension help. All books are paperback.

SAT Subject Test Biology E/M, 6th Ed., w/online tests
Deborah T. Goldberg, M.S.
ISBN 978-1-4380-0960-5, $24.99, *Can$31.50*

SAT Subject Test Chemistry, 14th Ed. w/online tests
Joseph A. Mascetta, M.S. and Mark Kernion, M.A.
ISBN 978-1-4380-1113-4, $22.99, *Can$28.99*

SAT Subject Test French, 4th Ed., w/online tests
Reneé White and Sylvie Bouvier
ISBN 978-1-4380-7767-3, $24.99, *Can$31.50*

SAT Subject Test Literature, 7th Ed., w/online tests
Christina Myers-Shaffer, M.Ed.
ISBN 978-1-4380-0956-8, $21.99, *Can$27.50*

SAT Subject Test Math Level 1, 7th Ed. w/online tests
Ira K. Wolf, Ph.D.
ISBN 978-1-4380-1133-2, $22.99, *Can$28.99*

SAT Subject Test Math Level 2, 13th Ed. w/online tests
Richard Ku, M.A.
ISBN 978-1-4380-1114-1, $22.99, *Can$28.99*

SAT Subject Test Physics, 2nd Ed.
Robert Jansen, M.A. and Greg Young, M.S. Ed.
ISBN 978-1-4380-0789-2, $18.99, *Can$22.99*

SAT Subject Test Spanish, 4th Ed., w/MP3 CD
José M. Díaz, M.A.
ISBN 978-1-4380-7561-7, $21.99, *Can$26.50*

SAT Subject Test U.S. History, 4th Ed., w/online tests
Kenneth R. Senter, M.S. and Eugene V. Resnick.
ISBN 978-1-4380-1074-8, $22.99, *Can$28.99*

SAT Subject Test World History, 2nd Ed., w/online tests
William Melega
ISBN 978-1-4380-1000-7, $21.99, *Can$27.50*

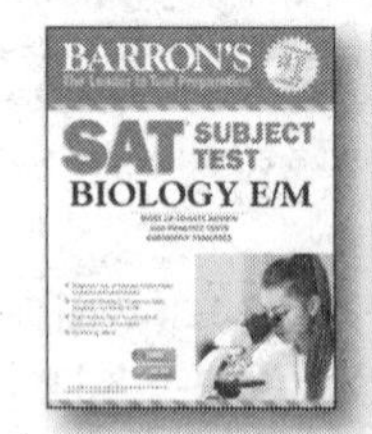

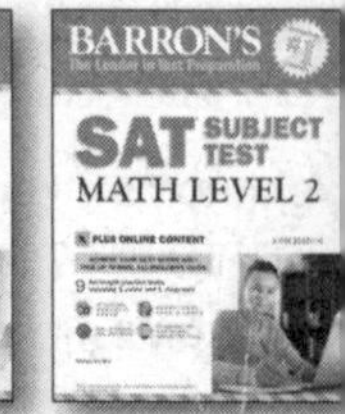

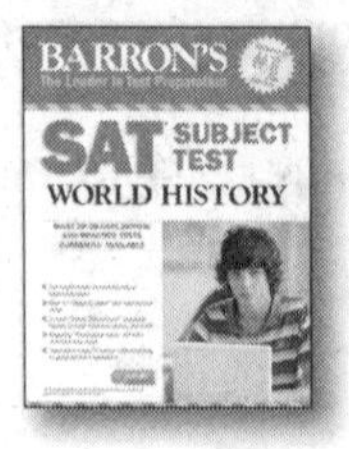

Available at your local bookstore or visit **www.barronseduc.com**

Barron's Educational Series, Inc.
250 Wireless Blvd., Hauppauge, NY 11788
Order toll free: 1-800-645-3476
Order by fax: 1-631-434-3217

In Canada: Georgetown Book Warehouse
34 Armstrong Ave., Georgetown, Ont. L7G 4R
Order toll free: 1-800-247-7160
Order by fax: 1-800-887-1594

Prices subject to change without notice.

(#87) R2/18